D1391095

OBSERVATIONAL ASTRONOMY
FOR AMATEURS

4

by the same author

AMATEUR ASTRONOMER'S HANDBOOK
INTRODUCING ASTRONOMY

OBSERVATIONAL

ASTRONOMY

FOR AMATEURS

J. B. SIDGWICK
F.R.A.S.

SECOND EDITION
prepared by
Dr. GILBERT FIELDER

THIRD EDITION
prepared by
R. C. GAMBLE
B. Sc., F.R.A.S.

FOURTH EDITION
prepared by
JAMES MUIRDEN

PELHAM BOOKS
London

First published in Great Britain by
Faber and Faber Ltd
1957
Second edition 1961
Third edition 1971

Fourth edition published by
Pelham Books Ltd
44 Bedford Square
London WC1B 3DU
1982

Sidgwick, J.B.
    Observations astronomy for amateurs.—4th ed.
    1. Astronomy—Observers' manuals
    I. Title
    II. Muirden, James
    520   QB44.2

ISBN 0 7207 1378 1

Typesetting by Cambrian Typesetters,
Farnborough, Hants.

Printed and bound in Great Britain
by Billing & Sons Ltd,
Guildford and Worcester.

CONTENTS

CONTENTS

CONTENTS

16. COMETS

CONTENTS

TABLE OF ABBREVIATIONS

The following commonly occurring abbreviations and symbols are used in the sense quoted unless the contrary is specified or is obvious from the context.

Å	Ångström unit = 10^{-8} cm
A.A.	*Astronomical Almanac*
A.A.H.	*Amateur Astronomer's Handbook*
AT	Apparent Time
A.U.	Astronomical Unit
α	Right Ascension
B	Latitude
B.	(followed by number): reference to Bibliography, section 20
B.A.A.	British Astronomical Association
Bibliographical abbreviations: see section 20	
c/i	colour index
CM	central meridian
D	aperture of objective in cm
Dec	Declination
d/v	direct vision
δ	Declination; or, diameter of the pupillary aperture of the eye
ET	Ephemeris Time
E/T	Equation of Time
F	focal length of objective
f	focal length of ocular; or, following
f/	focal ratio (F/D). $f/10$ is said to be a larger focal ratio and a smaller relative aperture (D/F) than, e.g. $f/5$
$\mathcal{F}$	equivalent focal length
GMAT	Greenwich Mean Astronomical Time
H	hour angle
HP	high power (of oculars)
HR	hourly rate
I	intensity (of radiation)

I.A.U.	International Astronomical Union
JD	Julian Date
L	longitude
LP	low power (of oculars)
LPV	long-period variable
λ	wavelength
M	magnification; or, integrated magnitude; or, absolute magnitude
M'	minimum useful magnification
M''	maximum useful magnification
M_f	minimum magnification for full resolution
m	metre
m	apparent stellar magnitude
m_b	bolometric magnitude
m_p	photographic magnitude
m_v	visual or photovisual magnitude
MSL	mean sea level
MT	Mean Time (prefixed by L = Local, or G = Greenwich)
μ	micron = 10^{-4} cm
N	refractive index
NCP	North Celestial Pole
nm	nanometre = 10^{-7} cm
NPS	North Polar Sequence
ν	constringence
OG	object glass
ω	deviation
p	preceding
p.a.	position angle
π	stellar parallax; or, transmission factor of an optical train
ϕ	latitude
R	theoretical resolution threshold
R'	empirical resolution threshold
r	angular radius of a given diffraction ring or interspace; or, angular separation of components of a double star
r'	linear radius of a given diffraction ring or interspace
RA	Right Ascension
SPV	short-period variable
ST	Sidereal Time (prefixed by L = Local, or G = Greenwich)
t	time; or, turbulence factor
UT	Universal Time
ZD	Zenith Distance
ZHR	zenithal hourly rate

EDITOR'S PREFACE TO SECOND EDITION

A book of this kind does not change greatly with the passage of time yet progress in Solar System astronomy has been remarkably rapid, recently, and many revisions have been made in order to bring the text and the Bibliography up to date. A note must be given here about the new ephemerides now in use, which introduce **Ephemeris Time**.

Following upon recommendations of the International Astronomical Union, various changes have been made to the content of the *Nautical Almanac* and the tables for 1960 *et seq* will be published in London and Washington under the respective titles *The Astronomical Ephemeris* and *The American Ephemeris*. Although these titles are different, the contents will be identical, and, in this book, both publications will be referred to as '*A.E.*'

The changes which have been made have arisen because of the need for greater accuracy in the predictions of astronomical events. Predictions tabulated for years prior to 1960 were based on the rotation of the Earth. However, the Earth's axial angular velocity is not quite uniform and hence equal intervals of Universal Time (UT), based on the Earth's rotation, are not equal by absolute standards. In an attempt to overcome this difficulty, accurate predictions will in future be referred to a new time scale, called Ephemeris Time (ET), based on the orbital motions of the Earth and Moon calculated from gravitational theory. These calculated motions must be compared with positional observations made over an extended period, and ET may then be determined accurately from the relationship

$$ET = UT + \Delta T$$

only after the correction ΔT has been determined with the requisite accuracy.

At the time of writing (1960), ΔT has not been well-determined, and astronomers wishing to compare their observations—which should, as always, be referred to UT—with predictions given in ET may add in the appropriate (at present, approximate) correction ΔT to the Universal Time. Values of ΔT are tabulated in the *A.A.* For 1960, for example, the correction is of the order of +35 seconds of time.

For completeness, the available *A.A.* data have been added, where appropriate, at the end of chapters in this book. The *N.A.* data have been retained and refer, of course, to years prior to 1960.

I should like to thank R. G. Andrews (Director of the Variable Star

Section of the British Astronomical Association), J. Heywood (Director of the Radio and Electronics Section of the B.A.A.), and J. Paton (Director of the Aurora and Zodiacal Light Section of the B.A.A.) for answering questions which arose during the revision.

G. FIELDER, Editor

EDITOR'S PREFACE TO THIRD EDITION

Since the second edition of this book appeared, in 1961, little revision to the text has been found necessary. As Dr. Fielder has said, in his preface to the second edition, a book of this kind does not change greatly with the passage of time. Careful attention has, however, been given to the data, which have been revised and brought up to date, and references to data from the *Nautical Almanac*, available only up to 1960, have been omitted.

R. C. GAMBLE, Editor

EDITOR'S PREFACE TO FOURTH EDITION

This new edition has been completely re-set, and the opportunity has been taken of bringing the text up to date. About one-fifth of the book has been re-written, and in addition the Bibliography has been updated.

The amateur astronomer of the 1980s faces rather different prospects from those confronting John Sidgwick when he wrote the first edition of his great work. Solar, lunar and meteor research in particular have changed in emphasis, and the chapters dealing with these fields would probably have been very different had he written them now. Other departments have been less affected by the passage of time, but the careful reader will find, both here and in the companion volume, *Amateur Astronomer's Handbook*, few pages that have not been amended in some way. It should be noted that *The Astronomical Ephemeris* and *The American Ephemeris* have now been replaced by a single volume, *The Astronomical Almanac*, to which frequent reference is made in this work.

Sidgwick laid great emphasis on the role of the B.A.A. Observing Sections in assisting the specialist observer. I have followed his example in seeking the advice of Directors or ex-Directors of Sections, and I wish to record here my gratitude to the following people for their ready assistance: Prof. V. Barocas (Sun), G. W. Amery (Moon), E. H. Collinson (Mars), W. E. Fox (Jupiter), A. W. Heath (Saturn), R. J. Livesey (Aurora and Zodiacal Light), S. W. Milbourn (Comets), and J. E. Isles (Variable Stars).

xvii

I am also grateful to the B.A.A. Council for allowing me to use information contained both in the *Handbook* and in individual section publications.

JAMES MUIRDEN, *Editor*

FOREWORD

This book is intended as a sequel or companion volume to my *Amateur Astronomer's Handbook*. The latter concerned itself with the instrumental and theoretical background of practical astronomy, and its plan and intention were those of a reference handbook. The present volume is devoted to the observational techniques employed in the various fields of amateur work. Though conceived as a single unit, the two books are individually self-contained and are independent of one another; cross-references between them have been kept to a minimum.*

In the Foreword to *A.A.H.* I stressed the value of the British Astronomical Association to all amateur astronomers in this country, and the importance to the practical telescope-user of joining one of its Observing Sections. The following Section Directors have read the relevant sections of this book in MS and have given me the benefit of their specialist knowledge in these fields: D. W. Dewhirst (Sun), H. P. Wilkins (Moon), H. McEwen (Mercury and Venus), A. F. O'D. Alexander (Jupiter), M. B. B. Heath (Saturn), James Paton (Aurorae), G. Merton (Comets), J. P. M. Prentice (Meteors), W. M. Lindley (Variable Stars) and W. H. Steavenson (Methods of Observation). To them I am deeply indebted, well realising how greatly the book has benefited from their generous expenditure of time and trouble on its behalf. Finally I should like to acknowledge my appreciation of the labours of Sarita Gordon, who drew the 230-odd diagrams for this book and for *Amateur Astronomer's Handbook*.

J. B. SIDGWICK
London
June 1954

---

*Where such occur, *Amateur Astronomer's Handbook* is referred to as *A.A.H.*

xix

SECTION 1

SOLAR OBSERVATION

1.1 Introduction

It being generally true that the amateur is plagued by, above all else, limitation of light grasp, the Sun offers the outstanding advantages of large angular size combined with an abundance of light. It is therefore particularly suited to observation with small instruments, and is one of the few professionally-studied objects on which valuable work (as distinct from work which is primarily of interest to the observer) can be done with amateur equipment. In spectroscopy, particularly, it is the one field in the whole of astronomy where the amateur can really go to town.

Within reason, focal length is more important than aperture in a telescope designed for solar work exclusively. But because light grasp is no longer a factor to worry about, except in the negative sense, it is sometimes forgotten that the degree of resolution of solar detail that is obtained is, to some extent, dependent upon D; although, since daytime atmospheric conditions usually limit resolution to something of the order of a $1''$ arc at best, there is little to be gained from an increase of aperture beyond about 15 cm, and routine observations of spots, faculae and prominences can be undertaken with an aperture of about 10 cm.

The usefulness of uncoated mirrors, as a means of reducing light grasp without at the same time reducing resolution, is emphasised elsewhere. It should be borne in mind that an inferior objective will perform just as badly on the Sun as on more difficult objects: all branches of observation demand objectives of the highest quality, and the Sun is no exception.

Solar observation is immensely facilitated, and its accuracy improved, by an equatorial mounting, especially if clock-driven. Visual observation by projection and the study of the chromosphere and prominences, in particular, are so difficult with an altazimuth that the equatorial may be regarded as virtually essential.

1

1.2 Observation by projection

The Sun is the only astronomical object bright enough to be observed in this way, and for general views it has the advantage that a number of people can observe the image simultaneously. Furthermore, the positions of the spots on the disc can readily be transferred to paper. Even for the detailed examination of small areas under high magnification, it may be preferable to direct observation provided certain precautions are taken (see section 1.3) though the contrary view—that such detail as pores and the granulation of the photosphere are best seen by direct vision—has been held by many experienced observers. It is certainly true to say that unless the screen is almost totally enclosed, so that extraneous daylight is excluded, low-contrast details will be seen only poorly or not at all.

Using a refractor, in the northern hemisphere, the projected image has the orientation shown in Figure 1 (*a*) when the Sun is on or near the meridian; but it must be remembered that the orientation with respect to the vertical depends upon the hour angle of the Sun, and near sunrise and sunset the image will appear markedly inclined. If a Newtonian reflector is used (eyepiece and projection screen on the eastern side of the tube, observer viewing from the western side), the orientation of the southerly Sun will appear somewhat as indicated in Figure 1 (*b*), the angle α corresponding to the Sun's altitude.

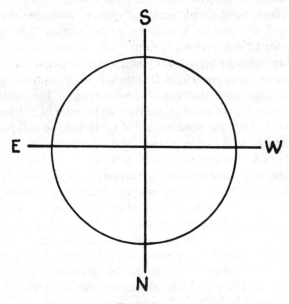

FIGURE 1a

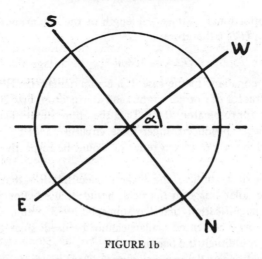

FIGURE 1b

The simplest way of establishing orientation is to allow the image to drift across the field of view after first of all positioning it centrally; that part of the image which first touches the edge of the field is the W (*p*) limb. If there is any subsequent doubt as to the N and S orientation, the tube can be raised or lowered slightly, and the corresponding motion of the image noted.

(i) Objective only:

Image scale being a function of F, a specially constructed Sun telescope (see section 1.15) is required to give an image of sufficient linear diameter in the focal plane. Taking the Sun's mean angular semi-diameter as 16′, the linear diameter of the primary image is given by

$$d=0.0094F$$

An easily rembered form of this relationship is that the linear diameter of the Sun's focal plane image is 11 mm for every 100 cm focal length.

(ii) Objective and ocular:

For observation by projection with ordinary telescopes, therefore, an ocular must be used to give an image of observable linear size. The diameter, d, of the projected image is directly proportional to the focal length of the objective, and to the distance, s, of the screen from the ocular, and

3

inversely proportional to the focal length of the ocular; putting M for the magnification (F/f), therefore,

$$d=k.M.s$$

where k is a constant. In the case of a 6 mm ocular (i.e. HP–x180– with a 75 mm refractor of normal f/ratio) and a projection box 20 cm long, the value of k is approximately 0.07. Thus the approximate diameters of the projected images produced under these circumstances by a f/15 75 mm refractor and by a f/6 15 cm reflector would be respectively 32 cm and 25 cm.

For whole-disc studies, the highest magnification that includes the whole of the solar image in the field should be used. For detailed work, the highest magnification that the atmosphere will stand. Ordinary Huyghenians are best; cemented oculars cannot be used, since the concentration of heat would melt the balsam.

The applications of the projection method are twofold:

(*a*) Whole-disc record of the spots and more obvious faculae. In this work, great attention should be paid to the correct representation of their sizes, positions, and orientations, and none to the detail of their structure.

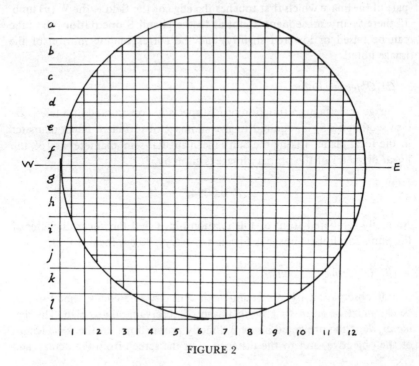

FIGURE 2

Pores can often be brought to visibility by gently tapping the telescope tube so that the image vibrates for a second or two.

Recommended size and reticulation of the projection blank: diameter 15 cm; horizontal and vertical diameters divided into 12 primary and 24 secondary parts, giving the smallest squares of the grid a side of 6.25 mm.

To facilitate the quick and correct identification of the positions of spots, the primary divisions may be lettered and numbered. It is essential that the grid be lightly and finely drawn (very sharp 2H pencil, e.g.), or small details such as pores will tend to be missed or even obscured. Some observers also mark in the grid diagonals. This certainly helps the correct delineation of the spot outlines, but has the disadvantage of cluttering up the blank and therefore tending to conceal small features.

Bristol board, or bromide photographic paper (unexposed, fixed, and washed) gives a good surface on which to draw the grid. Plaster of Paris, cast on plate glass and stripped off when set, has also been recommended as a projection surface.

(b) Laying down an accurately drawn and oriented outline of the umbra and penumbra of a spot or spot group, the detail to be filled in either by direct observation with a solar diagonal or from the projected image.

Unless the projected image is protected from daylight falling on it (see section 1.3) the finest detail of the HP image is better seen by direct vision than by projection. Drawing, however, is very much more difficult, owing to (a) alternating monocular and binocular vision, (b) discrepancy between the brightness of the alternately presented objects—solar image and drawing-board. If not even the outlines are laid down from the projected image, the orientation and scale of the drawing are also difficult to establish.

A suitable scale to work on is 45 cm to 60 cm to the solar diameter. A note of the diameter of the solar image must be made on every drawing which does not include the whole disc; this can be determined by comparing any convenient inter-spot distance with the same distance on the 15 cm projection.

The types of projection-screen holder suitable to the conditions of (a) and (b) above are different (see section 1.3), but the observational technique in the two cases is the same. A gridded circle, identical with that on to which the image is projected, but more heavily drawn, is pinned to a drawing-board beneath a sheet of tracing-paper. By means of the underlying grid, details of the image, seen projected upon the identical grid of the projection card, are transferred to the tracing-paper. The screen is observed, as much of the detail of a facula or spot as can be memorised is noted in accurate relation to the grid, and this portion of the image transferred to the drawing-board. The procedure is slow—though practice speeds it up to a surprising degree—since accuracy is the vital consideration.

The screen carrying the projection blank is oriented by allowing the Sun's image to trail across it: it is rotated about the optical axis until the limb or a spot (perferably small and well defined) trails one of the horizontal grid lines. The image is then centred by the Dec slow motion, when the horizontal diameter will be the EW median line, and the N and S points of the image will be the ends of the diameter perpendicular to it.

With an equatorial the image, having once been oriented, remains so, though both the orientation and the coincidence of the image with the projection disc should be checked periodically during observation. With an altazimuth the orientation must be readjusted at intervals whose length depends upon the magnification employed and the distance of the Sun from the meridian; this must be discovered by trial.

The date, UT, seeing scale number, E and W points, and heliographic coordinates of the centre of the disc (from *B.A.A.H.* or *A.A.*) must be added to every drawing, whether of the whole disc or of a single spot or spot group.

1.3 Projection apparatus

(a) For whole-disc observation with low magnification and small disc scale

The problem is to construct a frame for the support of the projection blank in a plane perpendicular to the optical axis, at such a distance (found by trial) from the highest-power ocular which includes the whole solar disc in the field as to give a solar image of the required size; the frame at the same time being rigid enough to maintain these adjustments and light enough to be attached to the drawtube without grossly upsetting the balance of the instrument and affecting its steadiness.

(*i*) The split hardwood collar *A* (Figure 3) is lined along its inner surfaces with felt, and clamped to the drawtube by means of the screws *S, S*. The image is oriented by rotating the drawtube. The bars *B, B* are clamped in a position parallel to the optical axis, *oa*, by the same tensioning screws. The framework *C*, which supports the projection card, is morticed and glued to the ends of these bars. Screen *D* excluded direct sunlight from the projected image; a circular screen with a central hole, to fit over the upper end of the telescope tube, achieves the same end and also helps to counteract the extra weight of the projection attachment. Adjustment of the size of the image is obtained by sliding the bars *B, B*, through the collar.

This design has the advantage that the distance of the screen from the

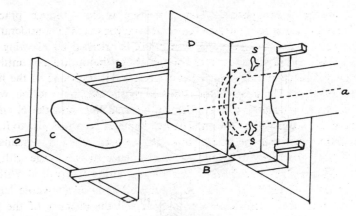

FIGURE 3

ocular is variable, and that the image can therefore be made to fill the 15 cm blank irrespective of the Sun's angular diameter.* It is, however, less steady than the design described below, and the accurate setting of plane of the projection perpendicular to the optical axis more difficult to obtain and maintain.

(*ii*) Sellers' open framework,† which was designed to be secured by a threaded eyepiece of the type used before push-in eyepiece fittings became common, can still be modified for clamping to the drawtube. The front end of the framework (Figure 4) consists of a 10 cm square aluminium sheet with a central hole just large enough for the threaded part of the ocular to pass through it, and a 12.5 cm square of plywood with a central hole just large enough to clear the drawtube collar; this is bolted to the aluminium sheet so that the two holes are concentric. Then when the collar is slipped through the hole in the aluminium rectangle and screwed into the drawtube, the combined aluminium and plywood sheet is clamped firmly between the ocular and the collar of the drawtube.

The framework is built up of aluminium slats on this forward end, as illustrated. Dimensions are omitted, since they depend on focal lengths of objective and ocular, and the correct distance of the projection screen from the ocular must be found by trial. The cross-pieces of the frame, supporting the projection card, are also omitted from the diagram. Orientation of the image is obtained by rotating the drawtube. Finally, the top,

* The Sun's angular diameter varies from 31'30".8 to 32'35".2 during the course of the year. At mean distance its diameter is 32'2".36.
† B. 3.89 (reference to Bibliography, section 20).

7

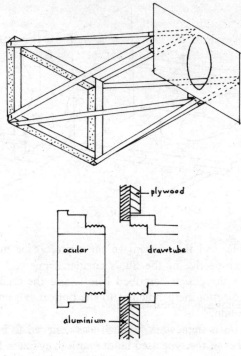

FIGURE 4

bottom, and one side of the framework are covered with black paper or cloth.

(iii) Many small (60–75 mm) refractors come complete with a single-arm, fully-adjustable projection screen, often designed to be used in combination with a totally reflecting diagonal. This can give satisfactory views of the whole disc if fitted with a light shade, but neither they nor the instruments themselves may be robust enough for serious positional work.

(b) For observation of small areas of the solar disc, with high magnifications and large disc scale

Owing to the dilution of the image with increased magnification, contrast will be reduced below the useful threshold unless daylight, as well as direct sunlight, is excluded from the image. The grid is therefore enclosed, and constitutes the rear wall of a nearly light-tight box.

To obtain the required distance from screen to ocular: let m' be the magnification of the ocular used for the projection of the 15 cm image; let m be the magnification of the highest-power ocular available for the large-scale projection; let d' be the distance of the screen in the 15 cm projection; then d, the required distance, is given by

$$d = \frac{nd'm'}{m}$$

where $n = 3$ for an image scale of 45 cm to the solar diameter,

$n = 4$ for an image scale of 60 cm to the solar diameter.

Make a plywood box of length d, and height and width equal to about one-quarter or one-third of the solar diameter—say 15 cm. The front wall of the box, by which it is attached to the telescope, is constructed as in para (a) (ii) above. The rear half of one side of the box is omitted, to allow examination of the image; the interior of the box should be given a matt black finish.

1.4 Determination of spot positions

Although accurate spot positions are derived by a number of professional observatories, with a precision far exceeding that attainable by any normally-equipped amateur, the absolute positions of the spots on the Sun's surface and their proper motions, particularly in longitude, form an interesting study. For this work, measures of whole-disc photographs made on a large scale, or the use of some kind of micrometer (see below), will be found necessary.

The heliographic data that are required are P, B_o and L_o (see sections 1.20, 1.21).

1.4.1 Position angle and distance from the centre of the disc by direct observation: Simple and convenient, though not of a high order of accuracy. It requires an equatorial driven at the solar rate for ease of manipulation.

Project the image on to a screen on which a grid of squares of convenient size has previously been drawn, and determine the scale by timing the drift of a small, well-defined spot over, say, ten squares. If, then, the time taken is t seconds, and the Dec of the Sun's centre is δ, the length of the side of one square corresponds to

$$\frac{t}{40 \cos \delta}' \text{ arc.}$$

9

Orient the grid so that a spot, or the solar limb, trails one of the sets of lines, and then set the solar image so that its centre coincides with the intersection of two grid lines. Count the number of squares, and estimate to 0.2 or 0.1 of a square, NS and EW from the centre of the grid to the spot. Suppose it lies x squares from the central EW web and y squares from the central NS web, and that the predetermined angular length of the side of each square is z. Then if θ' is the angle between the NS web and the line joining the centre of the spot and the centre of the disc, and p' the apparent angular distance of the spot from the centre of the disc,

$$\tan \theta' = \frac{y}{x}$$

$$p' = zx.\sec \theta'$$

and the position angle (always measured from the N point eastward) of the spot is:

$$\theta' \quad \text{if the spot lies in the N} f \text{ quadrant,}$$
$$180-\theta' \quad ''\quad ''\quad \text{S} f \quad ''$$
$$180+\theta' \quad ''\quad ''\quad \text{S} p \quad ''$$
$$360-\theta' \quad ''\quad ''\quad \text{N} p \quad ''$$

By applying P to θ' we obtain θ, the position angle of the spot relative, not to the N point, but to the Sun's central meridian.

1.4.2 Position angle and distance from the disc centre by projection: Project the solar image on to a 15 cm grid. Orient the latter so that the horizontal diameter coincides with the EW direction. Then θ', p' and θ can be derived as explained in section 1.4.1.

1.4.3 Heliographic coordinates: Stonyhurst discs: During the course of the year the value of B_0 varies from approximately $+7°.2$ to $-7°.2$. For many years, Stonyhurst discs showing the heliographic appearance of the whole solar disc for the eight whole-degree values of B_0 from $0°$ to $\pm7°$ were used by B.A.A. observers. Although these are no longer obtainable, an account of their construction appears in B.3.67; they are similar to those used at Zurich and other observatories for the direct derivation of sunspot positions, and the following description of their use may be of value.

Apply the drawing, with its marked EW diameter, to the Stonyhurst disc whose value of B_0 is nearest to the value at the time of observation. By means of the $0°-30°$ protractors at either side of the disc, orient the drawing to the correct value of P: it follows from the rule of signs for P that if it is positive,

E end of horizontal diameter should be $P°$ above the zero mark on the protractor, if P is negative,

E end of the horizontal diameter should be $P°$ below the zero mark on the protractor.

Read off from the disc the position of the spot E or W of the central meridian, and N and S of the equator. Call the former l and the latter b. l is positive if the spot is W of the meridian, negative if it is E; b is positive if the spot is N of the equator, negative if it is S.

The reduction then falls into two stages:

(*i*) Correction of the measured heliographic latitude for the difference between the latitude of the centre of the disc for which the Stonyhurst disc is constructed (B_d) and that at the time of observation (B_o).

(*ii*) Derivation of L, the spot's heliographic longitude, from the measured quantity l.

(*i*) If we write B for the heliographic latitude of the spot, then at the limb

$$B = b$$

at the centre of the disc,

$$B = b \pm (B_o - B_d)$$

and, generally, at a distance l E or W of the central meridian,

$$B = b \pm [\cos l(B_o - B_d)]$$

where the rule of signs is:

if $B_o > B_d$: b positive, the sign of the last term is +,
b negative, the sign of the last term is −,
if $B_d > B_o$: b positive, the sign of the last term is −,
b negative, the sign of the last term is +.

(*ii*) The value of L_o decreases at an average rate of $13°.2$ per 24 hours. Hence to obtain the value of L_o at the time of observation, a certain quantity must be subtracted from L_o at the preceding midnight (tabulated in *A.A.* or *B.A.A.H.* at 4-day intervals) or added to the value of L_o at the following midnight. The Table at the top of page 12 will facilitate this correction.

11

Interval	Change in L_0	Interval	Change in L_0
1^m	$0°01$	50^m	$0°46$
2	0.02	55	0.50
3	0.03	1^h	0.55
4	0.04	2	1.10
5	0.05	3	1.65
10	0.09	4	2.20
15	0.14	5	2.75
20	0.18	6	3.30
25	0.23	7	3.85
30	0.28	8	4.40
35	0.32	9	4.95
40	0.37	10	5.50
45	0.41		

Having obtained the corrected value of L_0, L is derived from

$$L = L_0 \pm l$$

Stonyhurst discs are useful when referring to a particular spot in a written communication, since it is sufficient for its identification to read off the coordinates from the 'nearest' disc without any reduction for date.

1.4.4 Heliographic coordinates: Ravenstone discs: Similar in function to the Stonyhurst discs, their use is summarised in B.3.27.

1.4.5 Heliographic coordinates: Porter's disc: The spot's coordinates can be quickly and easily deduced from rectangular coordinates measured from a single Porter disc, which fulfils the function of the set of 8 Stonyhurst discs. Copies of this disc, which is also available in transparent overlay form, together with instructions, are available to members of the B.A.A.;

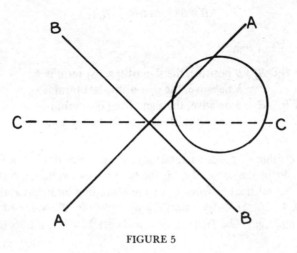

FIGURE 5

or the amateur can make his own copies from the prototype in B.3.83, and use them according to the instructions given under that reference. The disc is 15 cm in diameter, and the derived coordinates are correct to within a degree or two. The Porter disc undoubtedly provides the most convenient, if not the most accurate, method of deriving the heliographic coordinates of spots.

Two errata in the above-mentioned paper should be noted: (*a*) p. 63, lines 3 and 4 from foot, transpose the words *clockwise* and *anticlockwise*, (*b*) p. 65, line 4 of *Example*, for *anticlockwise* read *clockwise*.

1.4.6 Heliographic coordinates: polar coordinate method: This convenient method involves the measurement of θ, the spot's position angle, and r/R, its distance from the disc centre expressed as a decimal of the radius of the disc:

Derive an angle α, such that $\sin \alpha = r/R$.

Derive an angle N, such that $\tan N = \tan \alpha \cos (P - \theta)$, P, as before, being taken from *A.A.* or *B.A.A.H.*

If L = heliographic longitude of the spot,

   B = heliographic latitude of the spot,

   L_0 = heliographic longitude of the centre of the disc,

   B_0 = heliographic latitude of the centre of the disc,

Then $\tan (L - L_0) = \tan (P - \theta) \sec (B_0 + N) \sin N$,

$$\tan B = \tan (B_0 + N) \cos (L - L_0).$$

This method has the advantage that it can be applied to any correctly-orientated drawing, no apparatus beyond a ruler and protractor being needed.

1.4.7 Heliographic coordinates: Carrington's method: While this method gives positions of a high order of accuracy, it may be doubted whether the laborious reduction is justified by the type of observations that the average amateur can make. However, since the method has never been described in any readily accessible general text,* details of it are given below.

The method can be operated either by cross-wire micrometer or by a projection card, marked with two mutually perpendicular lines A and B, which are intersected by a third, fainter line C. Line C is adjusted accurately E-W by trailing a small spot along it. If a card is used, the adjustments to which attention must be paid are: cross-lines exactly perpendicular to one another, plane of the projection card perpendicular to the optical axis, and intersection of the lines located on the optical axis.

---

* It is described by Maunder in B.3.63, 3.64.

The image is allowed to trail across the micrometer webs or the cross on the projection card, such that its centre passes slightly above or below the intersection of the cross (Figure 5). Thus transits of the limb at the two webs will not be simultaneous. For the projection method it is convenient to use a 20 cm square of millimetre graph paper whose diagonals are clearly drawn in.

The observations consist of the timing of the transits of the E and W limbs and of the spot at each of the webs, A and B, to the nearest 0.1 sec. If the observer is working on his own, an audible signal such as that produced by a reliable mechanical or (preferably) electronic metronome, in conjunction with a tape-recorder, is probably the most satisfactory arrangement. Since the time intervals are arbitrary, it is unnecessary to work in MT seconds unless these happen to be the most convenient source of signals. The mean should be taken of at least three sets of transits. An example of a convenient layout for the record in the case of MT work is given below:

	Diagonal	Times of:			Mean time of transit	Interval between spot and disc centre	
		1st transit h m +	2nd transit h m +	3rd transit h m +			
		s	s	s		Over A	Over B
p limb	A				A_p		
	B				B_p		
spot	A				a	$a - A_c$	$b - B_c$
	B				b		
f limb	A				A_f		
	B				B_f		
solar centre	A				A_c		
	B				B_c		

A_c and B_c are derived from the transit times of the limbs:

$$A_c = \frac{A_p + A_f}{2}$$

$$B_c = \frac{B_p + B_f}{2}$$

The reduction falls into three stages:

(*i*) Derivation of position angle and distance from the centre of the disc, from the observational data.

(*ii*) Conversion of position angle and distance to heliographic co-ordinates.

(*iii*) Calculation of the area of the spot, if required (see section 1.5).

(*i*) Let θ = angle subtended at the centre of the disc by the line joining the spot and the N end of the solar axis,

P = position angle of the N end of the solar axis,

i = inclination of the Sun's path to the parallel of Declination passing through the centre of its disc.

P is taken from *A.A.* or *B.A.A.H.*; i from the Table on p. 16, θ is derived as follows:

$$\theta = \alpha + (\epsilon \pm i - P)$$

where

$$\tan \epsilon = \frac{A_f - A_p}{B_f - B_p}$$

$$\tan \alpha = \frac{a - A_c}{b - B_c} \; \frac{1}{\tan \epsilon}$$

The true (heliocentric) angular distance p of the spot from the centre of the disc is taken from the Table on p. 16 using the value of r/R (the distance expressed as a decimal of R, the solar radius) given by

$$\frac{r}{R} = 2 \sec \alpha \; \frac{b - B_c}{B_f - B_p}$$

(*ii*) Let B = heliographic latitude of the spot,

B_o = heliographic latitude of the centre of the disc.

Then
$$\sin B = \cos \theta \cos B_o \sin \rho + \sin B_o \cos \rho$$

Let L = heliographic longitude of the spot,

L_o = heliographic longitude of the centre of the disc,

l = longitude of the spot measured from the central meridian.

Then
$$\sin l = \sin \theta \sin \rho \sec B$$

and
$$L = L_o - l$$

15

Inclination of the solar path to the Dec circle passing through the centre of its disc:

Date		i	Date	
Feb	25	$+0°.06$	Apr	10
	5	$+0.05$	May	2
Jan	24	$+0.04$		16
	14	$+0.03$		27
	6	$+0.02$	Jun	5
Dec	30	$+0.01$		14
	21	0.00		22
	15	-0.01		30
	7	-0.02	Jul	9
Nov	29	-0.03		19
	19	-0.04		30
	7	-0.05	Aug	13
Oct	13	-0.06	Sep	8

Relationship between r/R and ρ:

r/R	ρ	r/R	ρ	r/R	ρ
0.100	$5°.7$	0.730	$46°.7$	0.910	$65°.3$
0.200	11.5	0.740	47.5	0.920	66.7
0.300	17.4	0.750	48.4	0.930	68.2
0.350	20.4	0.760	49.3	0.940	69.8
0.400	23.5	0.770	50.2	0.950	71.6
0.450	26.6	0.780	51.1	0.960	73.5
0.500	29.9	0.790	52.0	0.970	75.7
0.525	31.5	0.800	52.9	0.980	78.3
0.550	33.2	0.810	53.9	0.990	81.6
0.575	34.9	0.820	54.9	0.991	82.0
0.600	36.7	0.830	55.9	0.992	82.5
0.620	38.1	0.840	56.9	0.993	82.9
0.640	39.6	0.850	58.0	0.994	83.5
0.660	41.1	0.860	59.1	0.995	84.1
0.680	42.9	0.870	60.2	0.996	84.6
0.700	44.3	0.880	61.4	0.997	85.3
0.710	45.1	0.890	62.6	0.998	86.1
0.720	45.9	0.900	63.9	0.999	87.2

The amount of work involved in making and reducing the observations would not justify its application to more than the occasional interesting spot, such as one appearing in an exceptional latitude, or one suspected of exhibiting abnormal proper motion. When using an altazimuth instrument, the position of line C must be re-checked before each observing run is made, and this precaution might well be taken even when using an equatorial telescope.

1.5 Determination of spot areas

A graphical method of determining the areas of spots, whose accuracy is within the limits imposed by the projection method of recording the spot, is described by Sellers (B.3.88).

It is based upon the fact that the area of a square on the surface of a sphere, whose side subtends $1°$ at the sphere's centre, is very nearly 49 millionths of the surface of the hemisphere. If each side of such a square is divided into 7 parts, and the square further subdivided, each secondary square will have an area of one-millionth of the area of the hemisphere.

On a scale of 60 cm to the diameter, an unforeshortened $1°$ square will have a side of 5.25 mm; the side of each secondary square will then be 0.75 mm, the drawing of which is quite practicable with a very sharp 2H pencil.

The procedure is to project the Sun's image on to a gridded circle 60 cm in diameter. Copy the outline of the spots, with the greatest accuracy possible, on to a similar gridded circle. On this superimpose a grid of 5.25 mm primary and 0.75 mm secondary divisions—it need only be 5 cm or so across—drawn on tracing-paper. Count the secondary squares that are more than half covered by the umbra and penumbra (ignoring the remainder). This number will then be the area of the spot expressed in millionths of the solar hemisphere, uncorrected for foreshortening (ϕ).

If, for ease of drawing, a $2°$ -square grid is constructed (primary divisions 10.5 mm, secondary 1.5 mm), then ϕ will be four times the number of secondary squares. A still more time-saving simplification is to use 1-mm graph paper, in which case the number of squares counted must be multiplied by 1.72 to give the area in millionths of the Sun's hemisphere. In this case it would be the drawing of the spot that is made on a transparency.

To the area so derived a correction for foreshortening must be applied, unless the spot is at the centre of the disc. If the number of secondary squares of a $1°$ grid be ϕ, and the distance of the spot from the centre of the disc be r, the semidiameter of the disc being R, then the area corrected for foreshortening is given by

$$\Phi = \phi \sec \theta$$

where θ is an angle such that $\sin \theta = \dfrac{r}{R}$

17

Alternatively, Φ may be derived from direct measurement of the disc, using either the above expression and measuring θ, or

$$\Phi = \phi \cdot \frac{R}{x}$$

and measuring R and x. θ, R and x are the quantities shown in Figure 6.

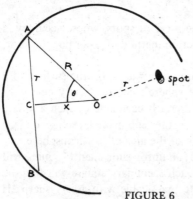

Given that the distance of the spot from the disc centre is r, from any point A on the circumference draw a a chord AB, of length $2r$. The distance from O to the midpoint, C, of this chord is x, and the angle subtended at O by the semichord AC is θ.

The precision of the method does not justify taking θ to more than the first decimal place. Since the inaccuracy increases very rapidly with sec θ, the area of a spot cannot be derived with sufficient accuracy to make the measurement worthwhile if $\theta > 60°$ approximately, except for provisional estimates of spots newly appeared on the E limb.

FIGURE 6

To save playing with protractors and compasses, the value of sec θ may be taken from the Table below, where it is tabulated against r (expressed in mm, for a 15 cm disc). Or these values of r may be used to construct concentric circles within a 15 cm disc on tracing-paper; when this is laid over the disc drawing, the appropriate value of sec θ for the spot may be read off without further calculation or geometrical work. The B.A.A. Solar Section has produced a transparent graticule for area measurement, designed to fit an image 15 cm in diameter.

r	sec θ	r	sec θ
10.5	1.01	45.0	1.25
14.8	1.02	48.0	1.30
18.0	1.03	50.5	1.35
20.5	1.04	52.5	1.40
22.8	1.05	54.2	1.45
24.8	1.06	56.0	1.50
26.8	1.07	58.5	1.60
28.2	1.08	60.8	1.70
29.8	1.09	62.5	1.80
31.2	1.10	63.8	1.90
37.0	1.15	65.0	2.00
41.5	1.20		

1.6 Direct visual observation

Adequate protection for the eye is the first consideration. The required intensity of the emergent pencil being in the region of 0.1% to 0.01% that of the incident radiation, reduction of aperture is useless; it is in any case undesirable owing to its reduction of the objective's resolving power. Available methods, which may be used singly or in combination, are (a) absorption, (b) fractional reflection at uncoated surfaces, (c) polarisation. When the image is viewed by projection, it will be remembered, no reduction of the incident intensity is required.

1.6.1 Absorption: Numerous materials have from time to time been recommended for rendering a telescope safe for direct solar observation. It is important to recognise that both the visual and thermal components of solar radiation are dangerous, and any effective absorbing filter must be capable of making a drastic reduction in both. The old-fashioned tinted-glass suncap is unsuitable and dangerous, because it is located at the worst possible place in the optical path, just at the exit pupil, where the combined radiation collected by the objective is concentrated. The intensity of the radiation at this point, compared with the direct solar radiation, corresponds approximately to M^2 for low and moderate magnifications, and will never be less than about 100 times normal, so that the filter will become fiercely hot and may even crack or melt. The correct place for a solar filter is in front of the objective, where it is not only free from concentrated heat but also excludes heat radiation from the interior of the tube.

When considering the effectiveness of any filter material, it should not be forgotten that blue and near-ultraviolet light can be harmful in only moderate overdoses; in fact, photochemical deterioration of the retina has been produced by blue light several hundred times less intense than infra-red radiation that caused equal, though different, damage.

Under normal conditions, a reduction of intensity of from 1,000 to 10,000 times (0.1–0.01%) is satisfactory, and it is convenient to reduce the intensity of the image to that acceptable under the dimmest feasible conditions of illumination, taking care of the last factor of X10 or so by using either light shade glasses or crossed polaroid filters in the eyepiece. If the objective is small (up to 75 mm D, say), a thin neutral photographic filter of density about 3.5 may be fitted into a removable cap and so placed over the objective. Totally fogged and developed photographic film has the advantage of absorbing heat radiation particularly well, but it cannot be recommended without precaution because of large variations from sample to sample. Mounted in removable caps, however, it is optically satisfactory for use with binoculars and very small hand telescopes.

19

A generally highly satisfactory material, which appears to be both safe and optically unimpeachable, is very thin aluminised mylar (plastic). It is available in sheets that can cover any aperture; a well-known example is sold under the trade name of Solar-Skreen. Although some observers have expressed reservations about the use of any filter whatsoever for solar observing, preferring to observe always by projection, it is fair to say that, during the past decade or so, thousands of observers have found it a cheap and convenient way of converting a telescope for solar use.

If D exceeds about 25 cm or so, some benefit may be gained by reducing the aperture with a stop, since conditions will rarely allow the resolving limit to be satisfied, or even approached. An eccentric stop of, say, 10 cm on a Newtonian of this size gives the benefit of an unobstructed light-path.

1.6.2 Fractional reflection: At every uncoated air/glass surface, the intensity of the reflected pencil is reduced to about 5% that of the incident. Hence, a single such reflection (e.g. solar diagonal) will have to be supplemented by a neutral filter of density about 2.0 to 2.5, or by further reduction of intensity by polarisation (see Table below). Two reflections will require a light filter (density in the region of 1.0 to 1.5), unless the Sun is already partially obscured by mist or cloud, when even a second uncoated reflection may reduce the brightness of the image too much.

Solar eyepieces: percentage transmissions

		1st unsilvered reflection	2nd unsilvered reflection	3rd unsilvered reflection
		5.0	0.25	0.0125
Neutral filter Density = 1.2	6.3	0.32	0.016	0.0008
Neutral filter Density = 2.5	0.32	0.016	0.0008	0.00004

Three reflections will bring the intensity of the emergent pencil within the required range of 0.1–0.01%, and only a narrow margin of reduction by polarisation need be allowed, for conditions of maximum solar brilliancy. More than three uncoated reflections will reduce the intensity below the required level under all conditions of observation. It should be remembered that an odd number of reflections reverses the image in one plane (L and R, or up or down), while an even number of reflections leaves the image as it is presented to the first reflecting surface.

Most solar diagonals employ a right-angled prism, arranged for external reflection at the hypotenuse, so that 95% of the radiation either passes through or is absorbed by the glass. Alternatively, the Herschel wedge, which is a first-surface reflecting prism of small angle (avoiding the undesirable second-surface reflections that would be returned to the field of view if a piece of parallel-faced glass were used), has been recommended. However, these and other more elaborate devices have been almost entirely superseded by the objective filters referred to above, and reference to them is mainly of historical interest.

A reflecting telescope designed exclusively for solar work can have an uncoated primary mirror, thus achieving the same effect as a solar diagonal. If an uncoated flat or an externally-reflecting prism is used in conjunction with such a mirror, little further reduction will be needed, and in the case of a partially-obscured Sun the reduction may already be excessive. In this case a coated diagonal is used in conjunction with an eyepiece filter. Uncoated flats must be of the wedge variety, a plane-parallel mirror giving trouble from secondary reflections in the field. On the whole, however, in view of the harm done to the definition by thermal air currents in the light path, the exclusion of as much heat (and dust) as possible by a full-aperture mylar filter would seem to be the best arrangement.

1.7 Instrumental: different heating, etc

The thermal inequalities that can ruin definition in solar work occur within the mirror itself (in the case of a reflector)—and also in the flat—within the remainder of the instrument (particularly in the column of air enclosed by the tube of a reflector or refractor), and within the immediate surroundings of the instrument.

Thermal motions within the tube can be minimised, as stated above, by filtering out most of the solar radiation before it enters the instrument, but problems may arise, particularly with metal-tubed refractors, from direct heating of the tube wall. An instrument left standing in the sunshine before observation commences will certainly not be in thermal equilibrium; therefore it should be shaded before use. During observation, a light screen fitted near the top of the tube will help to keep the instrument in shadow. Poor seeing is often experienced when a portable refractor is moved from a cool place, such as a shaded garage or shed, into the outside air. This appears to be due to stratification of the air within the tube, which is disturbed when the instrument is moved and produces a temporary condition of tube currents. Because of this, the telescope should be left in its new attitude, pointing at the Sun, for a few minutes before a final judgement of the seeing conditions is made. An ill-defined, woolly appearance of the sunspots is almost certainly due to tube currents, assuming of course that the normal performance of the instrument is satisfactory. The performance of a reflector, assuming that its optical components must be exposed to direct solar radiation, can be improved by blacking the interior of the tube; by using an open or lattice tube, or, if solid, providing it with numerous ventilation holes; and by using a tube of considerably larger diameter than that of the mirror, since air currents in the tube tend to stick to the outer parts.* Polishing the second surface of an uncoated mirror will also improve definition by reducing scattered light.

Refractors, though they suffer less than reflectors from unequal heating of the objective, are more subject to thermal currents in the tube.

Shading the observing site is an important factor in the improvement of solar definition. Movable canvas screens can be used to prevent direct sunlight striking the mounting—and the tube itself, when not in use—and the immediate surroundings of the instrument. If it is mounted in an observing hut, a large degree of thermal insulation during the daytime can be obtained by spreading a false roof of canvas, stretched on a light wooden frame, 30—40 cm above the true roof; the space between the two must be left open so that the air can circulate freely. All this has a direct bearing on the desirability of setting up a properly arranged solar observatory (see section 1.15).

The solar observer tends to be troubled much more than does the night-time astronomer by local or 'ground' seeing. After sunset the ground, though normally warmer than the air, is all the time cooling towards air temperature; during a sunny day, on the other hand, ground temperature is rising fast and the difference keeps on increasing, certainly until late afternoon. Not surprisingly, therefore, it will often be found that the best definition occurs fairly early in the morning—despite the greater thickness of atmosphere through which the Sun is then observed—and also, rather surprisingly (unless it is only then that an improperly shaded instrument settles down to thermal equilibrium), towards sunset.

This 'ground' seeing is surprisingly low-lying, and raising a telescope a few metres above the surrounding terrain can make a considerable improvement to image steadiness—hence the tower-telescope construction commonly employed in professional solar instruments. An alternative precaution, employed in the Big Bear Lake solar observatory in California, is to surround the telescope with water. Grassed ground is far superior in this respect to a heat absorber, and hence re-radiator, such as concrete.

* For a fuller discussion of tube currents, see *A.A.H.*, section 11.2.

1.8 Telescopic work

Although there is always the chance of observing an unusually brilliant flare (B.3.68), or of seeing exceptionally rapid sunspot development over a few hours, it must be admitted that white-light observation of the Sun is duplicating a field already very well covered by professional observatories. The following lines of research are partly observational and partly statistical, and the proposed 'Maunder minimum' of the late eighteenth century, to say nothing of recent investigations into possible secular changes in the solar diameter, indicate what may be achieved by re-analysis of existing records. Most periodical observation is confined to item 1 below, which does not preclude attention to items 3, 4 and 5 if circumstances warrant.

1. Mapping the spots daily (see section 1.4).

2. Drawing the detailed structure of spots. Attention should particularly be given to active spots and groups. Serial drawings of such spots, with sufficient aperture and magnification and on a large enough scale to show the fine detail, and made at intervals determined by the rate at which visible development is occurring, are of increasing value nowadays;* more and more spectrohelioscopes and spectroheliographs are coming into operation, and such drawings offer the possibility of correlating phenomena in mono-chromatic light with events visible in integrated light. The possibility of a repetition of the almost unique† Carrington-Hodgson observation of 1859 should also be kept in mind; the likeliest regions for a recurrence are complex spots at the height of their activity, and the remnants of great spots seen on a subsequent solar rotation.

Detail not otherwise visible is sometimes revealed by an extreme con-traction of the field, with its consequent reduction of photospheric background and concentration of the observer's attention. This is most conveniently achieved by means of a Dawes diaphragm in the focal plane of a positive ocular.

Accuracy—a matter of patience and experience more than of skill as a draughtsman—is essential; time spent making casual sketches is time wasted. *All* drawings, of this or any other type, must be labelled with the UT to the nearest minute—if necessary, the UT of the start as well as the completion of the drawing.

3. Besides the normal faculae there are two types that require special attention: (*a*) 'Dark faculae': vague, smudge-like markings, darker than the photosphere, about which very little is known and of which few obser-

* See also section 1.9.
† For references to other observations of flares in integrated light, see B.3.68,3.107.

vations have been recorded. They characterise the equatorial zone and are most readily seen near the centre of the disc; often contain a pore. Require moderate apertures (visible with 15 cm, but 20 cm is better). There is need for a considerable mass of observation of these objects, giving positions, appearance, and date and time of observation. (*b*) Polar faculae: occur within about 25° of the poles; fainter and smaller than the normal equatorial faculae; probably most numerous around spot minimum, but very little is known about them still. Best seen in large-scale projected images; recommended method of recording—dotted lines on a 15 cm disc. (See B.3.13).

4. Ill-defined spots. Occasionally spots appear which are unaccountably vague and ill-defined; that this is not an instrumental or atmospheric effect is shown by the fact that they have been seen simultaneously with spots of normal appearance. Visible with $75-100$ mm D. Record position, date, and time of observation.

5. Pores. Charting might give results of value, if continued over a long enough period. Pores should always be watched since they are often the precursors of spots, and more information is wanted about the earliest stages of spot formation. Pores appear to be formed by the increased separation of the solar granulation ('rice grain'); the normal separation of individual granules—which can be seen satisfactorily in the projected image provided extraneous light is excluded—is about 2″. Gently tapping the projection frame, so that the image vibrates slightly, often reveals small pores that would otherwise be missed. Very precise focusing is also essential.

6. Scope for more work on the longitude distribution of spots (see e.g. B.3.4), the correlation of longitude with other factors (such as size, longevity, type, the spot cycle, etc), and on the recurrence of spots in the same region of the solar surface. An example of the latter was provided by the great spot groups of February and July 1946; if the drift of the former is extrapolated to July it is found that the latter occurred in almost exactly the same position (latitude difference, 2°). It must be remembered that owing to the distribution of rotational velocity in zones, a given longitude loses its identity after 8 or 10 rotations.

7. Comparatively little study has been made of the laws governing the rate of growth and decay of spots, and of the correlations (if any) between this and factors such as spot type, latitude, spectrohelioscopic character, phase of the spot cycle, etc.

8. Scope for work on the rotation of spots. Pseudo-rotational effects which must be distinguished from true rotation spring from (*inter alia*) normal changes in the shape of the spot, changing orientation of the spot relative to the NS line (geocentric) during its passage across the disc, and

the common tendency of the leader in a pair, whose axis is inclined to the equator, to gain in longitude on the trailer.

As examples of what the amateur can do without even putting his eye to the telescope may be mentioned Dr Alexander's work on the longitude distribution of spots (B.3.4), including his discovery of a 400-day subcycle; and his analysis of the Greenwich photoheliographic results, throwing new light on the character of the 11-year cycle (B.18).

1.9 Photography

Uniquely among celestial objects, the Sun offers the photographer a super-abundance of light. The advantages that follow from this fact are notable: short exposures, no guiding, slow fine-grain emulsions, the opportunity to select moments of better-than-average seeing, and size of aperture only of importance in so far as resolution is concerned. The Sun alone offers scope for photography with a small altazimuth.

With most amateur telescopes, the prime-focal image is needlessly small and can be enlarged with a Barlow lens until it just fills a 35 mm frame (image diameter, about 20 mm), when quite small spots can be discerned if very fine-grain film is used. A reflector is more suitable than a refractor for this work, since the image is virtually colourless; to remove the bluish haze from a refractor's image, a yellow filter is required, which is particularly undesirable if colour film is being used. Mylar-type full-aperture filters are, or should be, nearly colourless, although with a tendency to suppress the red end of the spectrum and enhance the blue, so that the use of a light yellow filter may help to restore the colour balance.

Users of 35 mm black-and-white film who do their own processing will find materials such as Kodak Fine Grain Positive film, a blue-sensitive emulsion used for making transparencies from negatives, the most suitable for solar work. If cut film is used, Kodak Gravure Positive Film 4135 has similar characteristics. Both these films, being of relatively low sensitivity, will prove suitable when used with a lightly-filtered telescope. However, if the telescope is heavily filtered to visual standards, a faster emulsion will be required if exposure times are to be kept in the desired region of between 1/1000 and 1/100 second. The very fine photographs of F. Rouvière have been taken with a 20 cm Newtonian on Kodak Recordak Microfilm 5786, using a full-aperture aluminised filter passing 0.2% of the incident light, a Schott VG6 green filter, and exposures of between 1/125 and 1/500 second with the primary image enlarged to a solar diameter of about 13 cm (B.3.85). This particular film is panchromatic.

If the object is the detailed recording of a part of the solar surface, as against a whole-disc record, the observer needs to be sure that his image is (a) as steady as conditions will allow, and (b) capable of transferring all its details to the photographic emulsion. Failure to monitor the seeing conditions accounts for a very high percentage of rejected photographs. Normally, daytime seeing is relatively unstable, particularly near the ground and inside the tube, and although it is possible to take a series of exposures in the hope that some will coincide with an instant of good seeing, this is a wasteful process. The use of a single-lens reflex camera with a clear focusing screen makes it possible to expose only during moments of steadier seeing, as judged, typically, by the visibility of detail inside a sunspot. An auxiliary telescope, if used as a guiding instrument to direct the photographic telescope, can also be used as a seeing monitor, but will be less satisfactory than a direct-viewing arrangement because its own thermal characteristics may not match those of the main telescope. B.3.16 describes a somewhat elaborate automatic seeing monitor for a solar telescope, in which the shutter is triggered when the definition of the solar limb achieves a pre-set quality.

To ensure (b) requires some knowledge of the film being used, and this can be gained only by experience, since different processing techniques can effect the result. It might appear straightforward to match the film's quoted resolving power (in lines/mm) with that of the telescope objective, and to use an enlargement which makes the scale of optical resolution in the final focal plane coincide with or slightly exceed this value. But it must be remembered that the resolving power of a film is usually determined by tests on high-contrast images, whereas the fine detail on the solar surface (granulation, or the filamentary features of sunspot penumbrae) have relatively low contrast. This effectively lowers the resolving power of the film, and to record these objects satisfactorily means employing a larger image scale than theory might indicate. With such a superabundance of light as the Sun offers, this is not, of course, the hardship that it is in lunar and planetary work, and a factor of 6 or more between emulsion and telescopic resolution can and should be used. Taking, for example, a film whose laboratory resolving power is 200 lines/mm (typical for very slow emulsions), a factor of 6 would require the telescope's resolving power to be represented by two dots or lines .033 mm apart. Using this image scale, any loss of resolution must be attributed to factors other than the film itself. Whole-disc diameters corresponding to this order of resolution scale would be, for example:

Aperture (mm)	Solar disc (mm)
60	34
75	42
100	56
150	84
200	113

Because of the great range of brightness from photosphere to umbra, it is normally impossible to bring out the details both of the granulation and of the sunspots on a single print. If the negative is correctly exposed for the photosphere, the spots will be under-exposed (and vice versa, the photosphere will be over-exposed). A hard printing paper may be needed to reveal the granulation well, whereas the more contrasty details of a sunspot may respond better to a softer paper.

It will be clear from this discussion that serious solar photography demands an instrument specially designed for the purpose. A refractor, contrary to much popular opinion, has no particular advantage over a Newtonian reflector, provided that the latter's tube is sealed with a full-aperture (or eccentric-stop) filter. If the tube is solid, it should be made of a poor heat conductor and painted white; a jacket of the fibreglass material used for loft insulation may prove worthwhile. Exposures being instantaneous, an altazimuth mounting will work perfectly well unless the image requires to be orientated, although accurate orientation is not normally necessary in the case of close-up views of the surface.

Solar photography, as all other methods of observation, is much facilitated if a long-focus horizontal Sun telescope is available. The resolution of detail 1″ in diameter requires an aperture of only 125 mm. A long-focus Sun telescope with an aperture of 125–150 mm (as described in section 1.15) could therefore be extremely usefully employed. Such a layout permits the use of a much heavier camera than could possibly be attached to the telescope, it enormously simplifies the photography of the projected image, and it allows any type of shutter to be used under conditions where no vibration can be transmitted to the plate or image. Owing to the shortness of the exposures, a siderostat is not required (though anyone having gone to the trouble of laying out a solar observatory would certainly have installed one for visual, spectrohelioscopic, or other purposes), a silvered or aluminised flat mounted on a polar axis with Dec adjustment and RA slow-motion controls to the observing position being sufficient. Provision must be made for varying the position of the film along the optical axis —over a range (increase) of about $0.001F$—as the objective warms up. However, unless the aim is to record the whole disc on a fairly large scale,

27

using cut film (a rather expensive enterprise), it may be found that a horizontal solar telescope suffers seriously from ground seeing, due to the long light path between the plane mirrors and the image. Because of their relative freedom from distortion, compared with that to be expected in an eyepiece-magnified image, long-focus photographs are much more suitable for sunspot positional measurements.

It has already been emphasised several times that moments of minimum turbulence must be selected for the exposure of solar plates. For the recording of detail this often demands a lot of patience, and on some days is impossible of realisation. On the other hand, when turbulence is marked under magnification, and the recording of detail is impossible, fairly satisfactory whole-disc photographs may frequently be obtained. Turbulence, especially during the summer, is often comparatively slight when the Sun is low. Altitude, however, is a factor to be considered, since the Sun's actinic strength varies widely according to the length of the light-path in the atmosphere and the latter's water-vapour content. This is a less important consideration for the user of a visually corrected refractor than of a photovisual objective or reflector. A low Sun, or the presence of mist, must be countered either by longer exposures or, better, by a faster emulsion or a less dense filter. In so far as any general rule can be given, most satisfactory conditions are likely to be encountered around the middle of the day in winter and during the early morning or late afternoon in summer.

Whole-disc records of spot distribution are more quickly made with the camera than by eye and hand. Such work is largely a waste of time, however, being more than adequately dealt with by regular programmes at several large observatories. Furthermore a 75 mm image is liable to miss the smaller pores (for which, see below, a 10 cm image is the minimum); it is also too small for accurate measures of position to be made—these require 15 cm images, for which sheet film becomes very expensive if used regularly.

There is, on the other hand, plenty of scope for the photographic recording of small pores, the detail of spot structure, and the changes of active groups at frequent intervals, with full aperture, high magnification, and stringent selection of the moment of exposure. There is nothing like a continuous record of spot detail (unlike spot distribution), and in recent years—when at any moment a vital correlation with spectroheliographic observations may be wanted—it has become increasingly important.

If the smallest pores are taken to be of the order of 1″ in diameter, apertures not less than about 125 mm are desirable. Taking the limit of photographic resolution as .05 mm, the image-scale should not be less than about 80 mm to the solar diameter (see above); in fact, it would normally be considerably greater than this, since amplification would in any case be required unless the focal length were of the order of 25 metres.

Orientation of the image can most conveniently be established by stretching a thread across the field close to the film and orienting it EW by trailing the image.

Solar eclipses present perhaps the most fruitful opportunities for the solar photographer. According to the length of the exposures, and therefore to the phase of the eclipse, stationary short-focus cameras, clock-driven equatorials, or siderostat-fed long-focus telescopes may be used. Assuming that .05 mm is the limit of photographic resolution, we need be concerned with no movement of the image smaller than this during an exposure. Taking the Sun's angular motion as 15" per second, the maximum permissible exposures with a stationary camera for various focal lengths are as follows.

Focal length	Maximum permissible angular displacement of image (" arc)	Maximum permissible exposure (secs)
mm		
50	206	14
100	103	6.9
150	69	4.6
300	34	2.3
500	21	1.4
1000	10	0.7
2000	5.2	0.35
3000	3.4	0.23

Modern colour emulsions permit excellent coverage of all stages of a total solar eclipse. For general views of the landscape and sky at totality, exposures of the order of ½ sec at $f/4$ will be found suitable with a fast material such as Ektachrome 200, although the corona itself will be overexposed. To show the corona as it appears to the eye, using either a telephoto lens or a small refractor, exposures of around $\frac{1}{10}$ to $\frac{1}{2}$ sec at $f/16$, or their equivalent, will be suitable. The prominences, being much brighter than even the inner corona, will require a shorter exposure if they are not to be lost in the glare. A variety of exposures is essential for showing the form of an object like the corona, which has a considerable intensity gradient. A complete photographic programme, therefore, will require the use of at least two instruments, one for the prominences and inner corona, the other for the outer corona, which may be recorded out to a distance of 4 or 5 solar radii—each being used with various exposures, so that as complete a coverage as possible can be obtained. The coronal light being markedly polarised, the use of a polarising filter with its transmitting axis orientated successively NS and EW brings out this effect strongly.

Anyone intending to carry out photographic observations of a total solar eclipse is strongly advised to read accounts by observers of recent eclipses, such as those in *Sky & Telescope*. See also B.3.107.

1.10 Solar spectroscopes

The various types of spectroscope suitable for amateur use are described in *A.A.H.*., section 19.

1.11 Spectroscopic observation of the disc

Quite modest equipment will show the main features of the Fraunhofer spectrum, but for anything more than sightseeing a certain amount of dispersion and aperture is required. Spectroscopic attachments to telescopes require careful focusing if they are to reveal all that they are capable of showing. The general procedure is: (*a*) Bring the Sun's image to a sharp focus on the slit jaws (a magnifying glass is a help). (*b*) Adjust the spectroscope eyepiece to the position giving greatest sharpness to the edges of the spectrum. (*c*) Bring the Fraunhofer lines to maximum clarity by collimator adjustments. This will to some extent destroy (*b*), but simultaneous adjustment of collimator and eyepiece will quickly find the best combination. (*d*) Finally, by means of the telescope focusing rack, make the fine adjustment of (*a*); set the slit centrally and radially across the limb and bring the edge of the spectrum produced by the solar limb to the sharpest focus.

A small-scale map of the main features of the solar spectrum is given in Figure 7; see also Table opposite. B.3.1. shows the solar spectrum from 300 mm to 660 mm (total length about 400 cm).

The main characteristics of spot spectra are also visible with a very simple grating or single-prism spectroscope attached to a small telescope. With the slit set EW, sweep across the disc with the Dec slow motion until the spot band appears along the length of the photospheric spectrum; with this setting of the slit the spot's spectrum will remain visible while the spot trails along the slit (assuming no clock drive) or, in the case of a driven equatorial, with the minimum of adjustment in RA.

The edges of the band should be clearly defined, and the distinction between the central umbral spectrum and the bordering penumbral spectrum well marked (Figure 8). Points to watch for are, in particular:

30

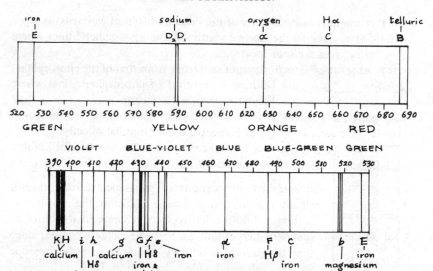

FIGURE 7

Table of the more important Fraunhofer lines

Designation of line	Rowland Revised Wavelength (nm)	Origin
K	393.4	calcium
H	396.8	calcium
h (Hδ)	410.2	hydrogen
g	422.7	calcium
G	430.8	iron, calcium
f (Hγ)	434.0	hyrdrogen
e	438.4	iron
d	466.8	iron
F (Hβ)	486.1	hydrogen
c	495.8	iron
b_4	516.7	{ magnesium / iron
b_3	516.9	iron
b_2	517.3	magnesium
b_1	517.8	magnesium
E	527.0	iron
D_3	587.6	helium
D_2	589.0	sodium
D_1	589.6	sodium
α	627.8	oxygen
C (Hα)	656.3	hydrogen
B	686.7−688.1	oxygen

31

(*a*) Variations in the density of the band in different wavelengths.

(*b*) Differences in the normal widths of the photospheric lines where they cross the spot spectrum.

(*c*) Absence of lines in the spot spectrum from that of the photosphere.

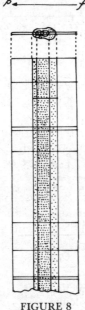

(*d*) Dilution or reversal of photospheric lines where they cross the spot spectrum.

(*e*) Distortion (noting in which direction) of lines in the spot spectrum due to radial velocity.

(*f*) Any differences between the spectra of the umbra and penumbra.

Records of observations of spot spectra must specify clearly the position of the slit relative to the solar image.

Flares—brilliant hydrogen outbursts in the chromosphere, which cannot properly be called eruptions since they have no radial velocity—require a spectrohelioscope for their full and detailed observation. They can however, be to some extent observed by the open-slit method providing the spectroscope is of a dispersion giving sufficient dilution of the integrated light when the slit is widened. Even if this condition is not fulfilled, some useful results can be obtained by the regular observation of the $H\alpha$ and $H\beta$* lines when the slit is lying over spots —particularly those in active development. Any widening or reversals of these lines should be noted—together with the time (beginning and end), position, and approximate intensity—and an attempt made to widen the slit slightly. Reversals on the disc can be seen with quite small instruments, with a narrow slit. In either case there is scope for regular patrolling of the spots. The widening of the hydrogen lines may be most impressive when the flare is intense, having been known to exceed 20 Å for short periods. Intense flares are of rare occurence, however, and typically last only a matter of minutes—exceptionally for several hours. All observations, however incomplete, are therefore of value, and may fill a gap in the spectrohelioscopic observations. Particularly required are observations of the early stages in the development of flares, and measures of area and intensity; but for these a spectroscope of considerable dispersion, if not a spectrohelioscope, is a *sine qua non*.

FIGURE 8

spectroscope of considerable dispersion, if not a spectrohelioscope, is a *sine qua non*.

---

*Flares have, at one time or another, produced strong emissions in D_1, D_2, D_3, H, K, and the *b* group, as well in the more usual $H\alpha$ and $H\beta$.

32

For a well-documented account of past observations of flares without a spectrohelioscope, see B.3.73.

1.12 Spectroscopic observation of chromosphere and prominences

Although the introduction of the narrow-band interference filter (section 1.13) has revolutionised monochromatic observation of the Sun, a prominence spectroscope can be made at home for a fraction of the cost of such a filter, and much interest and enjoyment will be derived from its construction and use.

Apertures of 10 cm and upwards are useful in this field, though the prominences can, of course, be seen with less. The dispersion of the spectroscope needs to be greater than that which will merely show the Fraunhofer spectrum, or insufficient weakening of the integrated light will be encountered when the slit is opened; a spectroscope to be used for observation of the prominences should be able to separate clearly the D_1 and D_2 lines in the yellow ($\Delta\lambda = 6\,\text{Å}$).

Given the requisite dispersion, increasing the magnification has the effect of increasing the apparent dimensions of the slit and hence of the prominence when the slit is opened. Given the dispersion, also (i.e. with a given spectroscope), a short-focus telescope has some advantages over one of longer F, since its primary image scale is smaller, whence a greater area of the prominence will be seen with any particular width of slit. The choice is between a larger fraction of the prominence being seen on a smaller scale, or a comparatively small part of it seen on a large scale; generally the former is preferable.

Owing to the fineness of the adjustment that holds the prominence visible, an equatorial accurately driven at the solar rate may be considered essential. With practice, the observation of prominences with an altazimuth, if it has good slow motions, is perfectly practicable. There are, however, enough unavoidable difficulties in the way of successful astronomical observation without indulging avoidable ones—and the cost of an equatorial (even of a motor-driven one) for instruments of moderate aperture is not nowadays prohibitive.

A prominence spectroscope* is one embodying the following features:

(*a*) Adequate dispersion.

(*b*) A prism or grating which is set permanently so that a prominent chromosphere line, usually $H\alpha$,† is central in the field.

---

* See also *A.A.H.*, section 19.

† $H\alpha$, $H\beta$, and D_3 (helium) are the most important prominence emissions visible to the eye.

(c) A slit which is offset from the optical axis of the telescope by an amount equal to $0.0047F*$ (the mean semidiameter of the solar image in the focal plane), so that rotation of the spectroscope about the telescope's optical axis will cause the slit, once it has been set tangentially on the limb, to travel round it. Alternatively, the solar image may be rotated on the slit jaws by means of an image-rotator.†

If the slit is placed radially across the limb at a point where there is a prominence, the $H\alpha$ line will be seen (given very precise focus and excellent seeing‡) to have a fine, bright prolongation which projects beyond the edge of the spectrum by an amount of depending on the height of the prominence. If the slit, while tangential to the limb and a few " of arc from it, lies centrally across a prominence, the centre of the $H\alpha$ will be reversed, while the rest of the field is dull; this is the slit position required for the observation of prominences. If the slit is slightly displaced towards the centre of the solar image from this position, so that it lies over the disc, the whole of the $H\alpha$ will be dark and the field bright; if displaced slightly too far outward from the solar limb, the $H\alpha$ will be dark and the rest of the field dim. If the slit is moved steadily outward from the limb position, the length of the hydrogen reversals will vary with the size and shape of the prominence, finally contracting to nothing as the slit moves off the prominence altogether. If the slit, while in the limb position, is opened slightly, each reversal will widen into a monochromatic image of the jagged profile of the chromosphere and the base of the prominence.

That the prominence should be visible when the widened slit is passing radiation which is only approximately monochromatic is explained by the fact that it is seen against a dark background. The loss of this contrast is, inevitably directly proportional to the width of the slit—and, for a given slit width, inversely proportional to the dispersion. Thus it is that a prominence is invisible by the open-slit method when it is projected on the disc, since the glare from adjacent wavelengths is then prohibitive. It is the job of the spectrohelioscope to overcome this.

The observational procedure is therefore as follows:

(i) Focus the solar image on the slit jaws.

(ii) Bring the $H\alpha$ line to a sharp focus in the centre of the field; slit nearly closed.

---

* The figures for maximum and minimum solar semidiameter are $0.00459F$ and $0.00474F$. Some slight readjustment of the slow motions will therefore nearly always be required while patrolling the limb.

† A.A.H., section 8.11, describes different forms.

‡ Visibility of the chromosphere and prominences is also much reduced by even a trace of haze or 'whitening' of the blueness of the sky.

(*iii*) Rotate the spectroscope till the slit is perpendicular to the Dec axis and tangential to the E limb (position angle 90°).

(*iv*) Rotate spectroscope clockwise,* so that the slit, maintaining its tangential orientation, travels round the limb through N (0°), W (270°), and S (180°) to E again. During this operation the image must be kept centrally in the field, which even with a clock drive will require touches on the slow motions.

(*v*) Whenever, during (*iv*), a bright reversed 'notch' appears on the *H*α line, continue the rotation of the spectroscope until it is at the centre of the line and open the slit to reveal the prominence.

(*vi*) With the slit 'open', make final focusing adjustments with the drawtube rack.

Records of prominence observations should include:

Date and UT.

Position angle (the sleeve of the drawtube should be graduated to show the p.a. setting of the slit).

Height (" arc).

Brightness.

Drawing, with scale. (In the case of active prominences a series of drawings should be made, the directions of motion being indicated by arrows.)

Any associated distortions of spectral lines.

Heights and other dimensions of prominences can be measured by means of an accurately drawn scale, reduced photographically on a glass diaphragm—both the number of graduations and the amount of reduction depending on the specifications of the instrument with which it is to be used—and mounted in the focal plane of the positive ocular of the view telescope.

The scale can be calibrated as follows: Orient spectroscope and scale so that both lie EW; the scale is then projected normally across the width of the spectrum. Adjust the telescope in Dec so that the Sun is central in the field; then clamp it p the Sun without altering the Dec setting. As the Sun transits the slit the spectrum will transit the scale. Time the transits of the p and f edges of the spectrum at any particular scale graduation ($t_2 - t_1 = T$). Also time the transits of either edge of the spectrum at two widely separated graduations, n divisions apart ($t'_2 - t'_1 = T'$). Let the Sun's angular semidiameter in " arc be R (taken from the *A.A.* for nearest

---

* If the spectroscope is mounted, not in the drawtube, but in a star diagonal, it must be rotated anticlockwise from 90° in order to follow the same sequence of position angles.

0^h UT). Then θ, the angular value of a single division of the scale, is given by

$$\theta = \frac{RT'}{\frac{1}{2}nT} \quad '' \text{ arc.}$$

There is great scope for systematic work on the prominences with adequate instrumental means. Many aspects of the relationship between spots and the overlying chromosphere are still imperfectly understood, largely owing to the former being too foreshortened for effective study when near the limb. The laws governing the motions of prominences,* again, will not be elucidated without the accumulation of many more observational data.

Since 1922 the daily appearance of the limb in $H\alpha$, as well as annual prominence statistics, have been published for the I.A.U. by Arcetri Observatory. The unit of profile area of a prominence is defined as the area contained between a $1°$ arc (heliocentric) of the Sun's limb and a concentric arc situated $1''$ arc above it:

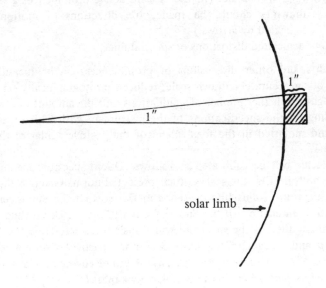

FIGURE 9

* B.3.33, 3.90.

1.13 Monochromatic solar telescopes

The development of the narrow-band interference filter now makes it possible for the amateur to observe both the prominences and photosphere with the equipment described in *A.A.H.*, section 22.5. The advantage of the filter-type telescope over the simple spectroscopic method described in section 1.12, or the spectrohelioscope (*A.A.H.*, section 22.6), is that it combines the best of both worlds, being as convenient as one and having the whole-disc coverage of the other. For observing prominences at the solar limb, a filter with a bandwidth of about 0.4 nm is adequate, the disc itself being occulted by an opaque stop in the focal plane. Observation of the disc features requires a narrower bandwidth, of the order of 0.05 nm. The latter type of filter is much more expensive and elaborate, since, being critically temperature-sensitive, it requires a heated jacket to keep it transmitting at exactly the right wavelength. Even the 0.4 nm filter requires some sort of temperature control.

Monochromatic filters for solar work are normally designed to pass the $H\alpha$ (C line, at 589.6 nm), since this is the hydrogen line to which the eye is most sensitive. Clearly, then, a filter telescope lacks the flexibility of the spectrohelioscope, which can be tuned to any desired wavelength and hence also used for the measurement of radial velocities of eruptive features; see, in particular, B.3.51. B.3.24, 3.25 and 3.37 describe monochromatic filter-type telescopes.

It is not easy to find commercial black-and-white films which are sensitised sufficiently far into the red to record $H\alpha$ light well. Kodak Plus-X is no longer as sensitive to this end of the spectrum as it used to be, and their special spectrographic film SO-392, available in 35 mm size, has been recommended. There is no difficulty in recording the C line using colour film.

1.14 Eclipse observations

1.14.1 Partial:

1. Timing of contacts. The first is necessarily less accurate than the last if observed visually. Observe the chromosphere with a curved slit nearly closed (if a straight slit is used, any slight misorientation will result in the point of contact being missed) and note the obtrusion of the lunar limb into the $H\alpha$ reversal. A good drive is desirable—and at least an equatorial with good slow motions is almost essential—in order to keep the slit accurately coincident with the limb. Last contract can be observed by the same method.

2. If observing visually, note and record any outstanding irregularities on the Moon's limb.

3. Record visibility of the Moon outside eclipse, with times.

4. A continuous plot of the intensity of the solar radio-frequency radiation as the lunar limb crosses the disc may give valuable information regarding the source locations of the radiation.

1.14.2 Total:

(*a*) *Visual*: The following notes are rather a suggested programme for a team of observers than for a single observer, who could not hope to cover more than a fraction of the ground.

1. Timing of contacts: second and third contacts are given by the disappearance of the last Baily's bead (white, cf. the red prominences) and reappearance of the first. Record: method used; time to nearest $1^s.0$ or better; longitude, latitude, and height above MSL of the observing station.

2. Shadow bands: white sheet laid on ground, with vertical rods on flat bases to indicate their direction. Record: times of appearance (usually $0^m.5$ to 2^m before second contact and after third contact); direction of motion (stating whether true or magnetic bearings); speed (both this and their direction are independent of the motion of the Moon's shadow).

3. Baily's beads: caused by combined effects of irradiation and the serrated limb; visible for a few seconds before second contact and after third. Of no scientific interest apart from their indication of the times of the contacts.

4. Prominences. Record: colour (been described as whitish, pink, red and yellow to the naked eye), identifying them by position angle from the N point of the disc.

5. Corona. Record: time of first visibility; extent and intensity; position and structure in relation to any prominences visible.

6. See section 1.14.1, paragraph 4.

7. Other phenomena: e.g. approach of the shadow (velocity up to 8000 km/h if oblique to the Earth's surface—i.e. near sunset or sunrise), often visible in the sky, or overland if the viewpoint is well above the surrounding terrain; degree of darkening during totality (faintest stars visible).

8. Conditions. Record: temperature and barometer at intervals; formation of dew, if any; direction and velocity of surface wind; wind deduced from movements of lower and upper clouds, if any.

(*b*) *Spectroscopic*: What work can be done is necessarily dictated by the equipment available. The following brief description shows what can be done with even a small slitless spectroscope. It should be so oriented that the length of the spectrum is normal to the line of cusps at the moment before totality.

1. Start observing about 10 minutes before totality.

2. Record time at which the curved Fraunhofer lines (arcs) become visible.

3. As the width of the continuous spectrum shrinks, bright arcs will be seen projecting beyond the continuous band.

4. Note if the prominent Fraunhofer arcs disappear simultaneously.

5. Instantaneous appearance of the flash spectrum at the E limb: second contact. Record time. The appearance and disappearance of the flash spectrum is an indicator of the moment of contact, which is sensitive to about $0^s.02$. Hence the possibility of a movie camera exposing 50 frames per second (normal speed, 24), with some method of synchronisation with radio time pulses. But to gain anything from such speeds, some form of sound-track time record would have to be used, and with typical amateur equipment synchronisation to the nearest 1^s would be nearer the limit possible (e.g. clock face reflected into the field of view).

6. Disappearance of the flash spectrum.

7. Disappearance of the chromosphere arcs and superimposed images of E limb prominences (during the short period of totality they will be complete rings).

8. Coronal rings (red and green, fainter than the chromosphere arcs or rings). Record: relative thickness and diameters of these rings; positions angles of any regions of abnormal intensity.

9. Appearance of chromospheric arcs at the W limb.

10. Appearance of flash spectrum at W limb.

11. Disappearance of flash spectrum and reappearance of continuous spectrum: third contact.

For the study of the coronal lines a spectroscope and view telescope of larger aperture, and correspondingly higher magnification, are desirable. For any detailed spectroscopic work, a slit spectroscope is, of course, necessary.

Anyone interested in the computation of eclipses should refer to B.3.26.

1.15 The radio astronomy of the Sun

Radio astronomical observations of the Sun are made at all wavelengths. In general four main types of instrument are used, the radiometer, the radio interferometer, the radio spectrograph, and the radio polarimeter. Observations with polarimeters yield information about magnetic fields when circular polarisation is observed. Amateurs in England have so far used interferometers in their expansion of 'low frequency radio astronomy' and radiometers for observations in the super high frequency region (centri-metric wavelengths).

Three distinct types of radio noise are observed from the Sun: (*i*) noise from the quiet Sun, due to a thermal mechanism; (*ii*) a slowly varying component having a period of weeks or months and observed in the range 3 to 60 cm, and (*iii*) bursts and noise storms. There are five main spectral types of solar noise at present and other types are under discussion. As in visual astronomy there is an important call for continuous observations of the Sun on a number of wavelengths. There is a need for spectroscopic observations by amateurs. The bursts and noise storms are best observed with a radio spectroscope using rhombic aerials. In general the noise storms which last for periods of the orders of hours or days are associated with large sunspots, whilst the outbursts which only last for minutes are associated with flares. In neither case can it be said that all flares and all sunspots will be associated with radio noise, and the *visible* spots are not the sources of these emissions.

In the same way the terrestrial effects of intense solar activity are not necessarily related to the number of sunspots. Observations during the International Geophysical Year showed that there was greater solar and terrestrial activity associated with the Sun in September, 1957, than in the following month, when the sunspot number was actually greater.

The following terrestrial effects have been observed to accompany solar flares:

(*a*) *Magnetic crotchets.* A disturbance of the Earth's magnetic field usually occurs simultaneously with the peak of the flare. These magnetic crotchets, as they are called, are restricted to the sunlit side of the Earth and last for only a short time.

(*b*) *Radio communication fade outs.* Radio communications are considerably disturbed when flares are generated. Apparently a new layer of ionisation is formed on the sunlit side of the Earth—just below the E region—forming a new D region.

(*i*) Short waves: these are disturbed and undergo absorption. The period of strained conditions is confined to the sunlit side of the Earth and lasts only for a matter of hours.

(*ii*) Long-wave phase anomaly: The ionisation of the D region is shown up by the fact that the phase between the ground- and sky-waves at the receiver changes. Much work has been done on this problem at 18,000 m.

(*iii*) Magnetic storms, as distinct from crotchets, affect the F region of the ionosphere.

(*iv*) The newly-formed D layer tends to improve long-wave propagation conditions. It also has the effect of increasing the atmospheric noise level at the receiver; particularly in the tropics.

(*c*) *Solar corpuscular streams.* After about 26 hours from the occurrence of a large flare there are often violent magnetic storms and aurorae. These are thought to be due to the ejection of particles at a velocity of 1600 km/sec from the Sun.

(*i*) The magnetic storms may last for several days and severely effect the F region of the ionosphere. Short-wave communication can be disrupted for several days.

(*ii*) Production of aurorae.

(*d*) *Cosmic radio noise.* At a wavelength of 16 m it has been discovered that the newly-formed D region causes the radio emissions from the cosmic sources to be absorbed.

(*e*) *Radio noise from the Sun.* Many flares are accompanied by outbursts which last for from several minutes up to an hour. They are many times the intensity of the base level of the solar radio noise.

Bursts and flares which are prevalent at metre wavelengths cause great difficulties in observing thermal radio emissions from the Sun. For this reason the thermal emission from the chromosphere and the inner corona is generally observed on much shorter wavelengths.

(*f*) *Sudden ionospheric disturbances (S.I.D's).* Of the above features, the increase in ultraviolet radiation from the Sun which increases the ionisation of the D region of the ionosphere can be classified with those effects which produce sudden disturbances in the ionosphere (S.I.D's). The measurement of S.I.D's is of importance and one of the methods used is to observe the fading undergone by cosmic radio waves in the region of 11 m. This is then a measure of the absorption from which the increase in free electrons can be computed.

The S.I.D's which have been discussed are clearly the:

(*a*) Magnetic Crotchet
(*b*) Short-wave Radio Fade Out
(*c*) Long-wave Phase Anomaly
(*d*) Fading of Cosmic Radio Noise
(*e*) Sudden Enhancement of Atmospherics (S.E.A.)

Of these one of the most interesting is the S.E.A. since amateurs can build a simple receiver to observe S.E.A's and this can be useful in maintaining a flare patrol. The D layer improves the reflecting properties of the long waves so that a receiver which is listening to thunderstorms which are generated from the tropics receives more noise from such storms when a solar flare occurs than it would otherwise. There is not however—and this is important—an increase in the number of atmospherics generated; it is only an increase in the integrated number which can be received. The apparatus required is simple, and can be built by the amateur, although a pen recording milliammeter is required. It is ideal for use in England, since the sky is often cloudy, and thus it provides an immediate comparison with other types of record. A simple receiver and recorder for monitoring S.E.A's on 27 KHz is described in B.3.56.

There is great scope for the amateur who wants only a simple aerial array in that a simple dipole array can be used to observe the Sun during the whole of its passage across the sky.

The literature devoted to radio astronomy is now vast. B.3.75 is a useful reference on the subject of magnetic storms. Special references to solar radio work will be found in the bibliography.

1.16 Horizontal Sun telescope

The amateur who decides to specialise in solar work should consider very carefully the advantages of a permanent solar telescope, into which a stationary image is fed by a coelostat. The difficulties involved are mainly non-astronomical: (*a*) possession of a suitable room or hut, or the space in which to build one, (*b*) expense—which, however, can be considerably reduced if he is prepared to do as much of the constructional work as possible himself.

The advantages are imposing:

(*a*) A much longer focal length than can be employed in an ordinary telescope, giving with ease a solar image of say, 50 mm diameter—which would be out of the question with a privately owned refractor or Newtonian.

(*b*) It permits the use of spectroscopes capable of doing justice to what even quite small apertures have to offer, when the object is the Sun. Spectroscopic attachments to a refractor or Newtonian must necessarily be light, and their dispersion small. No one who has ever seen the solar spectrum under wide dispersion (my own first view of it, at the focus of the Mt Wilson 150 ft tower telescope, was an unforgetable experience) will be prepared to go back to work with the type of spectroscope designed for attachment to small telescopes.

(*c*) Convenience of combining a direct-vision telescope, a solar-image projector, a camera, spectroscopes (prismatic or Littrow), and possibly a spectrohelioscope, mounted together under one roof in what is virtually a single instrument and instantly accessible and ready for use.

(*d*) Convenience of having the various instruments permanently mounted in comfortable observing positions.

(*e*) Convenience of the stationary image, obviating the necessity for a heavy and accurately driven equatorial, with continuously changing observing position.

(*f*) Comparative freedom from thermal currents, with a suitably designed installation.

(*g*) Finally, the unique scope that the Sun offers the amateur for valuable original research cannot be fully exploited with the ordinary telescope that is found in amateur hands.

Any number of different designs can be evolved, according to the nature and focal length of the objective and the ancillary instruments that are to be incorporated. The objective may be either an achromatic OG or a paraboloid. The specimen layout described below employs a 12.5 cm achromatic and is based upon the recommendations of F. J. Sellers, who himself built a Sun telescope of this type (Figure 10). See also B.3.32, 3.5.

The objective and the auxiliary instruments are housed in the oblong building oriented as nearly N and S as possible. (The scale given here is based upon a focal length of from 6 to 10 m; a shorter but wider room could equally well be used.)

A: masonry pier carrying the coelostat,* which consists of two 12.5 cm flats. The dotted area should be covered with a pent roof (no walls) to shade the window and the light-path from the coelostat into the hut; a sliding extension should be arranged so that the entire coelostat and pier are in shade when not being used; an additional shade, fixed to the pier itself, screens the collecting mirror from the Sun while in use. From the coelostat pier there should ideally be a clear view nearly to the horizon for $180°$ or more, centred on S.

O: objective (12.5 cm achromatic), mounted on a NS slide so that its position is adjustable between *v* and *w* to allow the image to fall at any required distance from *C*.

B: light-proof tunnel surrounding the travel of the OG.

C: 12.5 cm flat mounted with its centre on the optical axis of *O*. Its inclination is adjustable about a·vertical axis, so that the reflected pencil may be turned towards *V, W, X* or *Y* as required.

V, W, X, Y, Z: masonry piers or stout timber benches (according to the nature of the floor and foundations) carrying the auxiliary instru-

* *A.A.H.*, section 14.22.

ments; the collimating axis of each must pass through the centre of
C. Y carries the slit, grating, and ocular of a high-dispersion diffraction spectroscope; Z carries the collimating and image-forming
paraboloids (see below). If a Hale spectrohelioscope is to be
included in the equipment, this apparatus and arrangement is
already halfway to it, while an easily made rearrangement of the
piers would accommodate an Ellison autocollimating spectrohelioscope.* W takes the mounting for a projection lens (an
achromatic Barlow gives very good results); W' is a fixed
projection screen. V, X can be used for mounting cameras, prismatic
spectroscopes, etc.

D: movable screens with circular holes just large enough to allow the
the passage of the pencils.

Figure 11 shows in greater detail, and in elevation, a possible layout for
the grating spectroscope:

G: slit on which the solar image is formed by the objective after
reflection at C.

H: collimating paraboloid which reflects the light from the slit as a
parallel pencil to J. The centre of H and of the slit must, of course,
lie on an axis passing through the centre of C.

J: the grating (e.g. a 10 cm replica)†; being mounted so that it is
adjustable about a vertical axis, any region of the spectrum of any
order can be reflected to K. The grating cell should also be mounted
in such a way that the necessary adjustments to ensure that the
rulings are vertical can be made easily.

K: second paraboloid, which brings the spectrum to a focus at

L: eyepiece or plate-holder.

If a parabolic mirror is used instead of an OG, the layout must be
modified somewhat, though the principle remains the same. The flat,
C, is replaced by the paraboloid, which is fed direct from the coelostat.
Since tilting the mirror through wide angles (in order to bring the image
to instruments situated, say, near the side walls of the hut) would introduce
abaxial aberrations, the instrument piers must all be close to the optical
axis of the objective. This introduces some practical difficulties, and for
the amateur with limited resources the achromatic arrangement already
described is probably more satisfactory.

When planning a horizontal Sun telescope and deciding on the sizes of

---

* A.A.H., section 22.6.

† Replicas are not suitable for spectrohelioscopes, however (see A.A.H., section
22.6).

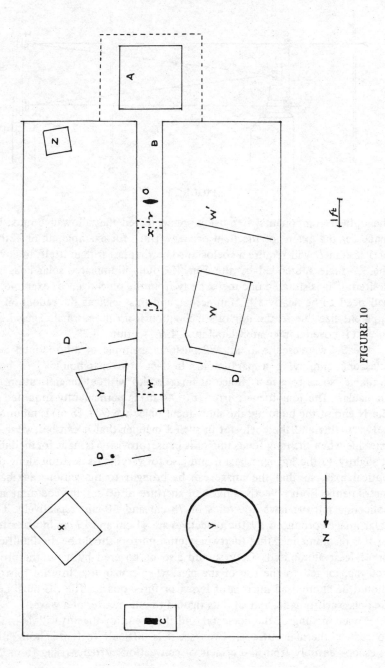

FIGURE 10

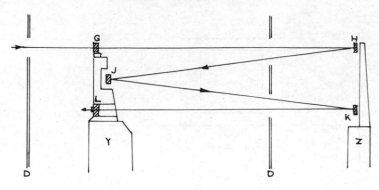

FIGURE 11

the optical components, it must be remembered that allowance must be made for the extent of the field of view. Thus, for example, an objective of 125 mm D will require a coelostat somewhat larger than itself, to cover the ½° angle subtended by the Sun, if a fully-illuminated solar image is desired; if it is situated one metre away from the objective, for example, it will need to be nearly 145 mm across, and this neglects the reduction in apparent size due to the mirror's inclination from the normal. The whole subject is covered in greater detail in *A.A.H.*, section 14.23.

The Snow telescope at Mt Wilson is an example of a horizontal Sun telescope employing a paraboloid. In fact, two paraboloids are here available, so as to give a choice of image-scales without magnification by an ocular. The long-focus mirror ($F = 44$ m) is permanently mounted at the N end of the building; the short-focus objective ($F = 18$ m) is mounted between this and the coelostat in such a manner that it can be swung to one side when the long-focus mirror is in use; provision is made for inclining it slightly to the incident beam, and also for varying its position along the optical axis, so that the image can be brought to the various auxiliary instruments. Both objectives have an aperture of 60 cm; the coelostat and collecting mirrors have apertures of 75 cm and 60 cm respectively. The solar images produced by the objectives are 41 cm and 17 cm in diameter.

It is recommended that the two exterior mirrors should be rhodiumised, or at least aluminised. They should also be covered by a weatherproof housing, locked to the top of the pier, when not in use. Interior mirrors should be aluminised mirrors of Pyrex or fused quartz. The OG must be a first-class achromatic, and all flats figured to one-quarter of a wave.

Proper shading of the coelostat and the S end of the hut will do much to reduce thermal currents. Even so, it is difficult to free a horizontal telescope entirely from the effects of convection currents rising from the heated ground. (Hence the two later Sun telescopes at Mt Wilson are vertical tower telescopes.) The outside of the building should be painted white, and its surroundings planted with shrubs to protect the ground from the direct rays of the Sun. Image distortion and change of focus due to heating of the flats and mirrors are materially reduced by using Pyrex or fused quartz, but even so an alteration of focal length of as much as about 1.5% during observation must be expected. The films should be renewed as soon as their reflectivity begins to fall off, or distortion will result.

1.17 Sunspot numbers and other solar data

Also known as Wolf Numbers, Wolfer Numbers, Zurich Numbers, and Sunspot Relative-Numbers. An arbitrary index of the daily spottedness of the disc. First devised by Wolf at Zurich, and continued by Wolfer.

The Wolf Number, R, is arrived at by counting the number of disturbed areas (both groups and isolated single spots) g, and the total number of individual spots f, giving the former 10 times the weight of the latter, adding the two together, and multiplying by a coefficient k, whose value depends on the aperture of the instrument with which f and g were counted, and also on the care and experience of the observer:

$$R = k \, (f + 10g)$$

The value of k is determined empirically: a long series of observations with a particular instrument by a particular observer are compared with the mass of data submitted to the Zurich clearing-house, and the coefficient selected which brings the two into best agreement. For a careful and experienced observer using the Zurich telescope (D = 10 cm, M = 64), k = 1.0; for smaller instruments, k will be slightly larger than 1.0; for larger instruments, slightly smaller.

Thus if a standard instrument, properly employed, shows 2 spot groups, one containing 2 spots and the other 3, as well as 4 isolated spots, the Wolf Number for the day will be 69:

$$g = 6 \text{ (i.e. } 2 + 4 \text{ disturbed regions)}$$
$$f = 9 \text{ (i.e. } 2 + 3 + 4)$$
$$\therefore \; R = 1 \, (9 + 60)$$
$$= 69$$

Daily Wolf Numbers and monthly means are published quarterly in B.3.105, which also includes for the three-monthly period:

(*i*) Chromospheric eruptions:

> date and time of observations and of maximum intensity,
> approximate coordinates,
> maximum width of $H\alpha$,
> maximum area,
> maximum intensity.

(*ii*) Active regions:

> rotation number,
> coordinates of centre,
> date of CM passage,
> age at CM passage,
> duration (solar rotations).

(*iii*) Diagram showing the hours during the three-monthly period during which the Sun was under spectrohelioscopic and spectroheliographic observation at the cooperating observatories.

(*iv*) Coronagraph measures of monochromatic (5303 Å and 6374 Å) coronal intensity at $5°$ intervals of position angle.

(*v*) Solar radio noise data.

(*vi*) Before 1945: character figures of calcium and hydrogen flocculi, on a scale from 0 (complete absence) to 5 (maximum encountered at the peak period of solar activity).

(*vii*) Before 1939: character figures for the Sun's ultraviolet radiation.

Since 1947 the Swiss Federal Observatory at Zurich has been broadcasting the daily Wolf Numbers on the short-wave channels of the Swiss Broadcasting Corporation. With the exception of those to South America (in Spanish) the broadcasts are in English. They are given on the 4th and 5th of each month as follows.

Date	Time (UT)	Wavelength (m)			Destination
	h m				
	07 20	25.39,	25.28		Australia
	15 05	19.60,	16.87		Far East
4th	21 50	19.59			S. America
	22 30	25.28			N. America
	23 40	31.46,	25.28,	19.59	S. America
5th	01 40	31.46,	25.28,	19.59	N. America
	03 05	31.46,	25.28,	19.59	N. America

1.18 Classification of spot types

Cortie's classification,* based upon the telescopic appearance of the spot, is simple and reasonably comprehensive:

1. Group of one or more small spots.

2. Two-spot formations:
    (*a*) With the leader larger than the trailer.
    (*b*) With the trailer larger than the leader.
    (*c*) With both members of approximately equal size.

3. Train of spots:
    (*a*) With the principal spots well defined.
    (*b*) With none well defined (patchy penumbrae and irregular umbrae).

4. Single spots:
    (*a*) Round and regular.
    (*b*) Round and regular, with small attendant spots or pores.
    (*c*) Irregular.
    (*d*) Irregular, with train of small attendant spots.
    (*e*) Irregular, with small companions irregularly distributed.

5. Irregular group of larger spots.

An alternative classification, based upon the magnetic properties of the spot, has been developed at Mt Wilson:

α : Single spot, or group of same polarity; flocculi symmetrical.
αp :   ”   ”   ”   centre of group *p* centre of flocculi.
αf :   ”   ”   ”   centre of group *f* centre of flocculi.
β : Bipolar pair; leader and trailer of about equal size.
βp :   ”   leader larger than trailer.
βf :   ”   trailer larger than leader.
$\beta\gamma$:   ”   leader and trailer accompanied by smaller spots of opposite polarity.
γ : Multipolar spots; rare, irregularly arranged spots of opposite polarity, not classifiable as β.

---

* Also sometimes referred to as the Stonyhurst classification (B.3.22).

1.19 Classification of flares

Ellison has proposed the following rough classification of flares on a basis of size only:

Class 1 : >100 ≤300 millionths of the Sun's hemisphere.
  "   2 : >300 ≤750   "   "   "
  "   3 : >750   "   "   "

This simple classification has not, however, been generally adopted, and the I.A.U. still recommends the older classification into four groups (1, 2, 3, 3+) based on a combination of area and brightness. Correct estimation of importance in these terms is unfortunately not easy and requires considerable experience on the part of the observer.

1.20 Classification of prominence types

The classic classification is that of Pettit (B.3.78), summarised below

I. Active prominences:

   Ia: Interactive; involving exchanges between two prominences.
   Ib: Common Active; material drawn down to a centre of attraction on the Sun's surface.
   Ic: Coronal Active; centre of attraction so strong that material is drawn down from the upper levels of the Sun's atmosphere.

II. Eruptive prominences:

   IIa: Quasi-eruptive; active, or active sunspot, prominence drawn bodily into a spot or centre of attraction.
   IIb_1: Common eruptive ⎫ active, or active sunspot, promi-
                               ⎪ nence which rises to great distances
                               ⎨ (100,000 to 500,000 km) from the
   IIb_2: Eruptive arches   ⎭ surface, and then disappears.

III. Sunspot prominences:

   IIIo: Cap prominences; resembles a low cloud over the spot.
   IIIa: Common coronal sunspot type; fan-like coronals converging on the spot.
   IIIb: Looped coronal sunspot type; as IIIa, but the coronals strongly curved or even looped.
   IIIc: Active sunspot type; similar to Ib pouring into the spot area.

IIId$_1$: Common surge; narrow filaments which rise, often to great heights, and then sink back along the same trajectory 'like an elastic band'.

IIId$_2$: Expanding surge; break into a spray, and begin to fade before sinking back into the spot.

IIIe: Ejection; small separate knots which are ejected and which do not return.

IIIf: Secondary eruption; spring from other sunspot prominences.

IIIg: Coronal clouds; form over a spot far above the chromospheric surface, and pour streamers down into it along curved trajectories.

IV. Tornado prominences:

IVa: Columnar ⎫ rotating columns of material; disappear
⎬ when the rotational velocity becomes
IVb: Skeleton ⎭ excessive.

V. Quiescent prominences: lack external streamers; characteristic 'palisaded' structure.

VI. Coronal prominences: form above the surface and pour material down to centres of attraction; when occurring as groups over spot areas, these are equivalent to IIIa and IIIb.

1.21 B.A.A.H. data

(a) RA ⎫
Dec ⎪
Angular diameter ⎪ at mean noon
P: Position angle of rotational axis ⎬ at 4-day inter-
B_0: Heliographic latitude of centre of disc ⎪ vals
L_0: Heliographic longitude of centre of disc ⎪
Time of transit every 4th day ⎭

(b) Dates of commencement of Carrington's (Greenwich Photo-Heliographic) series of synodic rotations

(c) Details of the year's eclipses

1.22 A.A. data

The following information will be of particular use to amateur observers:

(a) Synodic rotation numbers for the year
(b) Ecliptic and equatorial co-ordinates—daily ephemeris

(*c*) Heliographic co-ordinates, horizontal parallax, semidiameter, time of ephemeris transit—daily ephemeris

(*d*) Low-precision formulae for calculating co-ordinates and equation of time

(*e*) Details of the year's eclipses

1.23 Solar data*

Semidiameter at unit distance: $16' \; 1''.18 = 961''.18$.

Diameter: 1,392, 530 km.

Mass: 332,950⊕.

Volume: 1,300,000⊕.

Density: 0.26⊕ = 1.41 x water

Mean superficial gravity: 28.0⊕.

Stellar magnitude: −26.8.

Rotation number: Carrington's series of synodic rotations (also known as the Greenwich Photo-Heliographic Series). Rotation No. 1700 began on 1980 Sep. 25.49.

---

* The data in this section, and the corresponding sections under the various members of the Solar System that follow, are taken for the most part from *A.A.* and *B.A.A.H.*

SECTION 2

LUNAR OBSERVATION

2.1 Amateur work

It is unnecessary to dilate on the reasons for the Moon being an admirable object for study with small instruments.

Amateur work is mainly devoted to the following topics, some of which have in the past been given much more attention than others:

(a) Cartography.
(b) Lunar change.
(c) Colorimetric work.
(d) Occultations.
(e) Eclipses.
(f) Photometry.
(g) Photographic work.

Because of the catholic nature of what the Moon has to offer, almost any instrument will enable some sort of satisfactory work to be performed. Occultations of bright stars, to take an obvious case, can be timed as accurately with a small telescope as with a large one. The drawing of lunar detail, which must be treated more as an interesting and instructive pastime than scientifically useful or likely to lead to discovery, can be undertaken with any instrument. A reflector of 15 cm aperture will reveal more detail than any lunar observer could draw in a lifetime. For research into transient lunar phenomena (section 2.3) considerable magnification may be necessary, and the observer with a telescope of 25 to 30 cm aperture will have the advantage over users of smaller instruments.

Magnification must in general be suited to the type of observation. For worthwhile work on the detail of the lunar surface, instrument and atmosphere should be capable of permitting the use of x 300 or more.

Glare, which if unreduced may obscure the finer detail altogether, can be overcome in various ways: (a) increasing magnification: limited by

53

atmospheric turbulence, restriction of field, etc; (*b*) reducing aperture: consequent loss of resolving power; (*c*) neutral filter: the best of these three methods; (*d*) daytime observation: at first and third Quarters a polarising filter is useful for reducing the intensity of the background sky-light, whose maximum polarisation occurs along a great circle $90°$ from the Sun, while the sunlight reflected from the Moon is, on the average, polarised very little. (B.4.61).

In this field of observation it is impossible to overrate the importance of *any* technique (whether instrumental dodge or the organisation of the observations) to render the data as objective and impersonal, and as internally consistent and also comparable with those of other observers, as possible.

2.2 Cartography

The lunar surface has been surveyed in such detail by space-borne cameras that only the most foolishly optimistic observer could hope to discover any new permanent detail by Earth-based observation. The days of effective amateur mapping of the lunar surface have gone for good. Excepting the search for transient phenomena, then, lunar cartography must be viewed in the light of the personal benefits it may bring to the observer, rather than that of any addition to the bulk of scientific knowledge of the Moon.

These benefits are, nevertheless, considerable:

(*a*) Instrumental: experience will be gained of the effect of different magnification and seeing.

(*b*) Observational: to attempt to draw what one sees is the best way of training the eye to discriminate. It is not enough to stare at a view; this only cultivates a general impression. In putting pencil to paper, one is forcing the visual faculty to make decisions about discrete detail.

(*c*) Representational: few observers will be talented artists, but the object, or at any rate the prime object, is to produce less a convincing picture than an accurate representation of the detail seen. Relative positions and sizes of features are the first priority, and this is a skill which can and should be learned.

The following suggestions, aimed at observers undertaking lunar drawing for the first time, are freely adapted from the B.A.A. Lunar Section's *Guide for Observers of the Moon* (1979):

1. The magnification used should be the minimum necessary to show the required detail (this applies to all branches of observational astronomy).

2. The observer should be comfortably seated. It is very helpful to have the source of illumination fixed to the drawing board.

3. The outline of the formation, and the larger objects, must be sketched satisfactorily before attempting to draw fine detail. Outlines traced from photographs can be useful, but differences in illumination or libration may necessitate modifications.

4. Detail near the edge of a moving shadow may be either revealed (lunar sunrise, before Full) or obscured (after Full, lunar sunset). This must be borne in mind when deciding in what order to draw the different details; at local sunset, in particular, objects at the edge of the shadows should be drawn first.

5. Written notes about any interesting features should be made at the time, and a complete record of the times of starting and completion, conditions, instrument, etc., must be added to the drawing.

6. Drawings must never be 'touched up' after the session at the telescope has been completed.

2.3 Lunar change

Investigations into lunar change depends essentially upon consistent work carried out over long periods, since negative as well as positive observations are of value. They have not, therefore, been well covered by the highly intensive methods of space research, and, with the demise of serious amateur cartography, studies of albedo changes of discrete regions of the Moon, and searches for the so-called 'transient lunar phenomena' (T.L.P's), form perhaps the most active field of lunar work.

2.3.1 Albedo changes: It is impossible in theory, and often difficult in practice, to distinguish between changes of tone and changes of form (i.e. physical changes), since the latter can only be perceived in terms of the former unless the shadow provides a clue. No truly physical changes (of which Linné is the classical example) have been substantiated, although there have been innumerable suspected instances (see B.4.47, 4.56, the latter having a useful bibliography).

Albedo change investigations fall into two classes: the 'variable spots', whose tone appears to change regularly with the lunar day (these have by now been well studied, but their behaviour is still of interest); and the unpredictable, short-term changes. The majority of examples (the 'variable spots') fall into one or other of four categories: (*i*) those reaching maximum brightness between local noon and Full, (*ii*) those reaching minimum brightness between local noon and Full, (*iii*) those brightening continously throughout the day, (*iv*) those darkening continuously throughout the day.

In some cases it has been shown by the use of restricted fields that what had previously been accepted as a change in one area is in reality a contrast effect from changes in illumination in a contiguous area (Plato's floor being an instance). Polarising filters might also assist in distinguishing real from apparent tonal variations.

Many instances of non-periodic change have been recorded; see, for example, B.4.64, 4.67. The B.A.A. Lunar Section has claimed a number of variations in the brightness of Aristarchus, whose occasional very distinct visibility under earthshine conditions is well known. The difficulty in such work is to establish a scale of brightness by which the intensity of any feature may be measured unambigously. Extinction devices suffer from the serious drawbacks of all such photometers (see *A.A.H.*, section 21.2), while direct visual comparison is subject to numerous subjective errors. Photography, with subsequent measurement of the density of different parts of the negative, would seem to offer the best chance of success; B.4.53 describes a photoelectric device for scanning the Moon's surface.

2.3.2 Obscurations: A form of phenomenon in which a small area of the surface is claimed to appear hazy while the immediate surrounds remain clearly defined. Thorough knowledge of the terrain is necessary, as well as sufficient observing experience to be able to discern the merging of light and dark spots at the limit of telescopic resolution, an effect which can give the impression of local blurring. The crater Plato has acquired a reputation for producing 'obscurations', but many experienced observers have expressed doubts about these reports: tiny craterlets straddle a dark floor, and can create classic examples of the hazy semi-resolved effect. Obscurations near the terminator have also been reported. See, for example, B.4.63, 4.66 and 4.96.

2.3.3 Flashes: In the past, observers have occasionally reported seeing tiny pinpoints of light flashing out momentarily against shadows or the unilluminated hemisphere. There have been few recent reports of this phenomenon, once thought to be due to the impact of meteoroids.

2.3.4 Coloured glows: Local, usually reddish, coloration of the lunar surface. It is easy for the inexperienced observer to be misled by spurious colour either within the instrument itself (particularly if a refracting telescope is used) or due to dispersion in the Earth's atmosphere. Nevertheless, a number of independent sightings suggest that coloured glows do occur from time to time, the classic instance being the event seen by N. Kozyrev inside Alphonsus in 1958. Two well-known regions of permanent reddish coloration are the crater Fracastorius and a segment on the western wall of Plato.

Devices involving the rapid interchange of red and blue filters, to help in detecting any anomalous tints, have been used by British observers (B.4.78).

The prime targets adopted by the B.A.A. Lunar Section for T.L.P. studies are: Alphonsus, Aristarchus, Bullialdus, Cape Laplace, Copernicus, Gassendi, Plato, Proclus and Theophilus.

2.4 Colorimetric work

Though the Moon presents a pretty uniformly monochromatic appearance, browns, greens, and purplish tints do occur, as well as various tones of grey. A preliminary description of areas with most marked tints and variations of tint in 15 formations is given by Haas in B.4.55.

Usual method of observation has been to use the grey tint of a mare, at a uniform distance from the terminator, as a standard of reference. This is inevitably unsatisfactory, allowing personal idiosyncracies to run riot, as well as being intrinsically imprecise. The employment of colour filters to exaggerate or suppress tints seems to extend most chances of a useful technique, but much work is still needed in its development. There is scope here for the accumulation of a mass of observational data, prior to which no satisfactory discussion can be made.

Records should include: date, time, longitude of terminator (or, Sun's colongitude), atmospheric clarity, Moon's altitude, aperture and type of instrument, magnification, filter (if any, and giving the band-width whenever possible), method of comparison used.

Considerable work has been done, especially in America, on the colours of lunar shadows. With small apertures these invariably appear jet black, but with larger instruments a brownish tint can often be seen or imagined. Very faint at best, it is probably psychological—due to irradiation or the effect of contrast. But more work, as objective as possible, is needed. 'Coloured shadows' have been suspected in the case of (*inter alia*) Eudoxus, Klaproth, Petavius, Phocydides, Pythagoras, Stevinus, Victa, Zuchius. Such studies are now, effectively, a part of T.L.P. research. Examination of different maria and upland regions for mutual variation of hue is, because of the faintness of the colour and the large variations in visual colour perception, of subjective interest only.

2.5 Occultations

Since these are once again a popular field of amateur observation, there is always a need for more observers, particularly those with driven equatorial telescopes who can undertake the observation of reappearances (which at present are vastly outnumbered by observations of disappearances). The relative simplicity of the equipment, the ease with which the observations are made, and the small fraction of the observing time which they occupy, mean that they should appear on the regular programme of every serious observer.

Errors between the observed and calculated position of the Moon—occultations providing one of the most convenient and accurate methods of determining the Moon's position—may be due to a number of factors, error in mean longitude being the chief; the detection of these errors is a vital check on lunar theory. The steadily increasing mass of occultation observations can eventually be used to check on any secular changes in the Moon's motion.

Observations are useless unless made available.

All objects within $6°.5$ of the ecliptic are liable to be occulted; these include the planets (Mercury and Pluto only sometimes), Regulus, Spica, Pollux, Aldebaran and Antares (mag 1), β Tau, γ Gem, β and δ Sco, η Oph and σ Sgr (mag 2), the Pleiades and the Hyades.

Occultations tend to recur under similar conditions after a period of 6798^d ($18^y.6$), the time taken by the Moon's nodes to complete one revolution, though conditions may be reproduced more accurately after 6798 ± 27^d.

The length of time (z) over which a given star may be subject to successive monthly occultations depends upon its latitude. Assuming mean conditions, it is given, in days, by

$$\frac{z}{3°.17724} = x - y$$

where x, y are such that

$$\sin x = \frac{\tan (b + 1° \ 12' \ 36'')}{\tan 5° \ 9'}$$

$$\sin y = \frac{\tan (b - 1° \ 12' \ 36'')}{\tan 5° \ 9'}$$

where b = star's latitude,
$1° \ 12' \ 36''$ = Moon's mean semidiameter plus parallax,
$5° \ 9'$ = mean inclination of Moon's orbit,
$3°.17724$ = mean daily motion of the nodes.

When $b = 0°$, z (minimum) = 512^d (c. 20 successive occultations).
$b = 3° \ 56'$, z (maximum) = 2193^d (c. 80 successive occultations).
$b = 5° \ 9'$ (max), $z = 1517^d$ (c. 57 successive occultations).

The maximum time during which a star can be under occultation is $1^h 54^m$, the following conditions being observed:

(a) Occultation diametral, i.e. star, centre of Moon's disc, and observer in line at mid-occultation.

(b) Moon at apogee (minimum angular diameter, but also minimum velocity).

(c) Observer situated at equator (maximum displacement in same direction as the Moon, owing to the Earth's rotation).

(d) Moon on meridian at mid-occultation (maximum parallax in RA).

For a more detailed account of the conditions of occultations, see B.4.86.

2.5.1 Equipment and preparation for observation: A small refractor, mounted anyhow (since the telescope is most conveniently kept stationary), is perfectly adequate. Use moderate magnification and full aperture, and see that the observing position is perfectly comfortable and relaxed. The observable magnitude limit of a star near occultation will be considerably brighter than the normal limit of the telescope, even if the sky is transparent, because of the Moon's glare; the slightest haze can make even a naked-eye star difficult to see with a small telescope, particularly if the Moon is gibbous. Observations of faint occultations are facilitated if the telescope is driven, so that the star can be held at the centre of the field of view. A driven telescope is practically essential for the successful observation of reappearances.

If an undriven telescope is used, time the Moon's drift across the field diameter a few minutes before the event is due to occur. At half this time before the predicted time of occultation, set the telescope so that the star is at the f edge of the field. The occultation is then likely to occur at or near the centre of the field, which both aids observation and provides some degree of warning. This procedure is risky, however, if the observing station is far (say, more than 150 miles) from the tabulated station, and is impossible in the case of stars fainter than mag 7.5, whose occultations are not predicted. *B.A.A.H.* and *A.A.* predictions are normally accurate to within $0^m.3$, but may be inaccurate by $\pm 0^m.5$ or, exceptionally, $\pm 1^m$.

With a clock-driven equatorial, some degree of preparedness for the disappearance or reappearance is given by an accurately set wristwatch or clock visible from the telescope.

Predictions given in the *A.A.* and *B.A.A.H.* include coefficients for deriving the approximate time of disappearance or reappearance at any observing station other than those tabulated (see section 2.11).

B.A.A.H. gives predictions of dark-limb disappearances down to mag 7.5, bright-limb disappearances of the brighter stars, reappearances in the case

59

of brighter stars only, daylight occultations of stars of mag 1 or brighter, and all planetary occultations.

2.5.2 Timing: Accuracy required: for routine occultations, timings to the nearest $0^s.1$ should be attempted. To achieve something near this, the ordinary stopwatch measuring 60 seconds per revolution is not sufficiently precise, since its mechanism will probably work only to the nearest $0^s.3$ or so; one with 10 seconds per revolution is ideal, although, to be realistic about the accuracy attainable by visual means, a 30 seconds per revolution watch should be satisfactory. The watch must be set against a time signal of known precision, such as the telephone speaking clock (accuracy about $\pm 0^s.05$) or an international radio time signal. For methods of timing observations see, further, *A.A.H.*, sections 22.7, 28.19.

2.5.3 Observational record: The essential data required for the reduction of the observation are:

1. Designations of star (catalogue number); in the case of unpredicted occultation, sufficient data to ensure its correct identification.
2. Date.
3. Derived time of occultation.
4. Longitude and latitude (to nearest $''$arc, readily measurable from a 1:1250 Ordnance Survey map, if available), and height above MSL to nearest 30 m.

Additional data which should if possible be included are:
5. Atmospheric clarity.
6. Stellar mag.
7. Phase: Disappearance or Reappearance.
8. Whether at dark or bright limb; if latter, whether earthshine; if eclipsed.
9. Method employed.
10. Assigned weight, estimated accuracy, or degree of confidence.
11. Estimated personal equation, and whether its correction is incorporated in the derived time.
12. Any anomalous appearances: gradual or step disappearance,* pausing on limb, projection over the limb, etc.

It is worth tabulating the data—watch time, clock time, clock correction, watch correction (if any), personal equation, etc—neatly and logically in the Observing Book, to reduce the chances of arithmetical errors, such as the loss or gain of a whole number of seconds or minutes, the application of corrections with the wrong sign, etc.

---

* As is possible in the case of a close double.

Observations being submitted to the Nautical Almanac Office should be reported on the special forms that they distribute to regular observers.

2.5.4 Graze occultations: These are predicted, for stations in the British Isles, in the B.A.A. *Handbook*. Because of the uncertainty attached to these phenomena (there may be more than one occultation event, if the limb is irregular and the star reappears in a valley on the lunar limb), a tape recorder on which a commentary can be spoken against a background of reliable time signals enables the details to be documented later, at leisure. The Nautical Almanac Office issues, when known, a limb profile for each predicted graze occultation, so that the observer has at least some idea of what to expect. The normal procedure is to arrange for a group of observers to be spaced at various distances both inside and outside the predicted graze limit, covering a band several hundred metres wide.

Reports should include:

longitude and latitude of observing station,
instrumental details,
whether or not an occultation occurred,
whether passage behind lunar mountains was observed,
times of all observed disappearances and reappearances, to nearest $1\overset{s}{.}0$, or better if practicable.

2.5.5 Personal equation: *A.A.H.*, section 25, presents a general account of personal equation, and emphasises its dependence upon a variety of factors—notably the observational method employed.

In the particular case of occultation observations, the more important factors influencing personal equation are:

(a) disappearance or reappearance,
(b) method: eye-and-ear, chronograph, or stopwatch,
(c) bright limb, dark limb, or dark limb in earthshine,
(d) brightness of the star.

(a) Reappearances, as compared with disappearances are subject to a large systematic lag, irrespective of (b), (c) or (d), but they are particularly in demand because (1) far fewer reappearances than disappearances are observed—the ratio is about 1:10—and (2) they are more difficult to observe photoelectrically at professional observatories. Some 2,000 experimental observations by a single observer gave $0\overset{s}{.}10$ and $0\overset{s}{.}18$ as the average equation for disappearances and reappearances respectively, which suggests that the latter are not as difficult to observe as might be imagined.

Another reason for the general neglect of reappearances is that, since they occur after Full, they tend to be morning rather than evening phenomena.

(*b*) Experimental observations indicate that the chronograph method involves the largest but most uniform equation; the eye-and-ear method, the smallest equation but the greatest variations from the mean. That is, the chronograph is more subject to systematic error, the eye-and-ear method to accidental error. Available figures for the stopwatch are inconclusive; one observer found a constant lag of $0^s5 \pm 0^s1$ with an artificial star. The total lag in this case is compounded of the observer's personality and possible starting and stopping errors in the watch itself. Consequently the observer's equation must be redetermined if a new watch is brought into use. The results of various series of experimental observations of dark-limb disappearances, probably fairly representative of the equations of experienced observers generally, are quoted below:

	Mean	Limits	
Chronometer (mean of 10 observers) . .	0^s305	−0.1	−0.46
Eye-and-Ear (” 4 ”) . .	0.08	−0.03	+0.09
Stopwatch (” 2 ”) . .	0.055	−0.04	−0.07

(*c*) In observations of disappearances, the minimum equation is encountered at the dark limb illuminated by earthshine; at the invisible dark limb it is larger; and largest at the bright limb. The reason is probably that the greater the degree of warning received, the more the attention is focused on the muscular response rather than on the reception of the stimulus.* The following figures refer to 210 experimental observations with chronograph by two observers:

At visible dark limb	0^s25	0^s29
At invisible dark limb	0.29	0.32
At bright limb (favourable conditions)	0.33	0.32
At bright limb (excessive glare)	$\pm$ 2–3 secs	

(*d*) A small magnitude equation, inversely proportional to the brightness of the star, occurs in occultation observations. The following figures refer to three series of experimental observations, (*i*) by an observer using the eye-and-ear method, (*ii*) and (*iii*) by two observers using chronograph:

---

* See, further, *A.A.H.*, section 25.

	Magnitude	Mean equation
(i)	2.0−5.4	+0^{s}003
	5.9−7.0	0.035
	7.5−8.5	0.085
(ii)	4−5	−0.40
	6	0.47
	7	0.49
	8−9	0.59
(iii)	0−5	−0.35
	5−7	0.47
	7−10	0.65

Every observer intending to include occultations in his regular programme should determine his own personal equation under the different conditions described above. This can be done approximately (since neglecting the factor of limb nature) by observing a switch-controlled artificial star which is extinguished by an assistant working within sight or earshot of a clock.

Dunham has recently emphasised the great importance of visual occultation work, despite the fact that photoelectric measures are of superior accuracy. Since the profile of much of the lunar limb is not accurately known, and since the precise positions of many occulted stars have not been determined, the combined error may in many instances be equivalent to a timing uncertainty of about ±0^{s}3, rendering greater precision than this largely nugatory. Until the accumulating mass of photoelectric timings and improved astrometric positions refine our knowledge of the lunar limb, visual occultation timings are on average outweighed only 6:5 by photoelectric ones (*Sky & Tel.*, 54, No. 6, 478).

2.5.6 Prediction: Owing to the comprehensive data given in *B.A.A.H.* and *A.A.* (see sections 2.11, 2.12) it is not usually necessary for the amateur to undertake his own predictions. Nevertheless, the following notes on simple methods may be of interest:

1. Rigge's graphical method (B.4.75) quickly yields results which are accurate enough for the practical purpose of not missing the occultation.

2. Comrie's semi-graphical method, involving 3-figure work, also quickly gives results with a maximum probable error of the order of 1^m (B.4.49).

3. The parallax tables given in *B.A.A.H.*, 1929—designed for the reduction of the Moon's apparent position from its geocentric coordinates as given in the *A.A.*, thus facilitating the identification of stars involved in

unpredicted occultations—can also be used for approximate predictions of occultations.

2.6 Eclipses

Points to watch for include:

(*a*) Naked eye:
>   Definition of edge of shadow.
>   Colour and density of shadow throughout the eclipse.

(*b*) Telescopic:
>   Times of contacts.
>   Definition of edge of umbra (very variable and unpredictable); any irregularities in outline; width of 'edge' (region of noticeable density gradient).
>   Edge of shadow of uniform colour; whether same as inner region of umbra.
>   Visibility of limb and of surface features in different parts of the shadow:
>>   (*i*)   record of selected areas from bright to dark right through the passage of the shadow,
>>   (*ii*) any anomalies noticed elsewhere.
>
>   Any variation or colour between first and second half of the eclipse.

(*c*) Thermocouple observations.*
(*d*) Comparative visibility with different colour filters.
(*e*) Photographic record.

Besides details of the above, the record should include:
>   Longitude, latitude, and height above MSL of the observing station.
>   Atmospheric conditions, including details of any lunar halo.
>   Aperture and magnification (if not the same throughout, specify fully).

Other notes:
>   Use UT, to nearest $0^m.1$.
>   Distinguish between celestial and selenographic cardinal points, if used.

Use the faintest practicable red light for note-making. The visibility of the

* See also *A.A.H.*, sections 21.6, 21.7.

penumbra, when this alone lies on the lunar surface, can often be increased by the use of a neutral filter to reduce glare.

The total brightness of the Moon at different total eclipses varies considerably, and is of interest. In order to compare the Moon's light with that of a star or planet whose magnitude is known, the disc can be diminished effectively to stellar size by, e.g., virtual image reflected in a silver Christmas tree ball; real image formed by a short-focus lens held a metre or so from the eye; reversed hand telescope or binoculars. The Moon, and the comparison star or planet, are compared alternately, using whatever device is selected. Thus, at the total lunar eclipse of 18 November 1975, a mean magnitude of −2.9 at mid-eclipse was obtained by comparing the Moon with Jupiter, then of magnitude −2.4. On the other hand, the magnitude at the eclipse of 24 May 1975 was only 0.7.

2.7 Photography

Excellent lunar photographs can be taken with any standard camera attached to a reflector of 15 cm aperture or more. At the Newtonian focus of an $f/8$ instrument, using medium-speed film (200 ASA, say), the correct exposure will range from about 1/250 sec at Full to perhaps 1/25 sec in the crescent stage. A Barlow lens can be used to enlarge the prime-focus image to a diameter of about 20 mm, so that it will fit comfortably into the frame of a 35 mm camera. A 15 cm aperture instrument giving a lunar image of this size will effectively be working at $f/15$; hence, the exposures will now have to be about four times as long as those indicated above. The terminator regions being much dimmer than the rest of the disc, a certain amount of experimenting will be required to find the best all-round exposure for different phases. See, for an introduction to the subject, B.4.58.

A single negative or print cannot handle the whole range (something like 1000:1) of intensities encountered on the lunar surface. Nothing can be done to improve the rendering of a colour transparency, but a black-and-white print can be 'dodged' during enlarging, so that the terminator region receives less exposure than the rest of the disc and holds its detail better (B.4.73).

If it is required to do more than take general views of the whole surface, and to achieve the maximum resolution of which the telescope is capable (or, in practice, the maximum detail that the atmosphere is prepared to permit), considerable magnification of the primary image is necessary. Taking the limit of photographic resolution with medium-speed film as about 0.025 mm, it follows that the resolving power of the objective, represented in the focal plane, should be somewhat greater than this if the film's grain is not to set a limit on recordable detail. To achieve this sort of scale requires a focal ratio of at least 50. Taking $f/80$ as a commonly adopted ratio for high-resolution lunar and planetary work, the lunar image formed by a 15 cm instrument will be about 110 mm across, and for other apertures in proportion.

Because of its large initial focal ratio, necessitating an amplification of only x3 or x4, a Cassegrain is the most satisfactory instrument for this sort of work, but in practice some of the best amateur photographs have been taken with Newtonians of 25 to 30 cm aperture. The necessarily high amplification of about x10 is obtained by using a well-corrected eyepiece of between 6 and 12 mm f; the distance d between this enlarging eyepiece and the final focal plane, for an image enlargement of A, is given by

$$d = f(A + 1).$$

The relative faintness of the image, and the much longer exposure required (typically, from 0.2 to 2 secs), means that the lunar photographer cannot merely gauge moments of good seeing by watching the image in the viewfinder of an SLR camera, and then press the shutter: the image will probably be too faint for the effects of seeing to be obvious, while the mirror slam of the camera will almost certainly jar the telescope. The procedure commonly adopted by serious lunar and planetary photographers is to employ a leaved shutter of the Compur variety in a special purpose-built unit, and to take a series of photographs during periods of generally good seeing.

Approximate equivalent exposures at different lunar phases are as follows:

Age	$3\frac{1}{2}^d$	7^d	14^d	21^d	$24\frac{1}{2}^d$
Exposure factor	12	4	1	4	12

When the Moon is in apogee, the perigee exposure needs to be increased by a factor of about 1.25. The region of the disc in which maximum detail is required to be shown also has a bearing on the exposure. At Full the brightness of the disc is relatively uniform, but at other times the limb is markedly brighter than the terminator. Therefore somewhat longer exposures are required if the terminator is the area of interest (the limb being over-exposed) than when the limb is (the terminator then being under-exposed). In connexion with exposure times, reference may also be made to B.4.73.

Owing to the Moon's eastward motion lunar rate differs from sidereal rate. For exact following in RA, the clock needs to be retarded very slightly to counteract this eastward motion, which amounts on the average to about 0.023 diameters per minute. At accurate sidereal rate an amplified image of diameter 50 mm cannot be exposed for more than about 2^s5 (corresponding to a linear displacement of the image on the plate of approximately .05 mm). The Moon's motion in Dec amounts to about 0.0092 diameters per minute at maximum and zero at minimum, the latter occurring when the moon is in greatest N or S Dec. From these figures the maximum permissible exposure with a given equipment, driven at sidereal rate, can be calculated.

Anyone undertaking lunar photography is recommended to read the appropriate section in *A.A.H.*, as well as consulting, for example, B.5.10, 5.12.

2.8 Phase

The position of the lunar terminator is the most important single factor in the observation of the Moon; it is closely followed by libration and altitude, referred to in the next two sections.

The position of the terminator at a given time is specified in terms of selenographic longitude, which is reckoned E and W from the central meridian. The approximate longitude of the terminator throughout the lunation may be read from Figure 12. Or it may be derived from the

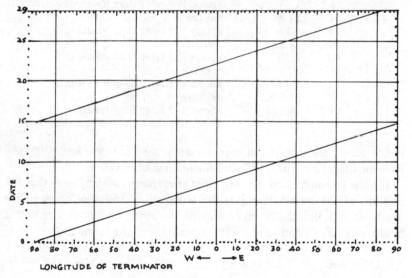

FIGURE 12

Sun's selenographic colongitude, whose daily value is tabulated in *A.A.* (E and W are used here in the classical sense):

67

Sun's colongitude C	Longitude of terminator
$0° - 90°$	$C°$ East
$90° - 180°$	$(180° - C)$ West
$180° - 270°$	$(C - 180°)$ East
$270° - 360°$	$(360° - C)$ West

Taking $29^d \; 12^h \; 44^m$ as the mean length of the lunation, a particular phase observed at night in a given lunation (lunation 0) will fall in daylight in the next lunation (lunation 1) and in subsequent odd-numbered lunations; in each even-numbered lunation from 2 onwards it will recur roughly 1½ hrs later. In a period of 15 lunations the day-occurring phase will again have worked round to the time of observation in lunation 0. Thus:

Lunation	Time of occurrence of a certain phase	
0	$0^h \; 0^m$ GMT:	midnight
1	12.44.	daytime
2	1.28:	about $1^h.5$ later than lunation 0
3	14.12:	daytime
4	2.56:	about $1^h.5$ later than lunation 2
5	15.40:	daytime
6	4.24:	about $1^h.5$ later than lunation 4
7	17.8:	daytime
8	5.52:	about $1^h.5$ later than lunation 6
9	18.36:	daytime
10	7.20:	about $1^h.5$ later than lunation 8
11	20.4:	evening
12	8.48:	about $1^h.5$ later than lunation 10
13	21.32:	evening
14	10.16:	about $1^h.5$ later than lunation 12
15	23.0:	about the same time as lunation 0

Hence phases are repeated at approximately the same time in the 2nd and 15th lunations, 59^d and 443^d after the original observation.

It must be emphasised, however, that phase means nothing more than the position of the terminator upon the visible disc. The appearance of the disc is altered, though the phase may be the same, by libration as well as by the varying inclination of the line of cusps with the terminator.

2.9 Libration

The second most important factor in lunar observation. It not only alters the inclination of the terrestrial observer's line of sight to every part of the visible lunar surface, but in addition is of paramount importance in the observation of the limb regions. In this work, full advantage must be taken of libration, which is continuously exposing a different region of the limb.

Libration in longitude—which can cause a maximum displacement of the disc's mean centre of $7°$ $45'$ (selenographic) in either direction—is the combined effect of uniform rotation and non-uniform revolution. Libration in latitude—which may displace the mean centre by $6°$ $44'$ on either side of the observed centre—is due to the non-coincidence of the equatorial and orbital planes. The mean centre may therefore be displaced by as much as $10°$ $16'$ on either side of the observed centre, owing to combined libration in longitude and latitude. Libration thus alters the appearance of the disc considerably—at extreme libration in longitude, e.g. Mare Crisium is almost on the limb—and is responsible for revealing about 60% of the lunar surface to us at one time or another. Diurnal libration, due to the displacement of the observer's viewpoint arising from the Earth's rotation, may amount to nearly $1°$ in addition to physical libration.

The following two tables show how libration is related to the displacement of the mean centre of the disc, to the shape of the lunar meridian, to the visibility of the limb, and to the optimum observing conditions for objects near the limb in different regions (E and W classical):

Libration	Displacement of mean centre of disc	Meridian convex towards	Exposed limb
Longit. +	E	E	W
—	W	W	E
Latit. +	S	S } equator	N
—	N	N	S

Region of disc	Optimum conditions of libration	
	Libration in Longit. (or Selenographic Longit. of Sun)	Libration in Latit. (or Selenographic Latit. of Sun)
N hemisphere	0	+
S hemisphere	0	—
E hemisphere	—	0
W hemisphere	+	0
NE quadrant	—	+
NW quadrant	+	+
SE quadrant	—	—
SW quadrant	+	—

Dates of maximum libration in longitude and latitude are given in *B.A.A.H.*; *A.A.* tabulates the displacement of the mean centre in longitude and latitude (to 0°.01) for every day of the year; the most exposed region of the limb, given the Earth's selenographic longitude and latitude (from *A.A.*), can be read off Figure 13.

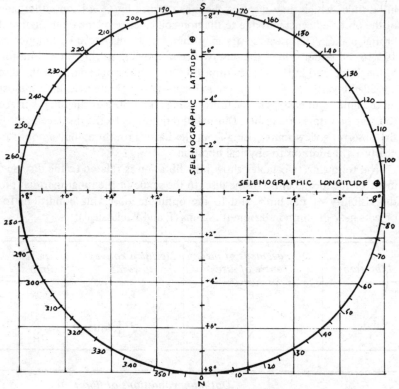

FIGURE 13
LIBRATION CHART

For the required date, take the Earth's selenographic longitude and latitude; plot the corresponding point against the axes. Connect the origin of the axes with this point and produce the line to intersect the circle. This point of intersection indicates the position angle of the region of the Moon's limb most favourably placed for observation. (Modified from R. E. Diggles, *J.B.A.A.*, **44**, No. 4, 144)

2.10 Altitude

The Moon's altitude at different phases varies with the time of year. Figure 14 representing the mean altitude of the Moon in its four phases throughout the year, as seen from a station in the Midlands, reveals the following relationship between phase, altitude, and season:

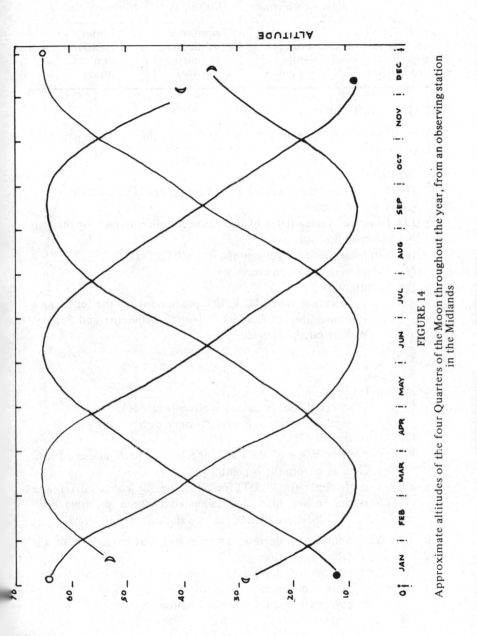

FIGURE 14

Approximate altitudes of the four Quarters of the Moon throughout the year, from an observing station in the Midlands

	Maximum altitude	Mean altitude	Minimum altitude
New	summer	equinoxes	winter
1st Quarter	spring	solstices	autumn
Full	winter	equinoxes	summer
3rd Quarter	autumn	solstices	spring

2.11 B.A.A.H. data

Vary slightly from time to time; in recent years the following data have been given:

(*a*) Details of the year's eclipses.

(*b*) The date and time (nearest 1^m) of the phases of all lunations.

(*c*) Table of apsides.

(*d*) Libration Table: dates of maximum libration in each of the four cardinal directions.

(*e*) Sun's selenographic colongitude for each day at 0^h.

(*f*) Tables of moonrise and moonset.

(*g*) Occultations:

Selection from the *N.A.O.* predictions for the following 6 stations: Greenwich and Edinburgh, Melbourne and Sydney, Wellington and Dunedin:

Date

Star

Magnitude of star

Ph: Phase, i.e. D(isappearance) or R(eappearance)

         Elongation of Moon, to nearest whole degree

UT: of the D or R

P: position angle of the star at the Moon's limb, measured anti-clockwise from the N point

$\left.\begin{array}{c} a \\ \\ b \end{array}\right\}$ coefficients (in ^{m} UT) for obtaining the corrected UT at a station other than that tabulated (with a probable error $<2^m$ if the separation of the two stations is <300 miles):

If $\pm\,\Delta\lambda=$ number of degrees the observing station is W/E of the tabulated station

      $\pm\,\Delta\phi=$ number of degrees the observing station is N/S of the tabulated station

        $T=$ corrected time for observer's station

        $T_o=$ tabulated UT

then $T = T_o + a\,(\pm\Delta\lambda) + b\,(\pm\Delta\phi)$

(*h*) Map showing the tracks of all grazing occultations visible from the British Isles.

2.12 A.A. data

The following information will be of particular use to amateur observers:

(*a*) Phases, perigee and apogee.
(*b*) Ecliptic and equatorial co-ordinates—daily ephemeris.
(*c*) Daily ephemeris for physical observations.
(*d*) Occultations.
(*e*) Tables of moonrise and moonset.
(*f*) Details of the year's eclipses.

2.13 Lunar data

Mean distance from Earth: 0.0025695 A.U.
Sidereal period: $27^{d}.321661$.
Synodic period: $29^{d}\ 12^{h}\ 44^{m}\ 2^{s}.9$.
Eccentricity: 0.05490.
Stellar magnitude of Full Moon at mean distance:$-$ 12.7.
Semidiameter at unit distance: $2''.40$.
Semidiameter at mean distance: $932''.58$.
Diameter: 3476 km.
Mass: $1/81.301 \oplus = 0.0123 \oplus = 3.674 \times 10^{-8} \odot$.
Volume: $0.0203 \oplus$.
Density: $0.60 \oplus = 3.34 \times$ water.
Mean superficial gravity: $0.16 \oplus$.
Lunation number: Brown's series of lunation numbers, starting at 1923 Jan 16. Convenient reference system for all lunar observations. Lunation No 680 began on 1977 Dec $10^{d}\ 17^{h}\ 33^{m}$.

SECTION 3

GENERAL NOTES ON PLANETARY OBSERVATION

3.1 Introduction

Planetary observation is at the same time one of the most popular branches of amateur work and one of the most exacting. The usefulness of larger than moderate apertures is much reduced by the atmospheric factor, which suits the amateur's purse; but the dividends paid by visual acuity are correspondingly high. The training of the eye and the mind to the particular requirements of planetary observation—such as the perception of infinitesimal tonal contrasts—is also a condition for extracting full value from the performance of the instrument.

The amateur intending to specialise in planetary observation is recommended to spend some time on the observation of drawings under similar conditions to those of the observation of the planet itself, and on the experimental investigation of his acuity by means of test drawings of 'planetary' detail observed from a distance both with and without the telescope. The advantage of being able subsequently to study close-up what he has just been observing under conditions similar to those obtaining in the actual observation of the planet will teach him more about visual acuity, thresholds of perception, and the observation and mis-observation of planetary detail than any amount of reading on the subject. Hargreaves' Presidential Address to the B.A.A. in 1944 (B. 5.4) should be read by all new planetary observers. See also *A.A.H.*, sections 2.6, 24.6.

3.2 Some requirements of planetary observation

(*a*) Experience, which is equivalent to the training of the eye, especially in the direction of sensitivity to detail near the threshold of size and contrast.

74

(*b*) Systematic observation: a homogeneous mass of observations, made over a large number of apparitions, is invaluable.

(*c*) Maximum possible reliability of all recorded detail, as regards shape, size, position, and intensity. All detail that is 'glimpsed', or otherwise uncertainly observed, must always be specifically recorded as such, and no drawing should exaggerate the clarity, definiteness, or certainty of any observed feature.

(*d*) Full instrumental details, a statement of the condition of the seeing, and the time at which the main outlines of every drawing were laid down must accompany all records.

(*e*) The 'working up' at a later time of more or less 'rough' notes made at the telescope is not a satisfactory practice. Factual accuracy, not the production of works of art, is the correct aim.

3.3 Instrumental

As regards aperture, there are two schools of thought. W. H. Pickering advocated moderate apertures for planetary work, i.e. from 30 to 40 cm, maintaining that owing to atmospheric factors, instruments of this size will show finer detail than larger telescopes on nine nights out of ten; even under the very favourable conditions in Jamaica, nothing was to be gained, in his opinion, by using more than 50 cm even on nights of the most superlative seeing. In support of this view he instances the fact that at the 1924 opposition of Mars he observed with 28 cm in Jamaica all that was visible with the Lick 36 in (91 cm) refractor. Owing to the comparatively poor seeing at Lick, however, it is possible that in both cases more would have been seen with apertures of about 50 cm. Jarry-Desloges, working under excellent conditions in Algeria, agreed with this general conclusion, reckoning that for planetary work 20–35 cm were superior to 50 cm on 95–98 nights out of 100.

Antoniadi disagreed with this, maintaining that increased light grasp (making accessible fainter tone and tint gradations) and resolution (making accessible finer structure in the image) will probably combine to beat the performance of a smaller telescope on average nights, will certainly do so on the best nights, and even on really bad nights can take advantage of the moments of improved seeing that occur intermittently.

H. P. Wilkins similarly emphasises the advantages of large apertures for planetary (as for lunar) work: with the Meudon 84 cm, for example, he was able to detect a mass of detail within lunar formations which his own 39 cm showed as comparatively flat and featureless even under superior magnification.

In section 1.4 of *A.A.H.*, which deals with the apparent brightness of extended (i.e. non-stellar) images, it is shown that

$$\frac{\text{telescopic brightness}}{\text{naked-eye brightness}} = \left(\frac{M'}{M}\right)^2$$

where M' is the minimum useful magnification,* and M (the magnification used) must be higher than M_r (the resolving magnification†) and should in fact be as much higher as the increasing dilution of the image and reduction of contrast will allow. But $\frac{\text{telescopic brightness}}{\text{naked-eye brightness}}$ is a function of aperture, and hence the same image brightness is given by a large aperture using a magnification M as by a smaller aperture using a lower magnification than M: e.g. by a 25 cm using ×800 as by a 75 mm using ×240. At the same time the larger aperture offers the further advantages of increased resolution and images of more than 3 times the angular size, although, in practice, magnifications exceeding ×400 are rarely usable in planetary work.

Planetary observations worth recording can be made with a 10 cm, though the desirable minimum is about 5 cm larger than this. With 15 cm, full participation in the observing programmes of the Planetary Sections of the B.A.A. is possible; for regular work on the inner planets rather more aperture is desirable—say, 20 cm as the useful minimum—though spasmodic valuable observations can of course be made with smaller instruments. Summarily: an aperture of 15 cm is sufficient for regular planetary work, although 30 cm increases the observer's range enormously, and anything larger will be more effective still on nights of good seeing.

Where aperture, without limit, has the undoubted advantage is in photographic work.

Regarding type of instrument, it is generally agreed that for critical definition of an extended image a refractor is superior to a reflector of equal aperture and optical excellence, owing to the different structure of the diffraction pattern in the two instruments.‡ The difference, however, is at least partially due to the difference in focal ratio, since a refractor at $f/20$ will be found superior to one at $f/15$, of equal aperture, where the delineation of fine detail is concerned. Where money is no object, the prospective planetary observer will invest in a refractor; where it is, he will be well advised to decide in favour of the larger reflector that the same outlay will secure.

* About 1.3 D (i.e. 1.3 × the aperture in cm).
† About 5 D at least.
‡ See *A.A.H.*, sections 2.2, 2.3, 2.6, 23.

Lowell's extreme view that for planetary work reflectors are virtually useless is not confirmed by the experience and records of innumerable amateurs. It is true, however, that the usefulness of a reflector for this type of work does not increase in direct proportion to aperture.

There is no doubt that a motor-driven equatorial eases the planetary observer's task. The very nature of the work compels him to spend long sessions (in the case of Jupiter, in particular, possibly several hours) at the eyepiece, and manual following with an undriven equatorial or an altazimuth, although it soon becomes virtually automatic, does distract the observer to some extent from the job in hand. On the other hand, it is not true to say that an automatic drive is essential, and the observer should certainly not be deterred from undertaking serious planetary observation because he has only an altazimuth telescope. What *is* indisputable is that all work involving critical definition demands objectives of the finest quality. A poorly figured mirror or an inferior OG will certainly fail to reveal detail that should be visible with the aperture concerned, and may even contribute spurious effects of its own.

3.4 Magnification

With given aperture, brightness of an extended image varies inversely with the magnification employed; given M, it varies directly as D. In planetary observation it is excess rather than lack of light that is, generally speaking, the embarrassment, and the value of aperture (up to the limit imposed by the atmosphere) lies primarily in its resolving power.

So far as generalisations can be made, glare is likely to appear as a factor tending to submerge fine detail present in the focal plane image when the magnification is reduced below about $8D$. Reduction of image brightness without reducing its resolution can be obtained by such means as neutral filters, partial reflection, polarising eyepieces, etc.

Saturn often appears to stand magnification better than either Mars or Jupiter. Mars, owing to its comparatively small disc, requires always the maximum M that the atmosphere will allow; it is comparatively seldom that the seeing will allow M as high as $20D$ or $25D$. The small disc of Uranus also necessitates use of the highest M possible. Venus naturally demands a rather wide range of M, and with small and moderate apertures can often be observed in the daytime with as much as $18D$ to $20D$. The magnification of Mercury's image is limited by want of illumination, unless D is considerable; amateur instruments will stand about $12D$ as the maximum that is normally useful.

In this connexion it is of interest to learn what sort of magnifications are employable with instruments outside the amateur range. Barnard considered that $15D–20D$ was much too high for planetary work with the Lick 36 in (91 cm) refractor: $8D–12D$ was nearer the mark, and in the case of Jupiter as little as $4D–6D$. Fine Martian and Jovian detail was invariably better seen with a magnification of $4D$ than $6D$, while only the coarsest detail was visible with $11D$. The latter was found suitable for the observation of Jupiter's satellites and some of the larger asteroids. Saturn, though benefiting more from magnification than its inner neighbours, was still seen better with $6D$ than $11D$.

The following figures summarise the discussion: they provide no more than a very rough indication of the sort of magnifications per cm aperture that will be found suitable for small (10 cm and less) and moderate (12–30 cm) apertures on nine nights out of ten. The corresponding figures for the Lick 36 in (91 cm) are added for comparison:

	Mars	Jupiter	Jupiter's satellites	Saturn	Uranus
Small apertures	15	15	25	20	25
Moderate apertures	6–12	6–12	15–18	12–16	15–18
Large apertures	4–6	4–6	12	8–12	12

Denning, who used a 32 cm reflector for many years, concluded that for planetary work about $10D$ was the most frequently applicable, $15D$ being superior only on rare occasions of superlative seeing.

Many observers who have had long experience of planetary work with a variety of instruments—Denning and Ellison among them—have been of the opinion that, with amateur instruments generally, from x200 to x400 is the best planetary magnification range, more or less irrespective of aperture; more than x400 involves considerable aperture, whilst even x200 requires reasonably good seeing. The whole matter of the atmosphere and seeing is discussed in some detail in *A.A.H.*, section 26.

W. H. Pickering's often repeated dictum that x400 is the minimum M with which original work can be carried out on the Moon and planets need not be taken too seriously, in the face of voluminous evidence to the contrary.

3.5 Oculars for planetary work

Of the characteristics exhibited by different designs of ocular,* the most important in planetary work are:

(*a*) Critical definition over a small field: for HP planetary (and lunar) work generally.

(*b*) Freedom from scattered light: especially in the observation of contiguous areas of greatly contrasted tone, and—even more important— for the visibility of small tonal contrasts near the threshold of visibility.

Providing that the focal ratio is reasonably large (as in the case of normally-constructed refractors and Cassegrains), ordinary Huyghenians give a satisfactory performance: they must be of first-class quality, however. Achromatic Ramsdens, orthoscopics and monocentrics have all proved satisfactory with Newtonians. In planetary observation small field of view, which is a feature of the monocentric type, is no drawback, since critical definition demands that the planet be fairly centrally located in any case. Freedom from scattered light is a major consideration, since any haze will obscure a low-contrast planetary detail, and in this connexion the solid Tolles ocular (a cheaper version of the monocentric) has much to recommend it.

Superlatively fine definition being the main prerequisite of the successful observation of planetary surfaces, suitably corrected oculars are absolutely essential with instruments of small focal ratio.

3.6 Colour observations

The estimations of colour are an important item in planetary observation. Long-term changes of tint, even if suspected, are virtually impossible to establish by existing techniques, owing to the absence of an objective standard against which to compare the estimates of different observers, or even those of the same observer made at different times. Colorimetry, in fact, is the most subjective branch of planetary work (cf. meridian transits, for instance).

Contributing to the unsatisfactory nature of colour observations are the following factors, *inter alia*:

(*a*) Difficulty of determining the personal equation for colour, and uncertainty as to whether it is subject to short-term and/or long-term variations. Most eyes have some degree of colour peculiarity (the two eyes of a single observer are not even necessarily the same), which can only be detected and estimated by comparison with simultaneous estimates of identical objects by other observers.

* Discussed in *A.A.H.*, section 8.1.

(*b*) The impossibility of treating the records of different observers as comparable, i.e. of being confident that discrepancies in the recorded observations indicate objective changes of tint, rather than different personal equations.

(*c*) The vague, unsaturated character of planetary (and lunar) tints generally.

(*d*) Factors affecting the colour estimates of the observer, the omission of which from the observational record further reduces the latter's value:

(*i*) Magnification: as M increases, the depth of all tints decreases. This effect starts to be operative at as little as $6D$ or $8D$, indicating the value of fairly large apertures for colour work.

(*ii*) Colour characteristics of the instrument. No colour estimates made with a refractor can be trusted implicitly; not only is secondary colour certainly present, but its evaluation is practically impossible. Reflectors, and achromatic eyepieces, are a *sine qua non* of satisfactory colorimetric work.

(*iii*) Relative darkness of the background, and atmospheric dispersion: high cloud, haze, and twilight modify tints as well as tones. Records should therefore state explicitly (α) whether the observation was made in daylight, twilight (within 2^h of the Sun on the horizon), or night, (β) altitude of object, (γ) atmospheric clarity in terms of the magnitude of the faintest zenithal star visible, or of the faintest star in the vicinity of the planet if the overall transparency is obviously not uniform.

How much false colour, due to atmospheric and instrumental dispersion, can be introduced into the image is illustrated by the following observational facts. The upper part of the image of a planet near the horizon will be tinted red, the lower part blue, owing to atmospheric dispersion. The colour introduced by a non-achromatic Ramsden varies from red at the centre of the field to blue at the periphery. Hence if the planetary image is placed in the upper half of the field, the two colour gradients will cancel each other, resulting in a more nearly colourless image. Definition may be improved in this way, but it is not difficult to appreciate that under conditions of this sort, objective estimates of colour are quite impossible. Colour estimates should always be made at or near culmination, and are worthless if the object is within $20°$ or so of the horizon.

Suggestions for colour observers:

(*a*) Continuous series of observations with a single telescope and magnification are required.

(*b*) Elimination of the personal equation might possibly be approximated to by the cooperation of a number of observers, working under identical conditions.

(c) Theoretically, every shade of colour is identifiable by the use of filters. In practice, the isolation of small patches of planetary colour by juggling with a large number of filters is out of the question. If, however, a filter is used, its number and/or name must of course be recorded.

(d) The record of each observation must, in view of the foregoing, be supplemented by the following data:

magnification employed,
date and time of observation,
state of the atmosphere,
altitude of the object,
filter (if any).

(e) Other things being equal, large apertures have the advantage over small, owing to the greater brightness of their images.

3.7 Radiometric observations

The measurement of planetary radiation requires apertures outside the range that the amateur can normally command—ideally not less than 75 cm. Reflectors are essential, since glass is opaque to the long-wave radiation re-emitted by the planetary surface after absorption. This re-emitted energy—constituting 56% of the total radiation of Mars, and 86% of that of the Moon—is almost completely absorbed by 0.1 mm of glass. Photoelectric cell windows have therefore to be made of rock salt, fluorite, or some similar material. *A.A.H.*, p. 442 may be referred to for the transmission curve of the atmosphere (whose absorption of these wavelengths is caused mainly by water vapour, carbon dioxide, oxygen, and ozone), and also for those of 0.165 mm of glass, 4 mm of fluorite, and 4 mm of rock salt. Anyone interested in such work should see B. 8.8, 8.9, 8.22.

3.8 Photography

With the exception of Venus, whose disc exhibits ultra-violet markings that can be photographed using suitable filters but are beyond the spectral sensitivity of most eyes, planetary photography lags behind visual observation in its perception of fine detail. The serious charting of planetary detail is a visual rather than a photographic field, at least as far as the amateur's equipment is concerned. The independent photographic confirmation of drawings and visual observations would certainly always be welcome; unfortunately, however, the details requiring such confirmation are usually those near the threshold of vision, which the camera is incapable of recording.

In a sense, then, planetary photography should not be confused with planetary observation. The observer who seriously wishes to study, say, the markings of Jupiter, their forms and rotation periods, will reach for a notebook rather than a camera. If he resorts to photography, it is because he finds the challenge arresting. Planetary photography, it must be admitted, is not the means to an end, but the end in itself. It is, essentially, an exercise in technique.

The characteristics of planetary images which are of particular concern to the photographer are their small angular diameter and their faintness. The first necessitates amplification of the prime-focal image; the second, together with the desire for improved resolution, demands the largest possible D. The demands are similar to those of lunar photography, and over and above these considerations hangs the permanent problem of atmospheric turbulence. The discussion in section 2.7 should therefore be referred to. Assuming a focal ratio of 80 and an emulsion speed of 200 ASA, the following exposures will probably be suitable: Mercury, 1^s; Venus, $0^s.1$; Mars, $0^s.5$; Jupiter, 1^s; Saturn, 4^s.

Planetary images being so small (in an $f/80$ system, $1''$ arc in the sky is represented by $0.00039D$), it is wasteful to obtain just one image on each 35 mm frame, yet most modern cameras have a shutter-locking mechanism that makes it impossible to obtain more than one exposure per frame. The serious planetary photographer will, therefore, have even more incentive than the lunar enthusiast to obtain or construct a special camera with a free winding mechanism that allows a continuous strip of film to be exposed. In this way, a hundred or more images can be obtained on a standard length of film, and the best selected for subsequent enlargement. Satisfactory cameras have been described by Rackham (B.4.73) and Dall (B.1.5), both of which have a retractable periscopic eyepiece to permit focusing and centring of the image just before an exposure is made.

To fix ideas, the following table (taken from *A.A.H.*, section 20.10) indicates the range of image sizes (in mm) of the Moon and planets in various $f/80$ systems:

D (cm)	*Moon*	*Mercury*	*Venus*	*Mars*	*Jupiter*	*Saturn*
10	68–78	.18–0.50	.38–2.5	.14–0.97	1.2–1.9	.57–0.80
15	103–117	.27–0.75	.58–3.7	.20–1.5	1.8–2.9	.86–1.2
20	137–156	.36–1.0	.77–5.0	.27–1.9	2.4–3.9	1.10–1.6
30	205–234	.55–1.5	1.20–7.4	.41–2.9	3.5–5.8	1.70–2.4
40	274–312	.73–2.0	1.50–9.9	.54–3.9	4.7–7.7	2.30–3.2
50	342–390	.91–2.5	1.90–12.4	.68–4.9	5.9–9.7	2.90–4.0

It is hardly necessary to add that the mounting for a photographic instrument must be exceptionally rigid, and that the driving mechanism must be capable of extremely accurate following over at least short periods of time. The latter is a different requirement from that for, e.g., stellar photography with a short-focus camera, when errors of a few " arc over the short term are permissible, provided the mean position of the image on the film does not change beyond this amount over several minutes.

3.9 Position of the ecliptic

The planets—as also the Moon, Zodiacal Light and Band, and Gegenschein —being confined to the zodiac, the varying altitude and inclination of the ecliptic are of practical interest to the observer.

Since the ecliptic intersects the equator at the equinoxes, its point of midnight culmination has an altitude of 37° for an observer in 53° N latitude (the Midlands) at the end of March and September. Its greatest and least altitudes are reached at the winter solstice (60½°) and the summer solstice (13½°) respectively. Hence winter oppositions of the outer plants are, in general, more favourably placed in the sky than those occurring during the summer.

At the spring equinox, that section of the ecliptic lying to the east of the Sun is inclined to the horizon at a larger angle than that lying to the west; at the autumn equinox this relationship is reversed. The relevance of these facts to the observation of the Zodiacal Light is obvious.

3.10 Sequence of apparitions

Superior planets:

As evening star: Opposition ⟶ E Quadrature ⟶ Conjunction.
As morning star: Conjunction ⟶ W Quadrature ⟶ Opposition.

Inferior planets:

As evening star: Superior Conjunction ⟶ E Elong. ⟶ Inf. Conjunction.
As morning star: Inferior Conjunction ⟶ W Elong. ⟶ Sup. Conjunction.

3.11 Data concerning the planets

B.A.A.H. gives the following general information about the planets each year, in addition to the specific data listed in sections 4–11:

(*a*) Graphical representation of the planets' visibility throughout the year, from which the times of rising and setting can be derived for stations in all parts of the British Isles.

(b) Scale drawings showing the changing appearances (phase and angular diameter) of the planets throughout the year.

(c) Elements of the planetary orbits.

(d) Dimensions of the Sun, Moon, and planets.

(e) Table of satellite data.

The *A.A.* contains a great deal of information which is vital to the planetary observer (also listed in sections 4–11).

The Annual Reports and Interim Reports of the various Observing Sections of the B.A.A. (published in *J.B.A.A.*) provide an almost uninterrupted review of planetary features and occurrences during the past 80 years or so.

SECTION 4

MERCURY

4.1 Apparitions

The following intervals between successive phenomena are approximate:

morning
- Inf. conjunction } 4 weeks (spring)–2 weeks (autumn)
- W elongation
- Sub. conjunction } 7 weeks (winter)–3½ weeks (summer)

evening
- E elongation } 7 weeks (autumn)–3½ weeks (spring)
- Inf. conjunction } 4 weeks (summer)–2 weeks (winter)

Elongations are always less than 28°–sometimes considerably so. In the N hemisphere, most favourable elongations E occur in the spring, most favourable elongations W in the autumn, since the planet is then at greatest altitude. In unfavourable years it happens that the spring and autumn elongations are small ones.

Mercury is observable telescopically for about 6 weeks at W elongation and 5 weeks at E elongation, during which time its angular diameter varies from about 5″ to 9″, and phase from nearly Full to a thick crescent. Its true brightness, as measured in stellar magnitudes, reaches a maximum around superior conjunction. At such times, however, it is so near the Sun as to be unobservable, and it appears brightest to the eye when seen against a darker sky at and near elongations.

4.2 Surface features

It is safe to say that Mercury will not offer the amateur any startling revelations–for many merely coaxing it into the field of view will be satisfying enough. Its markings are, for the most part, near the threshold of vision, and their observation and recording are correspondingly susceptible to personal visual idiosyncrasies. They consist of apparently permanent dusky markings, which, however, are of very variable visibility; and of whitish areas which characterise the limb regions and change from day to

85

day. There are in addition anomalous appearances of the disc itself: unequal cusps, undue thickening of the crescent phase, etc. To try to correlate what is seen in the telescope with the grosser features recorded by the 1974 *Mariner* probe is the most that can be hoped for. In the past, observers have reported anomalous appearances of the disc itself: unequal cusps, undue thickening of the crescent phase, etc. (see, for example, B.6.3, 6.4, 6.6).

The Reports of the former Mercury & Venus Section of the B.A.A. give a comprehensive summary of amateur observational work on the planet. Pre-*Mariner* charts of the planet will be found in B.6.1–3, 6.7, 6.9; for a modern description, see B.5.3.

4.3 Observational technique

Since the contrast in Mercury's markings in poor, low magnifications are indicated, provided they do not entail excessive glare. Glare will obscure faint markings more effectively than the reduced contrast dependent upon increased magnification. Hence the necessity of always observing Mercury (as also Venus) against a twilight or daylight sky. Observation by daylight also ensures a reasonable altitude above the horizon. McEwen found x135–x160 a satisfactory range of magnification with a 12.5 cm refractor.

The chief difficulties encountered in the observation of Mercury are the shortness of the apparitions, the smallness of the disc, and the planet's angular proximity to the Sun.

All drawings should be marked with the date, time (UT), type of instrument, aperture, magnification, and approximate altitude. A convenient scale for blanks is from 3 to 5 mm per " arc.

4.4 Finding by daylight

Never easy, because of Mercury's relative faintness (3–5 mags fainter than Venus, at similar phase) and its closeness to the Sun, and only under the clearest conditions is it likely to be picked up readily. In practice, only method (*a*) below is likely to produce consistent results. The remarks on observation of the Sun (section 1.7) also apply to daylight planetary work in general.

(*a*) With circles, taking RA and Dec from Almanac.

(*b*) When E of the Sun, take the difference between its Dec and that of the Sun from the Almanac. From known diameter of LP field, depress or elevate the telescope that amount from the Sun's centre; in the case of an altazimuth, let the Sun trail a web, and then depress or elevate the telescope in a direction perpendicular to the web. After an interval equal to the difference in RA between Mercury and Sun, the planet should be in the LP field; if not, very little sweeping will locate it.

(c) Starting from the Moon, on the day when the Almanac shows it and Mercury as being in conjunction, sweep in the direction and for the distance given.

When using either of the last two methods a wide-field finder (covering, say, 8° to 10°) is extremely helpful.

4.5 Solar transits

The remaining transits of Mercury to occur this century are as follows:

$$1986 \text{ Nov } 13^d \text{ } 4^h \qquad 1993 \text{ Nov } 6^d \text{ } 4^h \qquad 1999 \text{ Nov } 15^d \text{ } 21^h$$

(approximate times of mid-transit)

4.6 B.A.A.H. data

(a) Dates of elongations and conjunctions.

(b) RA
Dec
Phase (% of diameter normal to line
of cusps which is illuminated)
Elongation (° arc)
Distance

at 5-day intervals
throughout the
apparitions

4.7 A.A. data

The following information will be of particular use to amateur observers:

(a) Heliocentric longitude, latitude, radius vector — daily ephemeris
(b) Geocentric RA, Dec, distance — daily ephemeris
(c) Time of ephemeris transit — daily ephemeris
(d) Light time, magnitude, diameter, phase, phase angle, brilliancy, defect of illumination, meridian longitude — 2-day ephemeris

4.8 Data

Mean solar distance: 0.387099 A.U. = 5.791 x 10^7 km.
Perihelion distance: 4.600 x 10^7 km.
Aphelion distance: 6.982 x 10^7 km.
Sidereal period: 87.969^d
Mean synodic period: 115.88^d
Sidereal mean daily motion: 4°09.
Mean orbital velocity: 47.8 km/s.
Axial rotation period: 58.65^d

e: 0.2056244 (+ 0.0000002)

i : 7° 0′ 13″7 (+ 0″1)

Ω: 47° 44′ 18″9 (+ 42″7)

$\tilde{\omega}$: 76° 40′ 39″0 (+ 56″0)

L: 33° 10′ 6″07 (+53° 43′ 3″47)

Epoch 1950, Jan 1.5 UT (annual variations in brackets).

Angular diameter at unit distance: 6″68.

Angular diameter at mean inferior conjunction: 10″90.

Linear diameter: 4878 km = 0.38 ⊕.

Mass: 1.660 × 10^{-7}. = 0.0553 ⊕.

Volume: 0.06 ⊕.

Density: 5.43 × water = 0.98 ⊕.

Superficial gravity: 0.38 ⊕.

Mag at maximum brightness: about −1.8.

4.9 Elongations, 1980–2000

	Eastern				Western		
Date		Elong.	Mag	Date		Elong.	Mag.
1980 Feb	19	18°	−0.2	1980 Apr	2	28°	+0.6
Jun	14	24	+0.7	Aug	1	20	+0.4
Oct	11	25	+0.2	Nov	19	20	−0.2
1981 Feb	2	18	−0.3	1981 Mar	16	28	+0.4
May	27	23	+0.7	Jul	14	21	+0.6
Sep	23	26	+0.3	Nov	3	19	−0.3
1982 Jan	16	19	−0.3	1982 Feb	26	27	+0.3
May	8	21	+0.6	Jun	26	22	+0.7
Sep	6	27	+0.5	Oct	17	18	−0.3
Dec	30	20	−0.3	1983 Feb	8	26	+0.2
1983 Apr	21	20	0.0	Jun	8	24	+0.7
Aug	19	27	+0.6	Oct	1	18	−0.2
Dec	13	21	−0.3	1984 Jan	22	24	+0.1
1984 Apr	3	19	+0.2	May	19	26	+0.7
Jul	31	27	+0.6	Sep	14	18	−0.1
Nov	25	22	−0.1	1985 Jan	3	23	0.0
1985 Mar	17	18	0.0	May	1	27	+0.7
Jul	14	26	+0.7	Aug	28	18	+0.1
Nov	8	23	0.0	Dec	17	21	−0.2
1986 Feb	28	18	−0.1	1986 Apr	13	28	+0.6
Jun	25	25	+0.7	Aug	11	19	+0.2
Oct	21	24	+0.2	Nov	30	20	−0.3

Eastern				Western			
Date		*Elong.*	*Mag.*	*Date*		*Elong.*	*Mag.*
1987 Feb	12	18°	−0.3	1987 Mar	26	28°	+0.5
Jun	7	24	+0.7	Jul	25	20	+0.5
Oct	4	26	+0.2	Nov	13	19	−0.3
1988 Jan	26	18	−0.3	1988 Mar	8	27	+0.4
May	19	22	+0.6	Jul	6	21	+0.6
Sep	15	27	+0.4	Oct	26	18	−0.2
1989 Jan	9	19	−0.3	1989 Feb	18	26	+0.2
May	1	21	+0.5	Jun	18	23	+0.7
Aug	29	27	+0.5	Oct	10	18	−0.2
Dec	23	20	−0.2	1990 Feb	1	25	+0.1
1990 Apr	13	20	+0.3	May	31	25	+0.7
Aug	11	27	+0.6	Sep	24	18	−0.1
Dec	6	21	−0.2	1991 Jan	14	24	0.0
1991 Mar	27	19	+0.1	May	12	26	+0.7
Jul	25	27	+0.7	Sep	7	18	0.0
Nov	19	22	−0.1	Dec	27	22	−0.1
1992 Mar	9	18	0.0	1992 Apr	23	27	+0.6
Jul	6	26	+0.7	Aug	21	18	+0.1
Oct	31	24	+0.1	Dec	9	21	−0.2
1993 Feb	21	18	−0.3	1993 Apr	5	28	+0.5
Jun	17	25	+0.7	Aug	4	19	+0.3
Oct	14	25	+0.2	Nov	22	20	−0.3
1994 Feb	4	18	−0.3	1994 Mar	19	28	+0.5
May	30	23	+0.7	Jul	17	21	+0.5
Sep	26	26	+0.3	Nov	6	19	−0.3
1995 Jan	19	19	−0.3	1995 Mar	1	27	+0.3
May	12	22	+0.6	Jun	29	22	+0.7
Sep	9	27	+0.4	Oct	20	18	−0.3
1996 Jan	2	20	−0.3	1996 Feb	11	26	+0.2
Apr	23	20	+0.4	Jun	10	24	+0.7
Aug	21	27	+0.6	Oct	3	18	−0.2
Dec	15	20	−0.3	1997 Jan	24	24	0.0
1997 Apr	6	19	+0.2	May	22	25	+0.7
Aug	4	27	+0.6	Sep	16	18	−0.1
Nov	28	22	−0.1	1998 Jan	6	23	0.0
1998 Mar	20	18	0.0	May	4	27	+0.7
Jul	17	27	+0.7	Aug	31	18	+0.1
Nov	11	23	0.0	Dec	20	22	−0.2

Date			Elong.	Mag.	Date			Elong.	Mag.
1999	Mar	3	18°	−0.2	1999	Apr	16	28°	+0.6
	Jun	28	26	+0.7		Aug	14	19	+0.2
	Oct	24	24	+0.1		Dec	2	20	−0.3
2000	Feb	15	18	−0.2	2000	Mar	28	28	+0.5
	Jun	9	24	+0.7		Jul	27	20	+0.5
	Oct	6	26	+0.3		Nov	15	19	−0.3

SECTION 5

VENUS

5.1 Apparitions

The following intervals between successive phenomena are approximate:

morning	Inf. conjunction	}	10 weeks
	W elongation	}	31 weeks
	Sup. conjunction	}	
evening	E elongation	}	31 weeks
	Inf. conjunction	}	10 weeks

Elongations are never larger than about 47°. Venus is at maximum altitude (in the N hemisphere) at E elongations in the spring, when it may remain above the horizon till midnight. These particularly favourable apparitions recur at intervals of 8 years (= 13 Venus's years) from 1980.

Venus is observable for about 7 months at each elongation. It has been seen with the naked eye when only 5° from the Sun at superior conjunction (position taken from Almanac, found with binoculars, Sun then hidden by gable, and picked up with naked eye). Another recorded instance is of its naked-eye visibility 35^d after superior conjunction, when 9° from the Sun. It can be followed right through inferior conjunction when this occurs midway between the nodes: then up to 9° from the Sun, measured along the same hour circle, the crescent less than 1″ wide being visible with binoculars.

Maximum brightness occurs about 35 days after E elongation and 35 days before W elongation.

5.2 Surface features

The following are some of the points to look for:

(a) Irregularities in the cusps: widening or blunting; unequal sizes.

(b) Terminator shading present or absent; uniform from N to S, or varying in intensity; extent; depth of tone; colour; any lighter spots or areas within the shaded zone.

91

(*c*) Any irregularities in the terminator itself.

(*d*) Any difference between the observed form of the terminator and that which would be expected from the known phase.

(*e*) Visibility or otherwise of the unilluminated section of the disc; brightness compared with the day sky.

(*f*) Vague greyish areas; extent, colour intensity.

(*g*) Any relatively clearly defined markings should be recorded with the greatest possible precision.

(*h*) Width, brightness, and colour of the aureole.

See further B.7.1, 7.16.

5.3 Observational technique

For the detection of surface features the most favourable combination of large angular diameter and large phase occurs about midway between W elongation and superior conjunction, and again between superior conjunction and E elongation, when the planet's angular diameter is about 15″ and its phase gibbous.

The apparition for observational purposes, however, may be taken as extending from a little before W elongation to as near superior conjunction as the proximity of Venus to the Sun and its own decreasing diameter allow; and again from as soon after superior conjunction as the planet is observable, until the narrowing of the crescent makes observation worthless after E elongation.

At maximum brightness, glare will submerge the soft and uncontrasted detail of Venus's surface unless it is reduced in some way—such as:

(*a*) Observation in daylight.

(*b*) Ocular diaphragm or reduction of the objective's aperture.

(*c*) Herschel wedge, or other Sun diagonal.

(*d*) Neutral or colour filter.

(*e*) The presence of haze or high cloud often improves definition, probably by reducing glare.

Given observational precautions on the lines indicated above, faint markings, the darkening towards the terminator, and irregularities in the outline of the terminator may be seen with as little as 75 mm aperture. 20*D* or 25*D* can often be used when the planet is high in the sky.

Systematic observation of Venus calls for a sort of dogged perseverance. The value of negative observations should not be underrated. At the present time, much work is being directed to comparison of the apparent phase of the planet compared with that predicted; due to terminator shading, the phase always appears to be less than it should be (i.e., dichotomy occurs early at evening elongation, late at a morning one), the effect sometimes being equivalent to several days. It has yet to be established whether the atmosphere of the planet, or other factors, directly influence any variations in the apparent phase. Contrast undoubtedly plays a part, the phase appearing larger when a red filter is used, and much smaller when viewed through a blue filter. See, among many other references, B.7.5, 7.9, 7.11, 7.21.

Some observers have claimed that the use of a yellow filter, such as the Wratten 8 or 15, improves the visibility of markings, but there is no doubt that the most direct way of revealing them is by ultra-violet photography. It is upon the results of ultra-violet work that the relatively rapid rotation of the plant's atmosphere (about 4 days) is based (B.7.3). Successful photographs of the planet have been taken, in ultra-violet light, using a Chance-Pilkington OX9A filter, by Rackham (B.7.18; see also B.7.13).

Blanks for drawings should not be made too large, or there will be a tendency to elaborate the probably very small amount of detail that is actually seen. The standard size used by the B.A.A. is 50 mm to the planet's diameter. All drawings should be marked with the date, UT or GMAT, instrument (including details of any accessories), magnification, and approximate altitude of the planet at the time.

All observers of Venus are well advised to familiarise themselves with a useful paper on observational methods by P. Moore (B.7.17).

5.4 Finding by daylight

See sections 4.4, 5.1.

5.5 Solar transits

Of little practical interest at the present time: the next four transits occur on

$$\left.\begin{array}{l} \text{2004 June 7} \\ \text{2012 June 5} \end{array}\right\} \text{descending node}$$

$$\left.\begin{array}{l} \text{2117 December 10} \\ \text{2125 December 8} \end{array}\right\} \text{ascending node}$$

The last two transits occurred on 1882 December 6 and 1874 December 8.

5.6 B.A.A.H. data

(*a*) Dates of elongations, conjunctions, and greatest brightness.

(b) RA
Dec
Phase
Mag
Angular diameter
Distance
Elongation

} at 10-day intervals throughout the apparitions, except near conjunction.

5.7 A.A. data

The following information will be of particular use to amateur observers:

(a) Heliocentric longitude, latitude, radius vector—2-day ephemeris
(b) Geocentric RA, Dec, distance—daily ephemeris
(c) Time of ephemeris transit—daily ephemeris
(d) Light time, magnitude, diameter, phase, phase angle, brilliancy, defect of illumination, meridian longitude—4-day ephemeris

5.8 Data

Mean solar distance: 0.723332 A.U. = 1.082×10^8 km.
Perihelion distance: 1.075×10^8 km.
Aphelion distance: 1.089×10^8 km.
Sidereal period: 224.7^d.
Mean synodic period: 583.9^d
Sidereal mean daily motion: $1°.60$.
Mean orbital velocity: 34.9 km/s.
Axial rotation period: 243.0^d
e: 0.0067968 (−0.0000005)
i: 3° 23′ 38″.9 (+ 0″.04)
Ω: 76° 13′ 46″.8 (+ 32″.4)
$\tilde{\omega}$: 130° 52′ 3″.3 (+ 50″.6)
L: 81° 34′ 19″.20 (−135° 12′ 30″.3)

Epoch 1950, Jan 1.5 UT (annual variations in brackets).

Angular diameter at unit distance: 16″.82.
Angular diameter at mean inferior conjunction: 60″.80.
Linear diameter: 12,104 km = 0.96⊕.
Mass: 2.448×10^{-6} ⊙ = 0.826⊕.
Volume 0.86⊕.
Density: 5.24 x water = 0.95⊕.
Superficial gravity: 0.90⊕.
Mag at maximum brightness: about −4.4.

5.9 Elongations and inferior conjunctions, 1980–2000

E elongation	Inf. conjunction	W elongation
1980 Apr 5	1980 Jun 15	1980 Aug 24
1981 Nov 10	1982 Jan 21	1982 Apr 1
1983 Jun 16	1983 Aug 25	1983 Nov 4
1985 Jan 22	1985 Apr 3	1985 Jun 12
1986 Aug 27	1986 Nov 5	1987 Jan 15
1988 Apr 3	1988 Jun 12	1988 Aug 22
1989 Nov 8	1990 Jan 18	1990 Mar 30
1991 Jun 13	1991 Aug 22	1991 Nov 2
1993 Jan 19	1993 Apr 1	1993 Jun 10
1994 Aug 25	1994 Nov 2	1995 Jan 13
1996 Apr 1	1996 Jun 10	1996 Aug 19
1997 Nov 6	1998 Jan 16	1998 Mar 27
1999 Jun 11	1999 Aug 20	1999 Oct 31

The angular distance of Venus from the Sun at elongation varies only from $45°.3$ to $47°.2$, and its magnitude is about -4.1.

SECTION 6

MARS

6.1 Apparitions

Conjunctions and oppositions occur, usually, in alternate years: near perihelion, oppositions recur at intervals of about 800^d, near aphelion at intervals of about 765^d, the mean interval from opposition to opposition being 780^d. Thus each opposition occurs, on the average, 7 weeks later in the year than its predecessor.

Most favourable oppositions are those occurring at perihelion, in August (longitude of perihelion being $333°$). These recur at intervals of 15 and 17 years (7 and 8 synodic periods), and are less frequent than aphelic oppositions (February) in the ratio 2 : 3. Oppositions of Mars fall into two series, each lasting about 8 years. During one, each successive opposition is more distant than the last; during the other, each is nearer than its predecessor. Disregarding the factor of Declination, the most favourable for observation is the last opposition of the latter series.

Unfortunately for northern observers, Mars is always in S Dec at perihelic oppositions—Mars's perihelion lying in Aquarius—and the opposition following is more favourable for observers in Britain, the slightly reduced angular diameter being more than recompensed by the increased altitude.

Oppositions occurring during the early months of the year are unfavourable, since Mars is then near aphelion and its distance from the Earth is nearly twice that at a perihelic opposition. Favourable oppositions occur during the latter half of August and the early part of September.

The period during which Mars can be reasonably satisfactorily observed is limited to some 6 weeks on either side of opposition, and the practical advantage of perihelic operations lies less in the increased angular diameter of the disc than in the extended period over which observation is possible. This fleeting visibility of the planet makes it more necessary than ever that every opportunity for observation should be taken during the observing season.

The Martian S hemisphere is always turned towards the Earth at perihelic oppositions, the N at aphelic; British observers thus observe the northern hemisphere of Mars more favourably than the southern.

Angular diameter at aphelic opposition: about 14″,
  „   „   at perihelic opposition: about 25″,
  „   „   at mean opposition: about 17″.9.

At everage oppositions, 1″ is subtended by a linear distance of from 200 to 300 miles at the Martian surface; even at perihelic opposition it is always more than 175 miles.

Declination limits: ±25½° approximately. Observation is handicapped when the altitude of Mars is less than 30° or 40°; i.e. Mars in S Dec is not observable to full advantage from the British Isles.

Magnitude limits: at favourable opposition: about −2.8,
                  at conjunction: about +2.0.

6.2 Instrumental

Bare visibility of some of the principal surface features, and the polar caps, is possible with a telescope of only 75 mm aperture; but a 15 cm reflector is the smallest instrument with which even an experienced observer could record much of interest. At a favourable opposition, a small refractor will show the main features such as the M. Acidalium and Syrtis Major, and even some of the more prominent 'streaks' such as Thoth and Nilosyrtis.

The advantage of using apertures of the order of 30 cm is not confined merely to the improved views at opposition. They are equally valuable for allowing observations to be made over a longer period, when the disc is too small for adequate work with smaller instruments. The statement that apertures larger than 30 cm are rarely more effective than considerably smaller ones, because of tube and air currents having a prejudicial effect out of proportion to D, contains a grain of truth, but certainly should not discourage the observer from trying to get the largest instrument he possibly can; under good conditions, the large telescope will certainly prove its worth. The colours of the planet, in particular, come out much more vividly with a large aperture, because the intensity of the image raises the retinal perception from rod (predominantly grey) to cone (colour) vision. Comparison of what can be seen at a given time with different instruments (different sites, different observers, and different local circumstances) is worthless when trying to judge the effectiveness of different apertures.

For the amateur intending to observe Mars seriously, therefore, an aperture of between 20 and 30 cm is recommended. As with all planetary observation, a large focal ratio improves image clarity; therefore a refractor or a Cassegrain may give better results, aperture for aperture, than an ordinary Newtonian, unless the latter has an exceptionally long focal length. Equatorial motion, while a great convenience because of the freedom it gives the observer, is not essential.

Mars not only shows a small disc; its markings are often of low contrast. This raises conflicting demands, because a high magnification, while enlarging the disc, tends to reduce contrast. In general, powers of from ×200 to ×400 will be found most suitable. Accuracy of focus becomes increasingly important as the magnification is increased. It is often helpful to focus on the limb or the polar cap rather than on surface detail.

6.3 Longitude of the Central Meridian

The longitude of the CM at 0^h UT daily is given in *A.A.* The successive appearances cf the Martian surface are dependent upon the following facts:

Rotation period = $24^h\ 37^m\ 22^s.654$.
Same longitude transits the CM $37^m.4$ later each night.
The longitude system passes the CM at the rate of $350°.89202$ per day,

$$= 30° \text{ in } 2^h\ 3^m.1,$$
$$= 1° \text{ in } 4^m.1,$$
$$= 14°.62 \text{ per hour,}$$

from which it follows that the longitude of the CM at the same instant UT on successive nights will be progressively more westerly by about $14\frac{1}{2}°$ (i.e., features will appear to be displaced slightly to the right, as seen in an inverting telescope from the northern hemisphere). To put it another way, similar presentation of the disc will occur $37\frac{1}{2}$ minutes later on each successive night, until finally it passes into the daytime sky and is unobservable for a few weeks; after about 39 days, the original conditions will be restored.

The table below shows the time interval Δt after which a certain longitude will reappear on the CM, *n* nights after the original observation:

n	Δt	n	Δt	n	Δt	n	Δt
1	$0^h\ 37^m$	11	$6^h\ 50^m$	21	$13^h\ 3^m$	31	$19^h\ 16^m$
2	1 15	12	7 28	22	13 41	32	19 54
3	1 52	13	8 5	23	14 18	33	20 31
4	2 29	14	8 42	24	14 55	34	21 8
5	3 6	15	9 20	25	15 33	35	21 46
6	3 44	16	9 57	26	16 10	36	22 23
7	4 21	17	10 34	27	16 47	37	23 0
8	4 58	18	11 11	28	17 24	38	23 37
9	5 36	19	11 49	29	18 2	39	0 15
10	6 13	20	12 26	30	18 39		

In order to find the longitude of the CM at any time other than that given in the B.A.A. *Handbook* or *A.A.* tables, the following conversion table may be used:

Mins \ Hrs	0	1	2	3	4	5	6
0	$0°.00$	$14°.62$	$29°.24$	$43°.86$	$58°.48$	$73°.10$	$87°.72$
1	0.24	14.86	29.48	44.10	58.72	73.34	87.96
2	0.49	15.11	29.73	44.35	58.97	73.59	88.21
3	0.73	15.35	29.97	44.59	59.21	73.83	88.45
4	0.97	15.59	30.21	44.83	59.45	74.07	88.69
5	1.22	15.84	30.46	45.08	59.70	74.32	88.94
6	1.46	16.08	30.70	45.32	59.94	74.56	89.18
7	1.71	16.33	30.95	45.57	60.19	74.81	89.43
8	1.95	16.57	31.19	45.81	60.43	75.05	89.67
9	2.19	16.81	31.43	46.05	60.67	75.29	89.91
10	2.44	17.06	31.68	46.30	60.92	75.54	90.16
15	3.66	18.28	32.90	47.52	62.14	76.76	91.38
20	4.87	19.49	34.11	48.73	63.35	77.97	92.59
25	6.09	20.71	35.33	49.95	64.57	79.19	93.81
30	7.31	21.93	36.55	51.17	65.79	80.41	95.03
35	8.53	23.15	37.77	52.39	67.01	81.63	96.25
40	9.75	24.37	38.99	53.61	68.23	82.85	97.47
45	10.97	25.59	40.21	54.83	69.45	84.07	98.69
50	12.18	26.80	41.42	56.04	70.66	85.28	99.90
55	13.40	28.02	42.64	57.26	71.88	86.50	101.12

These values are based on the assumption of uniform rotation at the rate of $14°.62$ per hour.

6.4 Drawings

6.4.1 Preparing the blanks: Blanks should have been prepared before work at the telescope commences. The B.A.A. uses a diameter of 50 mm for both Venus and Mars drawings; the A.L.P.O. uses 42 mm. The phase of the planet, which near quadrature can be considerable, must have been derived from the ephemeris and allowed for. This phase refers to the width of the illuminated portion of the disc, measured along the line of greatest defect of illumination. For example, a phase of 93% indicates that a crescent whose greatest thickness is 7% of the disc diameter must be erased from the blank.

6.4.2 Execution: The correctness of the main outlines is of greater importance than niceties of shading, tone, and linear detail on the threshold of visibility. Therefore extreme care should be given to laying down the outlines, first of the polar cap, then of the main features in the central region of the disc, and finally of those near the limb. Beware of a systematic tendency (suffered by some observers) to place markings too far N or S, or E or W, on the blank.

The main features of the Martian surface are now thoroughly known, and the recording of the variations in their appearance constitutes the main body of the work possible with instruments of moderate aperture. Points for particular attention are variations in their size, shape, intensity, and colour; their occasional obscuration or even disappearance; and the occurrence of whitish areas at the limb and terminator, which fade and finally disappear as they approach the central meridian.

The main outlines completed, the finer detail is inserted. Only when the positions of the visible features are securely laid down should 'shading' be started. Its aim is to indicate the relative, not necessarily the absolute, intensities of the features whose outlines have been drawn. A change back to a LP ocular will often reveal tonal gradations within areas that appear uniform under high magnification. Written notes in amplification of the drawn features, intensities, etc, are often useful.

Finally, record the main colours of the image.

It is wise to omit all features and details that cannot be held steadily; at least, some distinctive form of notation should be used for details that are only glimpsed. It is a waste of time straining after microscopic detail with moderate apertures; there is plenty of macroscopic variation from opposition to opposition which is detectable with such equipment. It is better to be assured that in each drawing the shapes, sizes, relative positions, and relative tonal intensities of the main features are correctly stated.

The whole drawing should not take more than about half an hour; even so, rotation will have altered the appearance of the disc to a noticeable extent. It is always preferable to finish the drawing at the telescope, rather than to work it up from memory or from sketches: the dangers of the latter procedure are the tendency to 'finish' the drawing (i.e. to work it up to an extent sanctioned neither by memory nor, probably, by the eye in the first place) and the impossibility of comparing the drawing with the image for confirmation.

It should be possible to get one drawing per clear night throughout the apparition; two per night, as widely spaced as the planet's altitude will allow, are better. This routine is particularly desirable during the relatively scarce favourable oppositions. Any differences between a later and an earlier drawing of the same CM should be specifically noted, indicating whether the discrepancy is thought to be unintentional (inaccurate drawing), due to variation in the seeing, or accepted by the observer as objective.

At the end of the apparition a map of the whole Martian surface can, if it is thought worth while, be synthesised from the series of drawings.

6.4.3 Auxiliary data: The following data should finally be appended to each drawing:

(a) Date.

(b) Time (UT) of completion of main outlines.

(c) Longitude of CM at this time.

(d) Latitude of centre of disc.

(e) Aperture and magnification of telescope.

(f) Seeing conditions.

(g) Name of observer.

6.5 Surface features

The following brief and miscellaneous notes should be supplemented by B.8.6, 8.7, 8.15, 8.21, etc.

6.5.1 Dark areas: The standard I.A.U. map is now used generally by amateurs, because it forms a convenient reference for detecting changes in the 'normal' appearance of features, as well as supplying their identities for new observers. Work on the surface features should cover (a) changes in the shape and intensity of the dark areas; (b) the appearance of new features; (c) a special study of those regions known to be subject to seasonal or irregular changes. Amongst these may be included: Aethiopis, Araxes, Claritas-Daedelia, Cyclopia, Ganges, Hellas, Hellespontus, Thaumasia and Thoth-Nepenthes.

6.5.2 Atmospheric phenomena: Both yellow and white clouds are visible; the former—being dust raised from the surface by winds—are more visible by what they obscure than in themselves. The white clouds, which are true high-altitude ice clouds, are often made more conspicuous by using a blue filter (section 6.7).

6.5.3 Polar caps: To be distinguished from the whitish polar areas; irregularities in the outline; detached areas or temporary protrusions; apparent projection beyond the limb; general changes in shape and size.

6.5.4 Observing programmes: The following general areas of study have been recommended in a recent *Bulletin* of the B.A.A. Terrestrial Planets Section (April 1980):

(a) *D* less than 15 cm: observation of gross surface features and their overview contrast changes, bright clouds, limb brightenings, large polar cap details with light colour filters;

(b) *D* between 15 and 25 cm: studies of conspicuous surface detail are possible within 2–3 months of opposition. Seasonal intensity variations of surface features and cloud patches with the aid of colour filters can be studied successfully, especially if done in a continuous or routine manner. Positional micrometric work on the retreat of polar cap boundaries, dark feature boundaries, or transit times of features under change can be attempted. Successful photography with medium and fine-grain films at $f/130$–200 is possible;

(c) *D* of 30 cm and over: all of the work mentioned in (b) can be undertaken, with greater ease and accuracy. Critical visual and photographic observations of surface variations and atmospheric phenomena, on a professional level, are possible.

6.6 Intensity estimates

In an attempt at quantifying what must be, at best, highly subjective data, observers of Mars have agreed on a scale of intensity numbers to be applied to different features of the planet. Based on the recommendation of de Vaucouleurs, the gradations are numbered on a scale of 0–10, where 0 is the mean surface brightness of a polar cap, 2 the mean surface brightness of the light 'desert' areas near the centre of the disc, 6–7 represents the darkest markings near the centre of the disc, and 10 is the apparent depth of the night sky adjacent to the disc.

The use of intensity numbers forms a convenient way of indicating the different depths of tone on a drawing; but since the observer is using a self-adjusting basis on which to establish his intensity scale from night to night (rather like using a metre scale to measure an unknown length, and then graduating the length to check the metre scale), any temptation to treat this data as representing an objective standard must be resisted.

6.7 Colour and colour filters

To see colour other than the pinkish-ochre of the 'desert' areas requires considerable aperture, because only then is the intensity of light on the retina sufficient to bring the colour-sensitive cones into action. Also, careful choice of magnification is needed when making colour estimates. Generally speaking, too low a magnification yields a disc so small that no colour distinctions of any value can be seen at all; while over-magnification dilutes the image to such an extent that faint differences are suppressed.

The use of colour filters helps in the detection of clouds. The red Wratten 25 exaggerates the presence of yellow dust-clouds, while the blue 47B or 44A assists in the detection of white clouds.

By making a series of intensity estimates using different filters (red or orange, green and blue) the underlying hues of the different regions may be detected by their relative prominence or obscurity in the various colours (B.8.14).

6.8 B.A.A.H. data

RA
Dec
Magnitude
Angular diameter
P: p.a. of N pole measured E from N point of disc
Q: p.a. of point of greatest defect of illumination
Phase
Tilt: of the N pole towards or away from the Earth
Distance

at 10-day intervals throughout the apparition

Longitude of central meridian at 0^h UT daily throughout the year.

6.9 A.A. data

The following information will be of particular use to amateur observers:

(*a*) Heliocentric longitude, latitude, radius vector—4-day empheris
(*b*) Geocentric RA, Dec, distance—daily ephemeris
(*c*) Time of ephemeris transit—daily ephemeris

103

(d) Light time, magnitude, diameter, phase, phase angle, defect of illumination, meridian longitude, time of transit of planet's prime meridian—4-day ephemeris

(e) Ephemerides for Phobos and Deimos

6.10 Data

Mean solar distance: 1.523692 A.U. = 2.279 x 10^8 km.

Perihelion distance: 2.066 x 10^8 km.

Aphelion distance: 2.492 x 10^8 km.

Sidereal period: 686.98^d.

Mean synodic period: 779.94^d.

Sidereal mean daily motion: $0°.52$.

Mean orbital velocity: 24.1 km/s.

Axial rotation period: $24^h 37^m 22^s.654$.

$$
\left.
\begin{array}{l}
e: \ 0.0933589 \ (+ 0.0000009) \\
i: \quad 1° 51' \ (0".0 \ (\pm 0) \\
\Omega: \ 49° 10' 18".9 \ (+ 27".7) \\
\tilde{\omega}: 335° 8' 18".9 \ (+ 1' 6".2) \\
L: \ 144° 20' 7".08 \ (-168° 42' 50".52)
\end{array}
\right\}
\begin{array}{c}
\text{Epoch 1950, Jan 1.5 UT} \\
\text{(annual variations in} \\
\text{brackets).}
\end{array}
$$

Angular diameter at unit distance: $9".36$.

Angular diameter at mean opposition distance: $17".88$.

Linear diameter: 6794 km = 0.53⊕.

Mass: 3.227 x 10^{-7}⊙ = 0.106⊕.

Volume: 0.15⊕.

Density: 3.93 x water = 0.71⊕.

Superficial gravity: 0.38⊕.

Mag at maximum brightness: −2.8.

6.11 Satellite data

	I Phobos	II Deimos
Stellar mag at mean opposition distance	11.6	12.8
Mean distance from centre of Mars:		
A.U.	0.0000625	0.0001570
at mean opposn. dist.	$24".6$	$1' 1".8$
Sidereal period	$0^d318910$	$1^d262441$
	$= 7^h 39^m 13^s.8$	$= 30^h 17^m 54^s.9$
Synodic period	$7^h 39^m 26^s.6$	$1^d 6^h 21^m 15^s.7$
Eccentricity	0.0210	0.0028
Discoverer	Hall, 1877	Hall, 1877

6.12 Oppositions, 1980–2000

Date	Mag.	Disc diameter (" arc)
1980 Feb 25	−0.8	13.8
1982 Mar 31	−1.1	14.7
1984 May 11	−1.7	17.6
1986 Jul 10	−2.4	22.1
1988 Sep 28	−2.6	23.8
1990 Nov 27	−2.0	17.8
1993 Jan 7	−0.9	14.0
1995 Feb 12	−0.8	13.9
1997 Mar 17	−1.0	14.2
1999 Apr 24	−1.4	16.2

SECTION 7

JUPITER

7.1 Apparitions

Oppositions occur about one month later each year, the mean synodic period being $398^d.88$. Declination limits are approximately 25° N and S. Owing to the impossibility of wholly satisfactory observation when the planet is in S Dec, systematic observation is impossible in this country for about one-third of the revolution period (4 years in every 12).

The conditions regarding the visibility or otherwise of the same region of the surface on successive nights are given in the following Table. If a spot transits the CM at time t on night 0, then, throughout the following fortnight, assuming a rotation period of $9^h 55^m$:

Column (3) gives the times $\pm t$ of the transits of the same spot over the CM each night,

Column (4) gives the time interval that the spot is W or E of the CM at the same time t each night,

Column (5) gives the longitude difference between the spot and the CM at the same time t each night.

For example, if a spot is observed to transit the CM at $04^h 30^m$ UT on January 6, then the conditions for observing the spot on January 14 are as follows: it will transit the CM at $04^h 30^m + 6^h 20^m$ and $04^h 30^m - 3^h 35^m = 10^h 50^m$ and $00^h 55^m$ (col. 3), the latter alone being observable; at $04^h 30^m$ it will be $3^h 35^m$ (col. 4) or 130° (col. 5) west of the CM.

The chances of making repeat observations of the same region on or near the CM depend upon the length of the period during which Jupiter is observable each night (i.e. upon its Dec) and upon the time (early or late) during this period that the initial observation was made. Thus a spot observed early during the period of Jupiter's visibility on night 0 might be reobserved on nights 1, 2, 4, 6, 7 (rotations 3, 5, 10, 15, 17), etc; if late

106

in the night, on nights 1, 3, 5, 6, 7 (rotations, 2, 7, 12, 14, 16), etc; if approximately half-way between rising and setting, on nights 2, 3, 5, 7 (rotations 5, 7, 12, 17), etc. In other words, reobservation of a given region cannot be hoped for more frequently than at, generally, every other rotation, with gaps of as long as 5 rotations.

(1)	(2)	(3)	(4)	(5)
Night	Rotation	Times of CM transit $\pm t$	Time interval W/E (+/−) CM at time t	Longitude difference between spot and CM at time t
1	2	-4^h 10^m	$+4^h$ 10^m	151°
	3	+5 45		
2	4	−8 20		
	5	+1 35	−1 35	57
3	7	−2 35	+2 35	94
	8	+7 20		
	9	−6 45		
	10	+3 10	−3 10	115
5	12	−1 0	+1 0	36
	13	+8 55		
6	14	−5 10		
	15	+4 45	−4 45	172
7	16	−9 20		
	17	+0 35	−0 35	21
8	19	−3 35	+3 35	130
	20	+6 20		
9	21	−7 45		
	22	+2 10	−2 10	79
10	24	−2 0	+2 0	71
	25	+7 55		
11	26	−6 10		
	27	+3 45	−3 45	136
12	29	−0 25	+0 25	15
	30	+9 30		
13	31	−4 35	+4 35	166
	32	+5 20		
14	33	−8 45		
	34	+1 10	−1 10	42

7.2 Amateur work

Jupiter is, with variable stars, probably the favourite observational field with amateurs. The variety of its phenomena recommends it, and its angular size and brightness allow valuable work to be done with small instruments; neither an equatorial nor a motor drive is essential, though both are a convenience and will result in a larger number of individual observations being made. CM transit observations can be undertaken with even a 75 mm OG, but for regular work a 12.5 cm refractor or 15 cm reflector are the minimum desirable equipment. Transit measurements can be made almost as accurately with a 10 cm telescope as with a 20 cm, providing the spot is clearly seen, but there will be fewer of them.

The various branches of Jovian observation are:
(*a*) Longitude determinations by timing the CM transits of the features.
(*b*) Micrometer measures for the determination of latitude.
(*c*) General appearance of the disc.
(*d*) Colour and intensity observations.
(*e*) Photographic (see section 3.8).
(*f*) Satellite phenomena.
(*g*) Radio work.
(*h*) Other work.

7.3 Determination of longitude

The most powerful technique for unmasking the secrets of Jupiter's atmosphere that has yet been developed, and the main item in the programme of regular observers of the planet. It has the advantages of speed (the timing of the CM transits of 50–100 spots is often the work of a single session at the telescope); accuracy; producing quantitative data, with consequent ease and effectiveness of subsequent discussion; and finally, it goes straight to the core of the problem of conditions in the Jovian atmosphere.

Changes in the appearance of the visible surface are due primarily to the operation of the longitudinal currents: each latitude has its own 'normal' rate of drift, though the correlation of latitude with rotation period follows no known law (cf. the Sun); neither are the boundaries in latitude of the currents by any means fixed or permanent. The change of difference of longitude of adjacent spots or groups of spots, belonging to different currents can be extremely rapid. From the observed times of transit the rotation periods of the various currents are derived, and such measures,

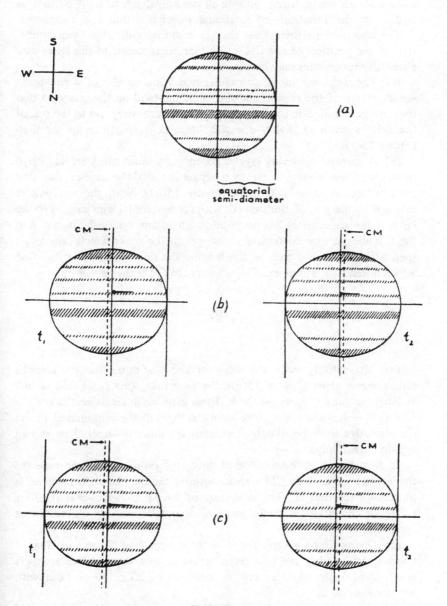

FIGURE 15

when collated and reduced, provide all our knowledge of these periods, as well as of the variations of rotational velocity within a given current.

The observations themselves may be made by radio, by visual estimation of the position of the CM, or by filar micrometer. In the latter case alternative procedures may be adopted:

(a) The webs are set to an angular separation equal to the equatorial semidiameter of the planet, and then superimposed on the image so that they are perpendicular to the belts (or, more accurately, set to the p.a. of the axis as given in *A.A.*), one web being kept tangential to the limb (Figure 15(a)).

(b) A method achieving greater accuracy by eliminating errors due to irradiation: the webs are set to a separation slightly smaller than the planet's equatorial semidiameter (Figure 15(b)). With the micrometer oriented to the p.a. of the axis (or with the horizontal web parallel to the belts, if the micrometer has no position circle) and one web tangential to the E limb, the time of transit of the spot at the other web is noted (t_1); then with the second web on the W limb, the time of transit at the first web is noted (t_2). The time of transit over the CM is then given by

$$t_1 + \frac{(t_2 - t_1)}{2}$$

(c) Alternatively, with the webs set at rather more than the planet's semidiameter apart (Figure 15(c)), the tangential web is set first on the W limb and subsequently on the E. These methods are least accurate when Jupiter is near quadrature, since similar settings at the illuminated and at the defective limb (markedly darkened) are more difficult than at two equally bright limbs.

In fact, however, the method of direct eye estimation of the time (to the nearest minute) of CM transit, without the help of a micrometer, is universally resorted to. The advantage of the micrometer—that it offers greater accuracy in a single measure—is outweighed by the following considerations:

(a) Owing to the greater speed of the eye method, a larger number of transits can be obtained in a given period of observation; when seeing is good, identifiable transits occur at the rate of 20 or 30 an hour with adequate apertures.

(b) Conversely, spots frequently come up to the CM in such rapid succession that only visual observation would obtain the transit times: there would simply be no time to readjust the webs on all of them, for it must be remembered that each recorded transit has to be supplemented with notes of the spot's position in latitude, shape, and intensity or colour, sufficient for its subsequent identification.

(c) Owing to Jupiter's high rotational speed, the eye can with practice become astonishingly accurate in its estimation of the moment at which CM transit occurs: one investigation of the accuracy of the method showed that practised observers agree consistently within 2^m, even though they may feel uncertain of the precise instant of transit by as much as 5^m; while errors of the order of 5^m are rare. The almost invariable tendency is to anticipate the transit, and once the sign and amount of the systematic error are determined (by comparisons with other observers, for example) a correction for personal equation can be applied. Though systematic, these personal errors are not necessarily permanent: it is not uncommon to find that their sign is reversed when Jupiter passes opposition (and the other limb then becomes defective), and that the right and left eyes have quite distinct systematic errors.

(d) The error in the derived rotation period becomes negligible when based on observations spread over a number of rotations, even though an individual transit may carry an uncertainty of from 2^m to 5^m. Rotation periods are not normally derived for features whose duration is less than about a month;* a total error of 5^m in the observations would produce an error in the derived period of only about $\pm 5^s$, and this error would decrease in proportion to the length of the interval between the first and last observations.

(e) The close juxtaposition in the field of an incisively defined web and an extremely ill-defined or poorly contrasted feature tends to suppress the latter altogether.

(f) The impossibility of repeating the micrometer measure, and the difficulty of keeping the limb web accurately tangential while the attention is directed to the meridian web (owing to their considerable angular separation).

(g) Visual determinations do not require a clock-driven equatorial.

Where there is scope for the micrometer is in the investigation of short-term variations of drift within a single current, since a single estimation by eye may give an error in the derived longitude of $\pm 3°$ or more. That such variations occur is well established, but little work has been done in this direction.

---

* Features surviving from one apparition to the next are comparatively rare; those lasting longer than one revolution period—e.g. the Red Spot and the South Tropical Disturbance—very much more so.

The observations should be recorded in the Observing Book in vertical columns:

Date. . . . .

(1)	(2)	(3)	(4) Longitude	
Serial No.	Description of feature	Transit time	λ_1	λ_2
1091	p end v.d. spot N edge NEB$_N$	$23^h\ 34^m$	—	177
1092	elong. proj. S edge STB	23 36	—	178
1093	consp. streak connecting NEB comps.	23 41	8	182
1094	ft. w. spot f RS	23 44	—	184

Column (1): serial numbers running consecutively throughout the apparition.

Column (2): as brief as is consistent with the certain identification of the feature. Recognised abbreviations (the observer can, and probably will, employ many more of his own devising) which are useful in this connection are:

N	North
S	South
p	preceding
f	following
d.	dark.
w.	white
v.	very
proj.	projection
comp.	component
elong.	elongation/elongated
consp.	conspicuous
indef.	indefinite/ill-defined
ft.	faint
RS	Red Spot
RSH	Hollow of the Red Spot
EZ	Equatorial Zone
SEB	South Equatorial Belt
SEB$_N$	North Component of the SEB
SEB$_S$	South Component of the SEB

NEB	North Equatorial Belt
NEB$_N$	North Component of the NEB
NEB$_S$	South Component of the NEB
S. Trop.Z.	South Tropical Zone
N.Trop.Z.	North Tropical Zone
STB	South Temperate Belt
NTB	North Temperate Belt
STZ	South Temperate Zone
NTZ	North Temperate Zone
SSTB	South South Temperate Belt
SSSTB	Southern component of SSTB (when the latter is divided into distinguishable components)
NNTB	North North Temperate Belt
NNNTB	Northern component of NNTB (when split into two)
SSTZ	South South Temperate Zone
NNTZ	North North Temperate Zone
SPR	South Polar Regions
NPR	North Polar Regions

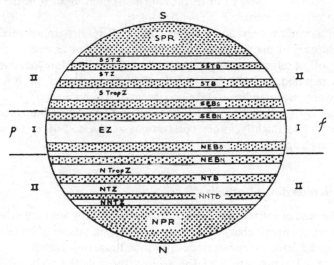

FIGURE 16

Column (3): To the nearest UT. Any clock or watch can be used that does not gain or lose more than 1 minute in 4 or 5 hours.

Column (4): The derived longitudes in Systems I or II respectively. Derived from *B.A.A.H.* or *A.A.* at the end of each observational session. Required data are:

113

(a) Longitude (L) of the CM at 0^h UT preceding or following the observation, in System I (between SEB_N and NEB_S) or II (N of NEB_S and S of SEB_N).

(b) Rate of change of L at intervals of MT in each System, from which the change of L (ΔL) during the interval between 0^h UT and the time of observation is derived.

Then the required longitude of the spot is

$$\lambda = L \pm \Delta L$$

according to whether the observation was made after or before 0^h UT. If $\Delta L > 360°$, add 360° to L; if the derived $\lambda > 360°$, reduce it by 360°; if in doubt as to which System the spot belongs to, enter up both λ_1, and λ_2.

Spots, though commonest in the equatorial and tropical regions, have been observed and measured right up to the edges of the SPR and NPR. Timed transits of the RS (p end, centre, and f end) are also extremely valuable, since its drift in longitude is considerable, and it is most desirable that as complete a record of its movements should be kept in the future as has been in the past.

In this work cooperation and the pooling of results are essential, so as to avoid gaps in the record of the apparition due to bad weather; a wide spread of observers in longitude is also desirable. More observers are always required, and the ranks of the Jupiter Section of the B.A.A. are open to anyone intending to take up systematic observing.

Records of CM transit observations should be sent to the Director of the Section fortnightly; other observations at the end of, or at intervals during, each apparition.

7.4 Determination of latitude

The changes of latitude of Jupiter's surface features are neither so large nor so rapid as their changes of longitude. Determinations of latitude need therefore be made much less frequently than those of longitude.

Figure 17 represents a median section through the body of Jupiter, taken in the plane which contains the line of sight from the terrestrial observer to its centre, O; P and E mark one pole and the equator respectively; PCE is thus a section of the Central Meridian; C is the centre of the apparent disc; P', where $P'O$ is normal to the tangent to the disc parallel to OC, is the N or S point of the apparent disc; X is a point on the CM whose latitude is required.

Two quantities have to be established by observation, the means of a number of micrometer measures being taken:

p', the polar semidiameter of the apparent disc (if the value of 1 turn of the micrometer screw is known, p' may be equated with the polar semidiameter—from *A.A.*—with negligible error in the final result).

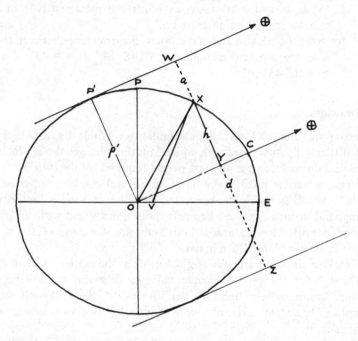

FIGURE 17

h, the angular distance of X north or south of the centre of the disc; it is derived by measuring XW (a) and XZ (d), the distance of X from each end of the CM,

whence $$h = \tfrac{1}{2}(d - a)$$

Then $b = \angle XOE =$ the required zenocentric latitude of X,

$\left.\begin{array}{l} h \\ p' \end{array}\right\}$ by observation,

$D_E = \angle COE$ (from *A.A.*),

$\phi = 21^\circ 2'.33$ (from $\sin \phi = 0.3590$, the eccentricity of the Jovian ellipsoid),

θ is such that $\sin \theta = h/p'$,

B is such that $\tan B = \sec \phi \tan D_E$,

115

and $$\tan b = \cos \phi \tan (\theta + B)$$
also b', the zenographic latitude of X, is given by

$$\tan b' = \sec \phi \tan (\theta + B)$$

where $b' = \angle XVE$, VX being normal to the tangent at X.

D, B, b, b' and θ are taken as positive if measured N from the equator, negative if measured S.

$b = b'$ at the equator and at the poles, the divergence between them reaching a maximum value of $3° 48' 38''$ at $b = 43° 5' 41''$, $b' = 46° 45' 19''$.

7.5 Drawings

Secondary in value to longitude determinations. Whole-disc drawings are made difficult by the speed with which rotation changes the appearance of the disc during even the shortest period required for the completion of a drawing which is justified by its accuracy and comprehensiveness. If very little detail is visible, however, a whole-disc drawing is better than nothing, and in such cases can be made quite quickly and with sufficient accuracy, provided the features at both limbs are attended to first and the central region of the disc filled in last.

Whole-disc drawings are also of great value at the extreme ends of each apparition; even though made under adverse observing conditions, and therefore comparatively unreliable or undetailed, they are well worth attempting, since any extension of the annual period of observation is highly desirable.

No completely satisfactory method of making drawings of Jupiter is available. Probably the best is the synthesis of a composite drawing from sketches and notes made at the telescope; but this must be compiled *immediately* after the observations are concluded, and a note as to its method of production always added to it. The consciously directed training of the visual memory is recommended.

Blanks of the correct degree of oblateness must be used; a suitable scale is about 50 mm to the equatorial diameter. The proportion of the lengths of the polar to the equatorial diameters is as $14:15$, giving an eccentricity of about 0.359. Suitable diameters would therefore be 46.5 mm and 50 mm; with the usual cotton-loop-and-pins method of constructing an ellipse, the pins would then be 18 mm apart, and the length of the cotton (equal to the major axis of the ellipse) 50 mm. Since so small an ellipse is difficult to construct accurately, it is recommended that a prototype blank be drawn, 5 times larger than the required size (major axis 250 mm, distance between foci 90 mm, minor axis 233 mm) and reduced photographically. The blanks actually used at the telescope can be traced or otherwise transferred from this reduced prototype.

Detailed drawings of individual areas, designed to reveal the nature of the changes to which they are subject, are both more valuable and easier to produce than whole-disc drawings. A convenient method of setting out the record is to make a scale of longitudes across the head of a large sheet of cartridge paper (the size of the scale depending upon the size of the area under review), and to make the drawings at each successive observed transit vertically below one another down the page, with a scale of dates and times at one side. Drifts in longitude can be revealed by taking a transit observation at the time each drawing is made.

Jupiter is a very suitable subject for the planetary photographer; see section 3.8.

7.6 Colour estimates

See also section 3.6.

There is great scope for colorimetric observations of Jupiter providing they are reliable, and despite the technical difficulties it is work that is well within the reach of amateurs. As an example of what can be done by observers without previous experience, and using modest equipment, the observations during 1951 of a B.A.A. member with a 15 cm reflector, ×240, may be mentioned. Confining his attention to the SEB_N and STB, he was able to establish the facts that their colours varied from brown to grey in different longitudes, and that the 'warm' and 'cool' sections of the belts were not static in longitude but (in the case of the SEB_N) were drifting eastward at a rate of approximately $28°$ per 30^d. The observations were mostly made without prior knowledge of the longitude of the CM, and were confined to nights when the atmospheric conditions were favourable.

What is required in this field, above all else, is a long series of strictly comparable observations, made with full precautions to guard against the intrusion of spurious colour effects. In this way it will ultimately be possible to clear up the present rather confused and tentative conclusions regarding suspected long-term colour variations in the equatorial regions: see B.9.17, 9.35–38; also B.9.5, 9.30 dealing with the colour of the polar regions.

7.7 Satellites

The four Galilean satellites (as opposed to the rest of the fourteen at present known) are visible in binoculars, and have even been reported with the naked eye (B.9.8). They form one of the most exciting discoveries for the newcomer to astronomy, and, although observation of their surface markings properly requires D of about 50 cm and over (and excellent seeing), some interesting work can be carried out using ordinary instruments. They are seen to be non-stellar with any aperture larger than about 10 cm.

7.7.1 Transits: Satellites and their shadows move in front of Jupiter from E (f) to W (p).

Before opposition: shadow precedes satellite, therefore falling on the disc before the transit begins.

After opposition: shadow follows satellite, therefore still being on the disc after the transit has ended.

The interval between the transit of satellite and shadow varies with the distance of the satellite from the planet and the time interval between the transit and opposition. It may be as long as several hours, often making the identification of shadow transits difficult, when nearer conjunction than opposition. IV may avoid transit altogether, passing above or below the planet. I (Io) is greyish in transit; limb shading may give it an elliptical appearance when seen with moderate aperture. II (Europa) has a very light tint, making it difficult to see when in transit against a zone. III and IV (Ganymede and Callisto) are both dark in transit, and may be confused with satellite shadows.

The timing of transits is of no great value, because of the indeterminate nature of ingress and egress.

7.7.2 Occultations and eclipses: Satellites move on the far side of Jupiter from W (p) to E (f).

From opposition to conjunction: satellites are occulted at p limb, reappearing from eclipse on the f side of the planet.

From conjunction to opposition: satellites enter eclipse before reaching the p limb, reappearing from occultation at the f limb.

I: Both entry into and exist from the same eclipse can never be observed owing to the satellite's nearness to the planet; it always passes from occultation into eclipse, or from eclipse to occultation. Duration of combined eclipse and occultation, about $2\frac{1}{4}^{\mathrm{h}}$.

II: Usually either entry into or exit from eclipse is alone observed, not both. Duration, about $2\frac{3}{4}^{\mathrm{h}}$.

III: Entry into and exit from eclipse can both be observed, except near opposition. Duration varies within wide limits, average about $3\frac{1}{4}^{\mathrm{h}}$.

IV: Immersion and emersion also individually visible, except near opposition. Duration varies within wide limits, average about 4^{h}. On the other hand, it may avoid the planet's shadow altogether; indeed, for two 3-year periods in every 12 years there are no phenomena for **IV**.

The satellites are not instantaneously extinguished on immersion, though the disappearance is rapid in the case of I and II. Timings of the first instant of visibility of a satellite emerging from eclipse, or (preferable, on the grounds of accuracy) the last visibility at entry into eclipse, can be of value in correcting the satellite ephemerides.

7.7.3 Mutual eclipses and occultations: These can occur only when the Earth is near the plane of the obits, i.e. for a few months every 5–6 years, the last occasion being in 1979. Important for the correction of their obits, and interesting to observe with apertures of about 15 cm and over.

7.7.4 Commensurabilities: I, II and III can never be experiencing the same phenomenon simultaneously, although any two of them may. This follows from the fact that their respective mean daily motions are related in such a manner that

$$n_I - 3n_{II} + 2n_{III} = 0$$

Jupiter's satellite phenomena are not merely interesting to witness, but are of considerable practical importance, for accurately timed observations are the data by whose means our knowledge of the satellite orbits is improved.

Jupiter without satellites: It occasionally happens that all four major satellites are simultaneously absent from the visible sky, being in transit (T), eclipse (E), or occultation (O). The future occasions up to A.D. 2000 when this will occur are listed below:*

Date	Duration	I	II	III	IV
1990 Jun 15	1^h5	OE	T	O	E
1991 Jan 2	1.2	T	OE	O	E
1997 Aug 27	0.5	O	T	OE	E

See also sections 7.11, 7.12.

7.8 Radio work

Jupiter can be one of the most active radio objects in the sky, emitting radiation across a large frequency range, although only the decametre radiation, from about 5 to 40 MHz, is of interest to the amateur. These emissions are presumably caused by Jupiter's atmosphere, since they

* From B.9.13

show strong correlation with certain System III longitudes and also with the longitude of satellite I. When active, Jupiter can be detected with very modest equipment (e.g., ordinary communications receivers), but the intermittent nature of its outbursts, means that there will be periods of days or weeks when nothing is received. In some ways, however, this is an incentive, since any positive observations become more valuable. Only long-term work can finally identify the nature of the disturbances that trigger such prodigious sources of energy.

The equipment for observing at a frequency of about 20 MHz is not complicated, but there are other problems. The Earth's ionosphere imposes difficulties of observation by limiting the angle at which radio observers can work. Another difficulty is that the observations are made in the communications band and it is very difficult to find free regions. The best way to approach this problem is to observe on a number of wavelengths for a week or more before deciding to erect an aerial system. One can then ascertain whether the wavelength is relatively clear from interference. Yet another problem is set by thunderstorms—these can interfere with the record and make it difficult to interpret *weak* Jovian bursts. In all cases care has to be taken to ensure that the noise bursts are not originating in the terrestrial atmosphere. This does necessitate practice in observing.

The previous remarks refer to non-thermal radiation. Thermal radiation has been detected from Jupiter using a 3.18 cm maser. This type of apparatus is quite beyond the pocket of the amateur. The apparent black-body temperature of the planet based on the mean diameter of the visible disc is $130°K$. Over a period of time the apparent black-body temperature does not appear to vary with rotation. (See B.9.2, 9.3.)

7.9 Other work

A notable example of what the enterprising amateur can achieve is provided by D. W. Millar's discovery, while collating 40 years' *Mem. B.A.A.*s, of a previously unsuspected and currently unexplained relationship between the latitude and rotation periods of belts that are not even contiguous; see B.9.19. And it is only necessary to glance through the literature that has accumulated around the perplexing S.Trop.Z 'Circulating Current' to appreciate that Jupiter still holds much of interest for the visual observer; see B.9.23, 9.42.

Other examples of areas of which observations abound in the literature and which for one reason or another have been intensively studied in the past include the Red Spot, possibly periodic (1919/20, 1928/29, 1938, 1943, and 1949) outbursts of spot formation and general activity in the SEB, similar suspected activity on the N edge of the NNTB, the S.Trop. Disturbance, etc. See, *inter alia*, B.153, 155–157. B.147, containing fully illustrated histories of the RS, S.Trop.Disturbance, Circulating Current, and SEB outbursts, should also be referred to for detailed descriptions of these features.

There is also room for a great deal more work on the detailed movements and behaviour of the spots, and of their interactions when passing closely to one another—as happens with spots situated near the edges of adjoining currents or longitude drifts, whose relative motion is often rapid: it has been noticed, for instance, that a bridge of lighter-toned material often joins such spots at about the time that their separation is minimal, beyond which bare fact little is known. That Jupiter can always spring new surprises is shown by the discovery in 1970 of the most rapidly-moving marking ever observed, a bright spot in the NTB associated with the North Temperate Current, having a rotation period of only $9^h 47^m 05^s$.

7.10 Daylight observation

When near quadrature, Jupiter must necessarily be observed in twilight, its highest altitude being attained when the Sun is above the horizon. Daylight observation with large apertures will, theoretically, extend the observational period very usefully; but the chances of good seeing coinciding with the necessary complete absence of haze (of itself, frequently, the sign of an unstable atmosphere) limits the effectiveness of the method. The glare must be reduced by using a red or orange filter to absorb as much of the skylight as possible. Polaroid is effective at and around quadrature, since maximum polarisation of sky-light occurs in a great circle 90° from the Sun, while reflected Jovian sunlight is unpolarised.

7.11 B.A.A.H. data

(*a*) RA, Dec, Mag, Polar and Equatorial diameters, and Distance: at 10-day intervals throughout the apparition.

(*b*) Longitude of the CM in System I (equatorial) and II (polar) at specified times throughout the apparition.

(*c*) Table of Change of Longitude at intervals of MT in the two Systems. (For more extended Table, see *A.A.*)

121

(d) Phenomena of Satellites I–IV throughout the apparition, in tabular form:
> times of beginning and end of eclipses,
> times of disappearance and reappearance from occultation,
> times of entry into and exit from transit,
> times of beginning and end of shadow transits,
> eclipse coordinates,
> configuration diagrams.

7.12 A.A. data

The following information will be of particular use to amateur observers:
(a) Heliocentric longitude, latitude, radius vector – 10-day ephemeris
(b) Geocentric RA, Dec, distance – daily ephemeris
(c) Time of ephemeris transit – daily ephemeris
(d) Light time, magnitude, diameter, phase angle, defect of illumination – 4-day ephemeris
(e) Meridian longitude for Systems, I, II and III
(f) Satellite ephemerides:
> I–IV: daily diagram, phenomena, superior conjunctions
> V: time of every 20th eastern elongation
> VI, VII: position at 4-day intervals
> VIII–XIII: position at 10-day intervals

7.13 Data

Mean solar distance: 5.202804 A.U. = 7.783×10^8 km.
Perihelion distance: 7.407×10^8 km.
Aphelion distance: 8.161×10^8 km.
Sidereal period: 4332.6^d (11.86^y).
Mean synodic period: 398.88^d.
Sidereal mean daily motion: $0°.08$.
Mean orbital velocity: 13.0 km/s.
Axial rotation period: System I: $9^h\ 50^m\ 30^s.003$.
System II: $9^h\ 55^m\ 40^s.632$.
Radio sources: $9^h\ 55^m\ 28^s.8$.
e : 0.0484190 (+ 0.0000016)
i : $1°\ 18'\ 21''.3$ $(-0''.2)$
Ω: $99°\ 56'\ 36''.0$ $(+36''.4)$
$\tilde{\omega}$: $13°\ 31'\ 1''.5$ $(+58''.0)$
L : $316°\ 9'\ 33''.57$ $(+30°\ 20'\ 32''.07)$

Epoch 1959, Jan 1.5 UT
(annual variations in brackets).

122

Angular diameter at unit distance: e 196″.94, p 183″.82.
Angular diameter at mean opposition distance: e 46″.86, p 43″.74.
Linear diameter e 142,700 km = 88,700 miles = 11.2$\oplus$.
$\quad\quad\quad\quad p$ 133,200 km = 82,800 miles = 10.4$\oplus$.
Mass: 9.55 x 10^{-4} $\odot$ = 317.9$\oplus$.
Volume: 1323$\oplus$.
Density: 1.32 x water = 0.25$\oplus$.
Mean superficial gravity: 2.69$\oplus$.
Mag at maximum brightness: -2.5.

7.14 Satellite data

(See Table on page 124.)

7.15 Oppositions, 1980–2000

Date	Mag	Disc diameter ($''$ arc)
1980 Feb 24	-2.1	44.7
1981 Mar 26	-2.0	44.2
1982 Apr 25	-2.0	44.4
1983 May 27	-2.1	45.5
1984 Jun 29	-2.2	46.8
1985 Aug 5	-2.3	48.5
1986 Sep 10	-2.4	49.6
1987 Oct 18	-2.5	49.8
1988 Nov 23	-2.4	48.7
1989 Dec 27	-2.3	47.5
1991 Jan 28	-2.2	45.7
1992 Feb 28	-2.1	44.7
1993 Mar 30	-2.0	44.2
1994 Apr 30	-2.0	44.5
1995 Jun 1	-2.1	45.7
1996 Jul 4	-2.2	47.0
1997 Aug 9	-2.3	48.7
1998 Sep 16	-2.4	49.6
1999 Oct 23	-2.5	49.6
2000 Nov 28	-2.4	48.4

	I Io	II Europa	III Ganymede	IV Callisto	V Amalthea	VI Himalia	VII Elara	VIII Pasiphae	IX Sinope	X Lysithea	XI Carme	XII Ananke
Mean distance from Jupiter: at mean opposition distance A.U. kilometres	2′18″.4 0·00281956 422,000	3′40″.2 0·00448620 669,000	5′51″.2 0·00715590 1,068,000	10′17″.7 0·0125865 1,879,000	59″.2 0·001207 118,000	1°2′50″ 0·076605 11,450,000	1°4′13″ 0·078516 11,730,000	2°8′35″ 0·15720 23,500,000	2°9′ 0·158 24,100,000	1°3′16″ 0·077334 11,570,000	2°3′24″ 0·1508336 22,570,000	c.2° c·0·148 20,900,000
Sidereal period: days days, hours, minutes	1·76913780 1.18.27½	3·55118108 3.13.13½	7·15455312 7.3.42½	16·68901805 16.16.32	0·49817923 0.11.57½	250·62 250.14.40	260·07 260.1.24	738·9 738.21.30	745 745.--.--	254·21 254.5.0	692·5 692.12.--	631 631
Mean synodic period (d. h, m, s)	1.18.28.35·95	3.13.17.53·74	7.3.59.35·86	16.18.5.6·92	0.11.57.27·6	266.0.--.--	276.16.--.--	631·2d	636d	270·01d	597d	551d
Orbital eccentricity	0·000	0·0001	0·0014	0·0074	0·003	0·1580	0·2072	0·41	0·275	0·1074	0·207	0·169
Diameter (km)	3652	2900	5000	4500	200?	100?	30?	20?	20?	20?	20?	20?
Mass ×10⁻⁵ Jupiter ×10⁻⁸ ⊙	4·5 4·294	2·5 2·421	8·0 7·627	4·5 4·300								
Stellar magnitude (at mean opposition distance)	4·8	5·2	4·5	5·5	13·0	13·7	16·0	18·8	18·3	18·6	18·1	18·8
Discoverer	Galileo / Mayer 1610	Galileo / Mayer 1610	Galileo / Mayer 1610	Galileo / Mayer 1610	Barnard 1892	Perrine 1904	Perrine 1905	Melotte 1908	Nicholson 1914	Nicholson 1938	Nicholson 1938	Nicholson 1951

Satellites XIII and XIV (about mag 20) are not included.

VIII, IX, XI, and XII are retrograde.
VI–X, and XII have highly inclined orbits.

SECTION 8

SATURN

8.1 Apparitions

Opposition occurs about 2 weeks later each year, the mean synodic period being $378^{d}.09$.

Dec limits: about $26°$ N and S.

As regards the reappearance of a given region, reobservation is possible at the 5th, 7th, 12th and 14th rotations, when a given longitude will be in the same position on the disc within about 3^{h} of the time of the initial observation.

The equatorial plane containing the rings is inclined to the orbital plane at about $27°$. Twice, therefore, in the planet's sidereal period of 29.46 years, the Earth and the Sun will be $27°$ 'above' or 'below' the plane of the rings, and 7.5 years later the rings will be viewed edge-on.

8.2 Instrumental

For regular work on Saturn, and the recording of observations of value, not less than 12.5 cm aperture is effective; if more than 25 cm, so much the better. The visibility of the projections on the N edge of the SEB (1943) with a 13 cm shows that at times smaller apertures than this can contribute useful observations; but apertures less than about 12.5 cm, though giving excellent views of the ring system, lack the resolving power to show significant detail.

For photographic work, see section 3.8.

8.3 Longitude determinations

Our knowledge of the rotation periods in different latitudes, and of the manner in which they are related, is still very inadequate owing to the rare occurrence of features having sharp enough definition to give good transit measures. No opportunity must be missed of timing the CM transit

125

of any sufficiently well defined spot (the edges of the equatorial belts seem particularly prone to spot production) or other features that may appear, and the position and estimated time of transit of any spot over the CM of the globe should be reported at once. Although irregularities in the main belts are the most likely to be noticed, observers should, nevertheless, keep an eye out for any dark or light markings in higher latitudes. *The Strolling Astronomer* (U.S.A.) publishes annually a table giving the longitude of the CM at 0^h UT daily, which includes correction for phase, light-time and the Earth's saturnicentric longitude. System I (analogous to that on Jupiter) covers the whole of the two Equatorial Belts and the EZ: period 10^h 14^m $13\overset{s}{.}08$ (rotation rate, 844°/day). System II includes the rest of the planet: period 10^h 38^m $25\overset{s}{.}42$ (rotation rate, 812°/day).

8.4 Latitude determinations

Of any features that appear, and, regularly, of the edges of the belts; those especially required by the Saturn Section are:

centre of the EZ,
edges of the SEB and NEB,
edges of the SP Band and the NP Band,
edge of the SP area and of the NP area,
centre of Ring C against the globe.

Moderate apertures and a micrometer are desirable; the measurement of drawings, however carefully made, involves a wider margin of uncertainty than direct measurement of the image; this is to some extent mitigated if the drawings or measures of a large number of observers are available for meaning and smoothing (hence the value of a clearing house, such as the Saturn Section).

The latitude of a surface feature is derived as follows:

h = its angular distance N (+) or S (−) of the centre of the disc; the quantities given by observation are a, b, its distances from the two ends of the CM, whence

$$h = \tfrac{1}{2}(a - b)$$

r = polar semidiameter (from *A.A.*),
B = saturnicentric latitude of the Earth (from *A.A.*),
B' is such that $\tan B' = 1.12 \tan B$,
B'' is such that $\sin (B'' - B') = h/r$.

Then the saturnicentric latitude, b, of the feature is given by

$$\tan b = \frac{\tan B''}{1.12}$$

126

and its saturnigraphic latitude, b', is derived from

$$\tan b' = 1.25 \tan b$$

8.5 Surface features

The importance of missing no opportunity of making longitude determinations of any surface feature has already been stressed. Reports of spots observed should always include estimated time of CM transit. Any changes in the general appearance of the belts, even though they may be useless for transit observations, should nevertheless be recorded with the greatest care.

Whole-disc drawings should preferably be made, and it is essential that the blanks be prepared beforehand; observing time is saved, greater accuracy is achieved, and a more consistent and therefore comparable set of records obtained. In constructing the blanks, the relevant dimensions need not be taken to a high degree of accuracy, and a single prototype will furnish blanks that will be available for several weeks. The following data are required:

Globe: p: polar semidiameter (from $A.A.$),

        e: equatorial semidiameter ($e = 1.12p$),

        Phase may be disregarded, since even at quadrature Ph is not less than about 99.7, but the globe should never be represented by a circular disc.

Rings: P: p.a. of N semi-minor axis of the rings (from $A.A.$),

        a_A : major axis of outer edge of Ring A (from $A.A.$),

        b_A: minor axis of outer edge of Ring A (from $A.A.$),

        a'_A: major axis of inner edge of Ring A ($= 0.88a_A$),

        b'_A: minor axis of inner edge of Ring A ($= 0.88b_A$),

        a_B : major axis of outer edge of Ring B ($= 0.86a_A$),

        b_B: minor axis of outer edge of Ring B ($= 0.86b_A$),

        a'_B : major axis of inner edge of Ring B ($= 0.66a_A$),

        b'_B: minor axis of inner edge of Ring B ($= 0.66b_A$),

        a'_C: major axis of inner edge of Ring C ($= 0.55a_A$),

        b'_C: minor axis of inner edge of Ring C ($= 0.55b_A$).

When the rings are not wide open it will be found unnecessary to compute all the above values, at any rate of the minor axis, owing to the exaggerated foreshortening: a_A and b_A, a'_A, a_B, a'_B, a'_C and b'_C will suffice.

The standard B.A.A. practice is to draw Saturn on a scale of a_A = 100 mm; the equatorial diameter then measures 45 mm. Partial drawings, to record the finer detail of any unusual feature, should be made on a larger scale and should be accompanied by a whole-disc drawing made at the same time. Each drawing (on a separate sheet) should be supplemented by the observer's name, aperture, magnification, date, and UT.

8.6 The rings

(a) Micrometer measures. Occasional apparent eccentricity of globe in relation to the rings.

(b) Degree of visibility and appearance of the rings near the times at which plane passes through the Earth (seen edge-on), through the Sun (illuminated edge-on)—as happens every 15 years approximately—and when the Sun and the Earth are on opposite sides of the ring plane (unilluminated side turned to the observer). Visibility on the two sides of the globe often unequal.

(c) Appearance and width of ring shadow on the globe.

(d) Appearance of globe shadow on the rings; the often-observed 'crotchet' in the outline of the shadow where it crosses Cassini's Division is probably an optical and/or instrumental effect; a white spot in the same position (often reported, particularly a few months on either side of opposition), concave shadow profiles, and other anomalies probably arise from similar causes.

(e) Markings or variations of tone on the rings.

(f) Divisions in the ansae (other than Cassini's) have often been suspected; but considerable aperture and moments of perfect seeing are needed for such observations to be of value. Any micrometer measures of such divisions are most valuable; failing this, estimate the position (expressed as a fraction of the ring width) by eye or from drawings.

(g) Variations of visibility of Ring C, or irregularities in its outline.

(h) Suspected existence of a faint ring outside Ring A.

8.7 Occultations

All occultations of stars by planets are of great interest and value for the light they may throw upon the density and extent of the atmosphere; in the case of Saturn they are doubly important for the similar information they can give regarding the ring system.

Each period of observation should begin with a quick survey of the sky slightly ahead of the planet (i.e. in the direction of its apparent orbital motion), and should it appear that any star is heading for occultation this should be given priority over all else, since they are rare occurrences. The record should show the times of the star's arrival at the edges of the ring system and at the planet's limb; for every change in the star's brightness or colour the time, its position, and the nature of the fluctuation should be recorded; also whether or not the disappearance at the limb was instantaneous, and if not, the length of the interval during which fading was observed.

8.8 Satellites

See also sections 7.7, 8.14.

There is great scope for accurate photometric work on Saturn's satellites, none of whose amplitudes or even mean magnitudes is well determined; the variation of III and IV is not even established beyond doubt (see B.10.14, 10.23).

I: Difficult, owing to nearness to the planet; mag about 12.0.

II: Also always a difficult object, though less so than I; mag about 11.7, variable.

III: Visible with a 10-cm, when far from the planet, mean mag about 10.3; suspected amplitude 0.25–0.5 mag, maximum at elongations, and W possibly brighter than E.

IV: Visible with a 10-cm, when far from the planet; mean mag about 10.4; suspected amplitude 0.25–0.5 mag; maximum at elongations, and E possibly brighter than W.

The mean daily motions of these 4 satellites are related in such a manner that

$$5n_I - 10n_{II} + n_{III} + 4n_{IV} = 0$$

V: Visible with 70-mm; easy with 85-mm when at elongation; mag ? 9–10; possible amplitude of about 0.5 mag.

VI: Readily seen with binoculars when near elongation; mag about 8.3 (see B.10.25).

VII: Difficult; mag about 14.

VIII: Largest amplitude of all the satellites: ? 9.5–11.0; maximum at W elongation, when visible with a 75-mm. See B.10.26.

IX: Difficult owing to its faintness and distance from the planet.

X: Not within the scope of the amateur observer. Mag about 14, close to edge of ring system.

Thus quite modest apertures can show a number of the satellites; but for accurate photometric work considerably more aperture is required than the bare minimum which allows them to be glimpsed when at maximum. Observational difficulties are considerable: (a) intrinsic faintness combined with proximity to the brilliant planet, (b) small amplitudes, (c) visibility affected by degree of opening of the rings, (d) visibility affected by their distance from Saturn, (e) lack of comparison objects.

It is advisable always to mask the planet and rings by an adjustable obstruction in the ocular; otherwise, the times of disappearance of the rings are the most favourable for satellite observations. Regarding (e) there are several alternatives: (i) field stars, when suitable ones are available, (ii) using VI (mag 8.3) as the comparison object, (iii) making relative inter-comparisons of two or more satellites; even the order of brightness, with no attempt at absolute values or even relative values in terms of magnitudes or steps, is preferable to nothing, (iv) photometer projecting an artificial star into the field.

Records of brightness estimates should always include the following supporting data:

time of observation,
p.a. of satellite from planet's centre,*
angular elongation from the planet,
whether or not an occulting bar was used,
identity of comparison object.

8.9 Colour and intensity estimations

See also section 3.6.

Objective variations of colour have been suspected (Ring C, for example) but are extremely difficult to substantiate.

The recording of the relative intensities of the different regions of the globe and the rings is an important part of the regular observer's work, and should be carried out frequently so that good means may be obtained. Ring C, again, should be given particularly careful attention; similarly the polar regions, for which a far greater quantity of observations is required.

Recommended scale of intensities, by means of which to annotate rough sketches of the main regions, is based on the brightness of Ring B just inside Cassini's Division in the ansae (scale no. 1; fractions being employed for anything brighter than this), and the tone of the planet's shadow on the rings or the sky background (scale no. 10).

The use of colour filters has become a regular part of serious observing programmes; the Wratten 25 (red) and 44A (blue) are recommended as standard. Repeated intensity estimates in red, blue and 'white' light can indicate different relative intensities (hence colour tints), and the belts and zones may appear to be of different widths. A curious effect known as the bi-colour aspect of the rings has been noted, in which one side of the rings, when viewed through a given filter, appears to have a different intensity from that of the opposite side. Possibly more obvious when the

* Note that the East point (i.e. p.a. $90°$) is not the same thing as eastern elongation.

planet is near quadrature, it would seem to be due to the arrangement of the ring particles.

8.10 Observing procedure

The B.A.A. Saturn Section has suggested that the following routine observations should be carried out by all observers, regardless of aperture, on every occasion:
1. General inspection of the planet and its field, noting:
   (*a*) Shape of shadows of rings on globe, and of globe on rings.
   (*b*) Any changes in the general shape or position of the belts on the globe.
   (*c*) Any star that seems likely to be occulted by the rings or the globe.
   (*d*) Position of any bright or dark spots or patches on the globe or the rings, or any projections from a belt which appear to move with the planet's rotation.
   (*e*) Any changes in the visibility, or irregularity in the outline, of Ring C.
   (*f*) Any sign of a faint ring outside the bright rings.
   (*g*) Position of any dusky patches or small divisions (e.g. Encke's) on the ansae of the rings.
2. Drawings of the complete planet should be made whenever the seeing is good.
3. Estimates of relative intensity of rings, belts, zones and shadows.
4. Any colour, if seen.

Opposite is reproduced the printed pro-forma issued to members for reporting their observations in a convenient form for subsequent analysis and discussion.

8.11 B.A.A.H. data

(*a*) RA
Dec
Magnitude
Polar diameter
Major axis of ring system
Minor axis of ring system
Saturnicentric latitude of Earth
   referred to ring plane
Distance

} at 20-day intervals throughout the apparition.

(*b*) Satellite data

SATURN

OBSERVER ....

Instrument(s) .... *Year of Observations* ....

ESTIMATES OF *(Cross out those which do not apply)*	INTENSITY COLOUR LATITUDES OF EDGES OF BELTS DISTANCES OF EDGES OF RINGS FROM CENTRE OF SATURN

Month											
Day											
U.T.											
Inst. & Power											
Conditions											
Definition											
Class											
FEATURE											

(c) Dimensions (linear, and angular at unit distance and mean opposition distance) of the ring system.

8.12 A.A. data

The following information will be of particular use to amateur observers:
  (a) Heliocentric longitude, latitude, radius vector – 10-day ephemeris
  (b) Geocentric RA, Dec, distance – daily ephemeris
  (c) Time of ephemeris transit – daily ephemeris
  (d) Light time, magnitude, diameter, phase angle, defect of illumination – 4-day ephemeris
  (e) Presentation of rings – 4-day ephemeris
  (f) Satellite ephemerides:
      Mimas, Tethys, Enceladus, Dione, Rhea: time of every eastern elongation, with supplementary tables
      Titan, Hyperion, Iapetus: time of all elongations and conjunctions, with supplementary tables

8.13 Data

Mean solar distance: 9.538844 A.U. = 1.427 x 10^9 km.
Perihelion distance: 1.348 x 10^9 km.
Aphelion distance: 1.506 x 10^9 km.
Sidereal period: 10759^d (29.46^y).
Mean synodic period: 378.09 days.
Sidereal mean daily motion: $0°.03$.
Mean orbital velocity: 9.7 km/s.
Axial rotation period (equatorial): $10^h\ 14^m$.

e : 0.0557164 (–0.0000035)
i :   2° 29′ 25″.2 (–0″.1)
Ω: 113° 13′ 12″.6 (+31″.4)
$\tilde{\omega}$: 92° 4′ 6″.6 (+1 10″.5)
L : 158° 18′ 12″.89 (+12° 13′ 36″.2)

Epoch 1950, Jan 1.5 UT
(annual variations in brackets).

Angular diameter at unit distance: e 166″.66, p 149″.14.
Angular diameter at mean opposition distance: e 19″.52, p 17″.46.
Linear diameter: e 120,800 km = 9.5⊕.
               p 108,100 km = 8.5⊕.
Mass: 2.859 x 10^{-4} ☉ = 95.2⊕.
Volume: 752⊕.
Density: 0.70 x water = 0.12⊕.
Mean superficial gravity: 1.19⊕.
Mag at maximum brightness: –0.2.
Ring system:

133

8.14 Satellite data

	I Mimas	II Enceladus	III Tethys	IV Dione	V Rhea	VI Titan	VII Hyperion	VIII Iapetus	IX Phoebe†	X Janus
Mean distance from Saturn: at mean opposition distance A.U. kilometres	30".0 0·0012401 185,000	38".4 0·0015909 238,000	47".6 0·0019694 294,000	1'0".9 0·0025226 378,000	1'25".1 0·0035226 528,000	3'17".3 0·0081660 1,221,000	3'59".0 0·0098929 1,485,000	9'34".9 0·0237976 3,561,000	34'52".7 0·086593 12,957,000	25".5 0·00106 159,000
Sidereal Period (days)	0·9424219	1·3702178	1·8878025	2·7369159	4·5175026	15·945452	21·276665	79·33082	550·45	0·7490
Mean daily motion	381°.9	262°.7	190°.7	131°.5	79°.7	22°.6	16°.9	4°.5	39°.25	480°
Inclination of orbit to ring plane	1°31'	0°1'.4	1°5'.6	0°1'.4	0°21'	0°20'	0°10'.4	14°32'.7	150°3'.7	0°
Mean Synodic Period: d.h.m.s. d.h.	0.22.37.12·4 0.22·6	1.8.53.21·9 1.8·9	1.21.18.54·8 1.21·3	2.17.42.9·7 2.17·7	4.12.27.56·2 4.12·5	15.23.15.25 15.23·3	21.7.39.6 21.7·6	79.22.4.56 79.22·1	523.15.36.– 523.15·6	0.17.59.– 0.18·0
Orbital eccentricity	0·0201	0·0044	0·0000	0·0022	0·0010	0·0290	0·104	0·0283	0·1633	0·0
Diameter (km)	500	600	1040	820	1580	5830	500	1600	200	400
Mass: x10⁶ Saturn x10¹⁰ ⊙	0·07 0·201	0·12 0·351	1·14 3·268	1·76 5·028	4·04 11·56	241 688·8	0·54 1·558	1·7 5·0	— —	— —
Stellar magnitude (at mean opposition distance)	12.1	11·7	10·3	10·4	10·0	8·3	14·2	10·8	16·5	14
Discoverer	Herschel	Herschel	Cassini	Cassini	Cassini	Huyghens	Bond	Cassini	Pickering	Dollfus

Recent faint *Voyager* discoveries are not included.

† retrograde.

		Diameter		Ratio
		kilometres	*at mean opposition distance (" arc)*	
Ring A	outer	272,300	43.96	1.0000
	inner	239,600	38.69	0.8801
Ring B	outer	234,200	37.80	0.8599
	inner	181.100	29.24	0.6650
Ring C inner		149,300	24.12	0.5486
Saturn equatorial		120,800	19.52	0.4440

Inclination of plane of rings to ecliptic: $28°068$.
Ω: $168°815$ (1950 Jan 0) + $0°014$ annually.

8.15 Oppositions, 1980–2000

Date	Mag.	
1980 Mar 14	+0.8	Earth passes through ring plane
1981 Mar 27	+0.7	
1982 Apr 8	+0.5	
1983 Apr 21	+0.4	
1984 May 3	+0.3	
1985 May 15	+0.2	
1986 May 27	+0.2	
1987 Jun 9	+0.2	
1988 Jun 20	+0.2	Ring system at maximum pre-
1989 Jul 2	+0.2	sentation
1990 Jul 14	+0.3	
1991 Jul 26	+0.3	
1992 Aug 7	+0.4	
1993 Aug 19	+0.5	
1994 Sep 1	+0.7	
1995 Sep 14	+0.8	Earth passes through ring plane
1996 Sep 26	+0.7	
1997 Oct 10	+0.4	
1998 Oct 23	+0.2	
1999 Nov 6	0.0	
2000 Nov 19	−0.1	

The apparent equatorial diameter of Saturn at opposition ranges from 18″.4 at aphelion (June opposition) to 20″.7 at perihelion (December opposition). The large brightness variation is due principally to the varying presentation of the ring system.

SECTION 9

URANUS

9.1 General

Opposition occurs about 4 days later each year. Dec limits about 24° N and S.

With a mean opposition mag of about 5.7, and a diameter of nearly 4″, Uranus is easily enough located once its approximate position is known. It offers no scope for amateur work, however, no details being visible except faint belts; these require at least 25 cm aperture. Magnitude estimates may be of value, since its brightness may fluctuate in a random fashion by several tenths of a magnitude (B.11.1). The most appropriate instrument for such work is a pair of binoculars. Since the planet changes its position on the ecliptic at the rate of about 4½°/annum, new comparison stars will have to be selected (from, e.g., B.23.10, 23.12) from time to time. At present (1981) Uranus is well south of the equator, and will be rather poorly placed for north temperate observers for the rest of the twentieth century.

9.2 Satellites

I, II: Both of the inner satellites are difficult objects; II requires apertures of the order of 50 cm, although I has been claimed with 30.5 cm.

III, IV: Should be seen without much difficulty with 20 cm; have been frequently glimpsed with 15 cm when Uranus was high in the sky, though not easily. Isaac Ward's astonishing feat of glimpsing them with only a 10.9 cm refractor must be counted a freak of vision.

V: Too faint for visual observation.

The observation of the Uranian satellites is facilitated if an occulting bar is used. A strip of red filter is cemented at the focus of the eyepiece and the planet's disc hidden behind it, so that the satellite shines in the clear portion of the field.

See also sections 7.7, 9.5.

9.3 B.A.A.H. data

Star chart showing path throughout the year.
Stellar Magnitude.

9.4 A.A. data

The following information will be of particular use to amateur observers:
- (a) Heliocentric longitude, latitude, radius vector—40-day ephemeris
- (b) Geocentric RA, Dec, distance—daily ephemeris
- (c) Time of ephemeris transit—daily ephemeris
- (d) Light time, magnitude, diameter—10-day ephemeris
- (e) Satellite ephemerides: times of northern elongation of all five satellites, with supplementary tables

9.5 Data

Mean solar distance: 19.18183 A.U. = 2.870×10^9 km.
Perihelion distance: 2.734×10^9 km.
Aphelion distance: 3.005×10^9 km.
Sidereal period: 30685^d (84.02^y).
Mean synodic period: 369.66^d
Sidereal mean daily motion: $0°.01$.
Mean orbital velocity: 6.8 km/s.
Axial rotation period: 16^h–28^h(?).

e : 0.0471842 (+ 0.0000027)
i :   0° 46′ 22″.8 (+ 0″.1)   Epoch 1950, Jan 1.5 UT
Ω: 73° 44′ 23″.7 (+ 18″.0)   (annual variations in
$\tilde{\omega}$: 169° 51′ 6″.1 (+ 58″.2)   brackets).
L : 98° 18′ 31″.03 (+ 4° 17′ 46″.13)

Angular diameter at unit distance: 68″.56.
Angular diameter at mean opposition distance: 3″.76.
Linear diameter: 52,000 km = 4.1⊕.
Mass: 4.37×10^{-5} = 14.6⊕.
Volume: 59⊕.
Density: 1.25 x water = 0.23⊕.
Mean superficial gravity: 0.93⊕.
Mag at maximum brightness: 5.5.

9.6 Satellite data

	I Ariel	II Umbriel	III Titania	IV Oberon	V Miranda
Mean distance from Uranus: at mean opposition distance A.U. kilometres	14″5 0·0012820 191,000	20″2 0·0017859 267,000	33″2 0·0029303 438,000	44″4 0·0039187 586,000	9″3 0·0008 119,000
Sidereal Period: days days, hours	2·520383 2.12·489	4·144183 4.3·460	8·705876 8.16·941	13·463262 13.11·118	1·41349 33h 56m
Mean Synodic Period (d.h.m.s.)	2.12.29.40	4.3.28.25	8.17.0.0	13.11.15.36	1.09.55.31
Stellar magnitude (at mean opposition distance)	14·4	15·3	14·0	14·2	17
Discoverer	Lassell	Lassell	Herschel	Herschel	Kuiper

SECTION 10

NEPTUNE

10.1 Apparitions

Opposition occurs about 2 days later each year. Declination limits, about ± 25°.

10.2 Telescopic appearance

Opinions differ as to the minimum aperture that will show the disc (diameter 2″.5 at mean opposition distance), and certainly the decision 'a disc or not a disc' when the object is at the threshold is one which is overwhelmingly influenced by such factors as the seeing and the experience of the observer.

It has been claimed to have been glimpsed with a 75 mm OG; doubtful with a 10 cm. Given good conditions, its appearance is certainly non-stellar with any aperture over 15 cm using adequate magnification, but under most circumstances the disc is probably invisible with about 12.5 cm or less, except when assisted by the eye of faith, since a magnification of about ×500 is the minimum desirable.

The best way of detecting its non-stellar character is to accustom the eye to the appearance of an accurately focused nearby star of about the same magnitude, and then switch the telescope quickly on to the planet.

Its opposition magnitude being about 7.8, it can be seen easily with binoculars. Magnitude estimates can be made as for Uranus, using a small telescope and a low power so as to include any convenient nearby comparison stars.

10.3 Satellites

I: Triton (retrograde and highly inclined) has been seen with a 15-cm, though an inexperienced observer will probably require at least 20 cm; here again the absence of atmospheric turbulence is almost the deciding factor.

140

If the body of the planet is hidden by an ocular bar, and the position of the satellite ascertained beforehand, it may be glimpsed with 12.5 cm, and should theoretically be visible with little more than 75 mm if the planet were high in the sky, which, unfortunately, will not be the case for British observers for many decades to come.

II: Nereid: discovered in 1949 May 1 by Kuiper with a 40-min exposure at the Cassegrain focus of the McDonald 82-in (208 cm) reflector, stopped down to 168 cm ($f/5$); it was then mag 19, and situated about 200″ from the planet. Period 360^d; mean distance from Neptune about 5½ million km; orbit inclined at about 28° to the ecliptic; its magnitude implies a diameter in the neighbourhood of 300 km, making assumptions regarding its albedo.

10.4 B.A.A.H. data

Star chart showing path throughout the year.
Stellar Magnitude.

10.5 A.A. data

The following information will be of particular use to amateur observers:
(a) Heliocentric longitude, latitude, radius vector—40-day ephemeris
(b) Geocentric RA, Dec, distance—daily ephemeris
(c) Time of ephemeris transit—daily ephemeris
(d) Light time, magnitude, diameter—10-day ephemeris
(e) Satellite ephemerides:
  Triton: times of eastern elongation, with supplementary tables
  Nereid: differential co-ordinates at 10-day intervals

10.6 Data

Mean solar distance 30.05802 A.U. = 4.497×10^9 km.
Perihelion distance: 4.458×10^9 km.
Aphelion distance: 4.535×10^9 km.
Sidereal period: 60190^d (164.79^y).
Mean synodic period: 367.49^d.
Sidereal mean daily motion: $0°.006$.
Mean orbital velocity: 5.5 km/s.
Axial rotation period: $18^h - 20^h$ (?).

141

e : 0.0085682 (+ 0.0000007)

i :   1° 46' 28".1 (−0".4)

Ω: 131° 13' 42".3 (+ 39".6)

$\tilde{\omega}$: 44° 9' 31".0 (+ 29".4)

L : 194° 57' 8".81 (+ 2° 11' 49".4)

Epoch 1950, Jan 1.5 UT (annual variations in brackets).

Angular diameter at unit distance: 73".12.

Angular diameter at mean opposition distance: 2".52.

Linear diameter: 48,400 km = 3.8⊕.

Mass: 5.177 x 10^{-5} ⊙ = 17.3⊕.

Volume: 54⊕.

Density: 1.77 x water = 0.32⊕.

Mean superficial gravity: 1.22⊕.

Mag at maximum brightness: 7.6.

10.7 Satellite I data

Mean distance from Neptune: 0.0023747 A.U.

   at mean opposition distance: 16".8.

Sidereal period: 5^d.876833 = 5^d 21^h.044.

Synodic period: 5^d 21^h 3^m 27^s.

Mass: 3.448 x 10^{-3} ψ = 1.8 x 10^{-8} ⊙.

Stellar mag at mean opposition distance: 13.

Discoverer: Lassell.

SECTION 11

PLUTO

11.1 General

At present passing through Virgo at a mean rate of $1°.5$ per annum. Declination limits about $±40°$, but for many years to come it will be in the equatorial regions.

Its opposition magnitude is about 13.8, so that under good conditions it can be seen with 15 cm aperture, although more may be necessary to ensure positive identification. The B.A.A. *Handbook* publishes annually a chart showing the movement of Pluto through its star field.

11.2 B.A.A.H. data

RA and Dec throughout the year.
Stellar Magnitude.
List of guide stars.

11.3 A.A. data

The following information will be of particular use to amateur observers:
- (*a*) Heliocentric longitude, latitude, radius vector—40-day ephemeris
- (*b*) Geocentric RA, Dec, distance—5-day ephemeris
- (*c*) Time of ephemeris transit—daily ephemeris
- (*d*) Light time—10-day ephemeris
- (*e*) Satellite ephemeris: times of northern elongation

11.4 Data

Mean solar distance: 39.45743 A.U. = 5.903×10^9 km.
Perihelion distance: 4.450×10^9 km.
Aphelion distance: 7.347×10^9 km.
Sidereal period: 248.4 tropical years.
Mean synodic period: 366.74 days.

143

Sidereal mean daily motion: $0°.004$.
Mean orbital velocity: 4.7 km/s.
Axial rotation period: 6.39^d.
e : 0.24852 ($\pm$0.0)
i : $17°$ $8'$ $34''.1$ $(-0''.2)$
Ω: $109°$ $38'$ $1''.4$ $(+49''.0)$
$\tilde{\omega}$: $223°$ $31'$ $20''.8$ $(+50''.3)$
L : $165°$ $36'$ $9''.2$ $(+1°$ $27'$ $59''.1)$
Linear diameter: 3,000–3,500 km.
Mass: approx. $0.002\oplus$.
Density: 0.8 x water $= 0.15\oplus$.

Epoch 1950, Jan 1.5 UT (annual variations in brackets).

11.5 Satellite data (Charon)

Mean distance from Pluto: 20,000 km.
    at mean opposition distance: $1''.0$.
Sidereal/Synodic period: 6.39^d.
Mass: $2 \times 10^{-4}\oplus$.
Diameter: 1,300 km?
Discoverer: James W. Christy.

SECTION 12

ASTEROIDS

12.1 General

Apart from the interest of tracking down and identifying the brighter asteroids, and as objects for photography, they offer little scope for amateur observation. Discovery is now carried on entirely by photography, and it is safe to say that none within the reach of small apertures (say, brighter than mag 12) remains undetected.

Vesta (4), the brightest, reaches mag 6 at opposition and is therefore at times just visible to the naked eye. Its brightness is variable, and there is perhaps scope for systematic magnitude estimations or photometric measures carried out over a long period.

Among the 25-odd which are brighter than mag 10 at opposition may be mentioned: Ceres (1), opposition mag 7.8; Pallas (2), mag 9.2; Juno (3), mag 9.4; Chaldaea (313), mag 9.0; Ilmatar (385), mag 9.7. Telescopically they are characteristically yellowish and indistinguishable from stars.

Work on their spectra and albedo (B.12.3, 12.8) has revealed three classes of asteroid: the carbonaceous or C-type, with a dark surface having an albedo of from 2% to 5%; the S-type (stony or silicate), albedo around 15%; and a small group of unusual bodies, such as Vesta, most of which have a very high albedo. The C-type asteroids appear to be the most common, although, because of their superior brightness, the S-type are easier to detect. Many well-observed asteroids show light variation due to their axial rotation; in the case of Vesta, for example, it is $5^h 55^m$. See also B.12.1.

About 2,300 asteroids have now had their orbits sufficiently well determined to be allotted a permanent number. Newly-discovered asteroids are given a provisional two-letter designation, indicating the half-month in which the discovery was made (January = A & B, February = C & D, etc.) and the order of discovery within that half-month; thus the second new object to be discovered in the first half of March 1981 would be designated 1981 EB.

145

The I.A.U. Minor Planet Centre, which issues circulars and ephemerides, is located at the Smithsonian Astrophysical Observatory. Orbits and other data are published periodically in the *Astronomische Nachrichten*. B.12.5, a monumental compendium summarising all available data for the first 1091 numbered asteroids, is indispensable for workers in this field. Ephemerides of the first four asteroids, Ceres, Pallas, Juno and Vesta, are published annually in *A.A.* The *B.A.A. Handbook* at present publishes ephemerides for asteroids reaching mag 11 or brighter at opposition.

12.2 Observation

Numerous asteroids are within the range of a 15 cm telescope, the difficulty being in distinguishing them from the stars, for which purpose an atlas showing all the stars at least to the faintness of the body being sought is desirable (e.g., for many purposes B.20.22; B.20.4 (*Eclipticalis*) will serve for the few asteroids brighter than mag 9). Alternatively, the field known to contain the body can be sketched, and re-compared the following night to detect the moving object, although this method is inappropriate if the asteroid is at or very near a stationary point.

An alternative way of detecting motion is to set a NS wire at the focus of the eyepiece and to time the transits of the field objects among which the asteroid is known to lie. A subsequent run of transits the following night should reveal a discrepant timing, due to the motion in RA of the planetary body. This method may aid the identification of slow-moving objects, for which a freehand chart may be prohibitively inaccurate.

Photography is, clearly, the most convenient way of recording asteroids, since a 35 mm camera, exposed and guided for ten minutes or so, will reach the 11th magnitude without difficulty; repeated exposures on subsequent nights will reveal the orbital motion. To show trailing, exposures of the order of an hour or more, and large F, are required. Manning has photographed asteroids down to mag 16, using a 26.5 cm Newtonian set to follow the asteroid's computed path.

Any predicted occultations of stars by asteroids are included in the B.A.A. *Handbook*, and every effort should be made to observe and time such events. Particular attention should be paid to possible additional occultation of the star by an invisible satellite or companion, as occurred in the case of Hebe in 1977, and Herculina and Melpomene in 1978.

12.3 B.A.A.H. data

For Ceres, Pallas, Juno, Vesta:

    (*a*) Date of opposition.

(b) RA
     Dec $\Big\}$ at 10-day intervals throughout
     Horizontal parallax   the apparition.
(c) Magnitude table.

For other asteroids reaching magnitudes brighter than about +11 during the year:
(a) Date of opposition.
     Magnitude at opposition.
(b) RA $\Big\}$ at 10-day intervals throughout
     Dec   the apparition.

12.4 A.A. data

For Ceres, Pallas, Juno, Vesta:
(a) RA (astrometric, and apparent−astrometric)
     Dec (astrometric, and apparent−astrometric) $\left.\rule{0pt}{5em}\right\}$ at 0^h ET daily
     Horizontal parallax   throughout the
     Distance from Earth   apparition.
     Ephemeris Transit time
     Photographic magnitude at selected dates.
(b) Stellar magnitude at 40-day intervals throughout the year.
(c) Dates of conjunction, opposition, stationary points, and of conjunctions with other members of the Solar System.

12.5 Data

Ceres
   Mean solar distance: 2.767 A.U. = 4.139 x 10^8 km.
   Sidereal period: 1.681 tropical years.
   Axial rotation period: $9^h\ 5^m$.
   Diameter: 1003 km.
   Mass: 60 x 10^{22} gm.
   e: 0.079.
   i : $10°.6$.

Pallas
   Mean solar distance: 2.767 A.U. = 4.139 x 10^8 km.
   Sidereal period: 1.684 tropical years.
   Axial rotation period: not known.
   Diameter: 608 km.
   Mass: 18 x 10^{22} gm.
   e: 0.235.
   i : $34°.8$.

147

Juno

Mean solar distance: 2.670 A.U. = 3.994×10^8 km.
Sidereal period: 1.594 tropical years.
Axial rotation period: $7^h\ 13^m$.
Diameter: 247 km.
Mass: 2×10^{22} gm.
e: 0.256.
i : 13°.0.

Vesta

Mean solar distance: 2.361 A.U. = 3.532×10^8 km.
Sidereal period: 1.325 tropical years.
Axial rotation period: $5^h\ 20^m$.
Diameter: 538 km.
Mass: 10×10^{22} gm.
e: 0.088.
i : 7°.1.

SECTION 13

ZODIACAL LIGHT, GEGENSCHEIN
AND ZODIACAL BAND

13.1 Introduction

It may be taken as a sign of the difficulty of making regular observations of the Zodiacal Light and its associated phenomena from temperate latitudes that the B.A.A. Observing Section, devoted at least partially to its observation, had to be terminated due to lack of interest. In the latitudes of Great Britain and the northern U.S.A, the Zodiacal objects appear as mere wan ghosts of their true selves, and that to observe them under anything approaching satisfactory conditions it is necessary to approach at least to within 35° of the equator. It is arguable that none of the numerous unsolved problems connected with these objects is soluble by observations made outside the tropics, and certainly it is true that half a century's observations made in this country (the majority of them being no more than bare statements of visibility) have produced no significant increase in our knowledge of their nature or even behaviour.

13.2 Appearance and nature

The Zodiacal Light consists of two cones, apparently centred on the Sun, and lying in or near the ecliptic. The apices of the cones are commonly 60° to 90° from the Sun in high latitudes; 100° or 110° in the tropics. In the tropics, too, they are visible to within about 20° of the Sun, while the evening cone remains visible at least to the end of the first lunar quarter.

The brightness of the cones is commonly referred to that of some part of the Galaxy, but this can be misleading if the latter is near the horizon and meaningless unless the particular region of the Galaxy is specified. Thus in the tropics the brightest parts of the cones are frequently brighter than the densest regions of the Milky Way in Sagittarius, as seen there; in this country, also, they are as bright as, or brighter than, the Sagittarius region as seen here, when they are observed under optimum conditions.

149

But that is not to say that the Zodiacal Light appears as bright in England as in the tropics. However, the brightness of the Light varies.

The Gegenschein, situated at or near the anti-solar point, is very much fainter than the cones, though normally brighter than the Zodiacal Band, when both are visible. Its characteristic shape is oval rather than circular, its axes being in the proportion 2 : 1 roughly. From high latitudes it is typically about $10° \times 20°$ in extent, though in the tropics it may be seen to extend over at least $30°$.

The Zodiacal Band is a very faint, parallel-sided extension of the apex of the visible cone (sometimes visible as 'wings' on either side of the Gegenschein), about $5°$ to $10°$ wide; its intensity falls off on either side of its median line, and also from the cone to about $135°$ from the Sun, whence it increases again until the Band merges with the Gegenschein at $180°$ from the Sun. In Britain it is never brighter than about $\frac{1}{3}$ G in Monoceros (usually much less), and even in the tropics is a difficult object near the threshold of vision.

The Light may be explained in terms of a disc of dust and some electrons in the plane of the Earth's orbit, centred on the Sun, and physically connected with the corona. It has been calculated that the observed intensity could be reproduced by particles 2 metres in diameter spaced 2000 km apart. Very high altitude investigations have been performed by Blackwell and he has concluded (B.13.1) that the Light is due mainly to the scattering of sunlight from interplanetary dust particles.

13.3 Observational conditions

In British latitudes the evening cone is best seen during the moonless periods from February to April, the morning cone during the moonless periods from August to October. It is at these times that the ecliptic makes its greatest angle with the W and E horizon respectively: during the summer and winter the ecliptic, in these latitudes, is inclined at about $40°$ to the horizon at sunset and sunrise; in the spring the angle at the W horizon is larger than this in the evening, and smaller at the E in the morning; in the autumn these conditions are reversed. In the tropics both cones are well seen throughout the year. In latitudes near $50°N$, from August – when the midnight twilight has ceased – until the end of September or a little later, the northern sky maintains a distinct brightness. This, also, may be a part of the Zodiacal Light, because low-latitude observations show that the light of the main cones can be traced up to ecliptic latitudes of at least $\pm 50°$. During the corresponding period before the summer solstice, the Milky Way interferes with the phenomenon.

The cones should be looked for about 2^h after sunset or before sunrise, when the Sun is some 20° vertically below the horizon. The requisite local conditions are absence of cloud about the horizon concerned, absence of twilight, absence of artificial lights (i.e. the effective observation of the Zodiacal Light is restricted to country districts—even in the wartime blackout the pollution of urban atmospheres was a hindrance), a minimum of 10 minutes' dark adaptation of the eyes, and a dim red light for recording the observations. Given these conditions the Light is frequently visible to normal eyesight in Britain, though the Gegenschein—and, still more, the Zodiacal Band—requires exceptional eyesight, and cannot be effectively observed outside the tropics. During the Second World War, both were indeed regularly observed as far N as Cumberland, but mere visibility yields little information of value. The Gegenschein can only be seen during moonless periods when it is projected on a relatively starless region of the sky; the other conditions must be even more rigidly observed than in the case of the Light. During December and January it is rendered invisible by the Milky Way; during February and March it is well placed in Leo; passing into Virgo it becomes too low for effective observation in these latitudes, though occupying relatively starless regions in Aquarius and Pisces during the summer and early autumn; during the autumn, conditions again become favourable, with the Gegenschein in Aries and Taurus.

The Gegenschein is best located as follows: determine its approximate position on a star atlas, 180° from the Sun. Then let the dark-adapted eye wander at random over the area. If spotted, concentrate first on its position, then upon its shape and extent, resting the eyes frequently and avoiding prolonged staring.

13.4 Work, and further data required

As has already been remarked, it is probable that none of the many unsolved problems connected with the Zodiacal phenomena is capable of solution by observations made in this country. Examination of the mass of observational material that has accumulated during the last fifty years forces upon one the conclusion that in this field, where the observing conditions are of paramount importance, English observers provide more information about these conditions than about the objective nature of the object they are observing.

It is, for example, quite impossible to detect objective variations in the brightness or extent of the Light when the observing conditions impose very much wider variations on its brightness and extent as observed. The outstanding need is of observational data that are objective, accurate, and strictly comparable. This is just what, regarding the first desideratum at least, the British observer cannot provide.

The main points at issue, and upon which further data are required, are:

13.4.1 Brightness: Is there any systematic difference of brightness between the two cones?

The brightness of the Zodiacal Light has been observed to increase by a factor of about 2 following a very large solar flare (B.13.14). At present there seems to be some, but not conclusive, evidence that both long-period variations—up to several months—and sudden fluctuations occur; the former are difficult to disentangle from seasonal meteorological causes; the latter—characteristically a sudden increase of brightness, followed by a slower fading over several nights—is more likely to be objective.

What is the luminosity distribution within the cones themselves? How does the brightness fall off from base to apex, and from central axis to edges?

Are reported 'ripples' and 'flickerings' of intensity (travelling along the cones with a period of a few seconds) objective? If so, are they related to auroral pulsations? Or due to some such factor as fatigue, or related to the glimpsing of an object near to the visual threshold?

There is a great need for a long series of objective and comparable photometric observations of the Light and the associated phenomena. Bousfield's photometric work in Queensland (B.13.2) has shown what can be done, given suitable observing conditions.

Photometric observations should always be made with the Sun at the same vertical distance below the horizon, to ensure consistency. If recording the brightness cf. the Milky Way, always specify the region that is being used as standard. Three photometric methods suggest themselves: visual estimations, compared with the Galaxy; instrumental;* elongation of the apex as a criterion of brightness—this, if valid, would at least be more precise than visual estimations.

13.4.2 Colour: It has at times been suspected that the evening cone is of a 'warmer' tint than the morning. Few records exist, however, owing to the difficulty of the observation. But it seems to be established that the Zodiacal Light is of longer wavelength than the Milky Way, being of a definitely yellowish tint.

13.4.3 Position: The axis of symmetry of the Light is very close to the ecliptic. Blackwell (B.13.1) believes that the plane of the Light is closer to the average plane of the planetary orbits than it is to the ecliptic.

* *A.A.H.*, section 21.3.9, describes a very simple photometer which is suitable for work of this nature.

Observations of position relative to the ecliptic are important, though decisive observations will probably come from observers in the subtropics, where the ecliptic can be perpendicular to the horizon.

It is not established whether or not the Zodiacal Band adheres to the ecliptic. Bousfield (Queensland) puts it in a plane inclined to the ecliptic, though a different one from that derived by Hoffmeister, which coincided with the orbital plane of Jupiter.

More data are also required concerning the position of the Gegenschein relative to the ecliptic.

13.4.4 Extent: Can the elongation of the apices of the cones be correlated with the sunspot cycle? As a result of a discussion of all published observations of the Light from 1668 to 1939 Thom (B.13.20) discovered an apparent, and as yet unconfirmed, correlation between the visible extent of the cones and the solar cycle, their size being maximal during or just before every second spot maximum; the suspected correlation is therefore with the 22-year period of the Sun's magnetic activity rather than with the spot cycle.

Can the elongation of the apices be correlated with the time of year— if so, is it independent of terrestrial meteorology? Thom also unearthed apparent maxima during January and June.

Can the width of the cones (which is variable by about 100%) be correlated with the time of year?

Are the observed variations of size of the Gegenschein objective? There is some evidence that it is more elongated at the solstices than at the equinoxes.

13.4.5 Analysis: From the individual observations (documented below) a series may be amassed with a view to detecting changes with time. The records may be examined to see if the physical extremities or the brightness of the various parts varies with, e.g., the seasons or the solar cycle. Comparison may also be made with solar flare activity, terrestrial magnetism, and auroral records. But to be useful such studies require many observations made over a long period by several observers for mutual corroboration, and the record of achievement from temperate latitudes is not encouraging. Tropical observers, especially at elevated sites, are much more favoured.

13.5 Observational records

It is convenient to keep both an Observing Book and a Permanent Record. In the latter the data recorded at the time of observation are entered in a slightly modified form, more suitable for discussion, at the conclusion of each period of observation.

For routine observation the following data should be included:

13.5.1 General:

Date.

Location of observing site.

Time (UT).

Interval ± sunset/sunrise.

Atmospheric clarity: either by minimum visible magnitude or by simple verbal description.

Moon: age, and interval from setting or rising.

Artificial lights.

Length of dark adaptation: should be not less than 10 mins, 20 mins being better before observation of the Gegenschein or Zodiacal Band.

13.5.2 Zodiacal Light:

The ill-defined quality of its boundaries as seen in this country does not usually justify the making of a chart of its position, or the specification of the whole boundary. It will suffice to give:

Position of apex, expressed as an angular distance from Sun (see also section 13.6).

Angular width at horizon i.e. azimuth difference of points on the horizon where the boundaries of the cone, if produced downward, would intersect it.

Brightness cf. Milky Way in Cygnus or Gemini (e.g. 2G, ½G, etc), the comparison region being specified.

In the Permanent Record the Position of the Apex is converted to Distance of Apex from Sun (if this has not been recorded direct), the RA and Dec obtained by observation being converted to Longitude, and the Sun's longitude being taken from the *A.A.* Further data, which should be recorded if the visibility of the Light justifies them, are:

Deviation of the apex from the ecliptic.

Relative definiteness of the N and S boundaries.

Notes on the intensity distribution within the cone.

Coordinates or bearing of the intersection of the axis of the cone and the horizon.

13.5.3 Gegenschein:

Position: drawn on a tracing of the naked-eye stars in the region.

Maximum and minimum diameters and their orientation.

Brightness.

Whether any connexion with the Zodiacal Band is detectable.

In the Permanent Record are added:

154

Distance of the centre N/S of the ecliptic.
Coordinates of the centre.

13.5.4 Zodiacal Band:
Position.
Angular width at specified points.
Brightness.
If visible on both sides of the Milky Way, where the latter crosses it. In the Permanent Record the position as expressed with reference to the stars in the Observing Book is converted to RA and Dec. Also added is: Distance N/S of the ecliptic at specified points.

13.6 Methods of recording position and extent of the Light

1. Location of the apex, or of sufficient points along its boundary, with reference to the stars, to enable the coordinates of the apex or the whole outline of the cone to be later plotted on a tracing from a star chart. A method somewhat similar to that employed for describing the position of a meteor path can be used. For example:

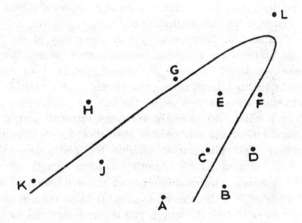

FIGURE 18

Position of apex would be specified as: *L ¼ F.*
Position of whole cone would be specified as: *A ½ B,*
C ¼ D,
E ¾ F,
L ¼ F,
G,
H ⅔ J,
inside (i.e. < 3°) *K.*

2.   Location of the apex, or of sufficient points to define the whole cone, in terms of altitudes and azimuths, with the time of observation.

3.   Graphical: Make tracings of a 60°-wide zodiacal zone, including all stars brighter than mag 4 or 5, but omitting all reference lines and systems of coordinates. Use these tracings for drawing in the outline of the Light, etc, if possible also its axis and some isophotic contours, and the horizon; this can conveniently be done on a ground-glass sheet covering a box containing a red light just bright enough for the tracing to be seen. If the stars are marked in ink and the observations recorded in pencil, the same tracings may be used over and over again; the Permanent Record is traced off at the conclusion of each observation, half a dozen of the brighter stars being included for purposes of orientation. This permanent tracing can then be laid over the original chart and the position of the ecliptic drawn in.

13.7 Instrumental

For routine observation of the Zodiacal Light all that is required is: a clearly graduated rule (1 cm at arm's length subtends about 1°*), a watch, a red desk light, a star atlas, and the *Astronomical Almanac*.

Photography: see section 13.8.

Photometry: Comparability and objectivity of the data are the vital considerations in assessing heterogeneous observational material, and these are largely lacking in existing records. Instrumental photometry eliminates to a large extent the effects of the variable sensitivity of the individual eye, as well as of the diverse sensitivities of different eyes; there remains the necessity for some form of control to allow for the application of the unavoidable correction for variable atmospheric transparency. One photometer suitable for Zodiacal work is described in *A.A.H.*, section 21.3.9. Another consists simply of an optical wedge with a coarse wire mounted across the diaphragm of a negative ocular. The telescope is directed at a starless field on the axis of the Light and the scale is read when the wire is just visible against the background; a second reading is taken with the telescope directed to another part of the sky at the same altitude. In each case (as, indeed, always) the mean of a series of readings is taken, the wedge being moved right away from its previous position between each setting. See also B.13.8.

13.8 Photography

In temperate latitudes it is unlikely to produce much of practical value

* See *A.A.H.*, section 36, for methods of estimating angular distances.

except under rare conditions of extraordinary clarity. The faintness, low contrast, and angular extent of the Light make it a difficult object, and the required homogeneous series of photographs is virtually unattainable: in addition to atmospheric transparency there must be neither twilight nor moonlight present. Under good conditions, using Tri-X film, it can be readily recorded with a 4-minute exposure at $f/3.5$ (B.13.3; see B.13.16 for photographs of the Zodiacal Light and Gegenschein). Apertures of $f/2$ fog prohibitively rapidly, and the best modern results have been obtained with wide-angle lenses (e.g., 28 mm F with 35 mm film) at between $f/3.5$ and $f/5.6$.

SECTION 14

AURORAE

14.1 General

The visibility of aurorae in both hemispheres is related to geomagnetic latitude and longitude, the axis of which passes through both magnetic poles and the centre of the Earth. Aurorae are to be seen most frequently in the auroral zone (geomagnetic latitude 67°), with a progressive diminution both polewards and equatorwards. The aurora itself forms in space around an oval about the magnetic poles, and this oval is offset from the magnetic axis towards the night side of the Earth, expanding or contracting according to the state of the interplanetary magnetic field due to electrified particle clouds. The auroral zone on the surface of the Earth is therefore the locus of the night side of the auroral oval at its point at local magnetic midnight, which latter is defined as the time when the Sun, the magnetic pole and the observer at that point lie in the same plane.

A small pocket spectroscope (which, with slit open, is capable of showing the characteristic green auroral line) is useful for establishing the auroral nature, or otherwise, of suspected sky glows.

The regular observation of aurorae requires no expensive equipment; what it does demand is an observing station in the country, well removed from artificial lights and urban 'smog', and with as nearly a 100% unobstructed horizon as possible. Though observers in the south see aurorae less frequently than those in the north, southerly stations are important in investigations of great aurorae and in latitude surveys or auroral frequency. Southern stations also have an advantage over northern during the summer months owing to the shortness of the northern nights.

If an assistant is available he should sit within earshot of the observer in a lighted room with a watch in front of him, recording the observations as they are called out and adding the time to each. If the observer is working alone, or if he is using star charts for the record, he will need to provide himself with a very subdued red light.

14.2 Frequency of auroral displays

The mean frequency with which aurorae could be seen per annum, assuming no cloud cover, and the comparative figure for actual sightings, are given below for various locations in the U.K.:

	Geomagnetic latitude	Mean annual frequency	Observed frequency
Lerwick	63°	100	30
Kirkwall	62°	80	20
Inverness	61°	63	15
Aberdeen	60°	45	10
Glasgow	59°	32	7
Newcastle	58°	22	5

Geomagnetic midnight in the U.K. occurs at about 22^h local time, with the centre of auroral activity at about 350° true. The annual frequency of occurrence varies with the solar cycle, peak activity taking place one or two years after sunspot maximum, depending upon the intensity of the solar cycle and the location of the observer.

Aurorae may be active and violent, due to flare activity, or may be of a quiet form lasting for several nights in succession and recurring about 27 days later, in step with the rotation of the Sun; these latter are due to holes (formerly called M regions) in the corona. Aurorae may be associated with magnetic storms and radio effects, but the one may be observed without necessarily being accompanied by the other.

Active aurorae may initially take the form of a glow at the horizon, developing into one or more arches of light together with shaft-like rays stretching upwards. An auroral storm may evolve further, with movement of the structures and flame-like changes in the brightness of the light. The peak of the storm may appear overhead to form a corona, with rays converging towards a central point. The storm may then dissolve into patches of cloudy light before finally disappearing; and there may be repetitions of the whole phenomenon on the same night. Figure 19 shows some typical auroral forms. Whether or not a climax of this kind is reached depends upon both the physical severity of the storm and the observer's location. The following table gives the relative frequency of peak activity of storms on auroral nights observed in central Scotland:

Year	No. of nights when storm peak of given intensity recorded						
	Glow	Arc	Rayed arc	Ray structures	Active movements	Corona	All sky
1976	0	1	0	0	0	0	0
1977	5	7	7	5	3	2	0
1978	10	7	4	2	9	1	4
1979	21	7	3	9	7	0	0

Aurorae occur more frequently at the equinoxes.

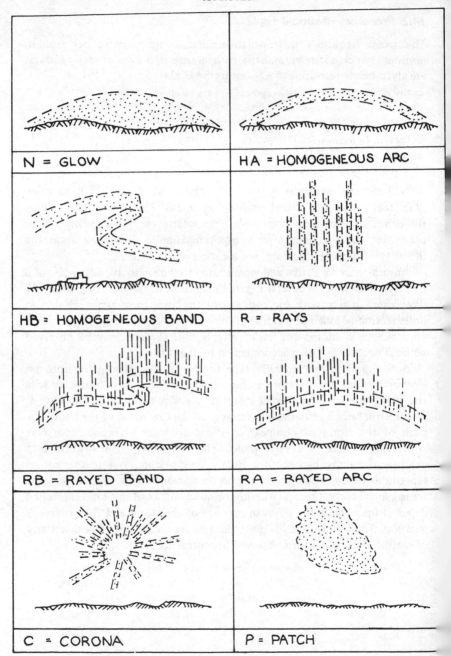

FIGURE 19

14.3 Visual observation

The predominant spectral line of the aurora is the green emission line of ionised oxygen (557.7 nm), and, in skies free from town lights, this may be detected in a pocket spectroscope with the slit open. Better still, in cloudy or hazy weather, especially if sodium lighting or moonlight is present, the green light may be effectively detected using an interference filter tuned to this wavelength. Through such a filter, the glow may be detected between clouds, which will appear black against the bright auroral background.

Under favourable conditions, merely noting that an aurora is occurring can be of value, giving the date, time, and location of the observer. The observation's value is enhanced by adding a description of the event if possible accompanied by a sketch. More useful data still may be recorded by making simple measurements of angular height and width of the auroral forms. Horizontal angles are measured eastwards from true N, using Polaris; vertical angles are measured from the true horizon. To obtain the latter accurately, a cloud alidade or some other device with a plumb bob or levelling bubble will be effective. A home-made rule marked to show units of, say, 5° when held at arm's length, is helpful.

Detailed observations of an auroral storm are usually made at about 5-minute intervals, with additional observations at each change of activity and form. There is an internationally accepted code for recording auroral activity, which was used in the International Geophysical Year (1957) and updated for the International Year of the Quiet Sun (1964). Details of the code are as follows:

Symbol	Meaning
Elevation	
Base	Angular height from horizon to bottom edge of form.
Top	Angular height from horizon to top edge of form.
Direction	
Azimuth	Compass bearing (true) to point being measured.
Condition	
q	Quiet. No movement of form, brightness or activity.
a	Active. Movement in structure or brightness. Undefined.
a_1	Active. Bands which are folding or unfolding.
a_2	Active. Rapid change of shape of lower form.
a_3	Active. Rapid horizontal movement of rays.
a_4	Active. Forms fade quickly, to be replaced by others.

p	Pulsating. Undefined.
p_1	Pulsating. Rhythmic change of form as a whole.
p_2	Pulsating. Flaming with light variations moving upwards.
p_3	Pulsating. Flickering with rapid irregular variations.
p_4	Pulsating. Streaming of irregular horizontal variations in homogeneous forms.

Qualifying symbol

m	Multiple. Several groups of the same type or form.
f	Fragmentary. A part only of an auroral form.
c	Coronal. Rays converging overhead at the magnetic zenith.

Structure

H	Homogeneous. Uniform in shape and intensity.
S	Striated. Lines of brighter and darker light.
R_1	Rayed. Short length rays.
R_2	Rayed. Medium length rays.
R_3	Rayed. Long rays.

Auroral form

N	Glow. Unspecified form. Unidentifiable form. Auroral light.
A	Arc. Uniformly curved arch of light.
RA	Rayed arc. Arc from which rays reach upwards.
R	Ray. A shaft of light reaching upwards.
B	Band. A partial arc or twisted band of light.
RB	Rayed band. Band from which rays reach upwards.
V	Veil. Curtain-like structure, which may be folded.
P	Patch. An isolated cloud or surface of auroral light.

Brightness

1	Weak. Barely visible.
2	As bright as moonlit cirrus cloud.
3	As bright as moonlit cumulus cloud.
4	Bright enough to cast a shadow.

(Note: Observers tend to over-estimate brightness)

Colour

a	Red in upper portion of form only.
b	Red in lower border only.
c	White, green or yellow.
d	Red in main region of form.

e   Red and green mixed.

f   Blue and purple.

The following example illustrates the use of the code:

$$a_3 \ c \ f \ R_2 \ B \ 3 \ c \ 25 \ 30 \ 330 \ 020$$

which would be interpreted as meaning a rapid, horizontally moving rayed band of medium length, fragmentary, converging to a corona, as bright as moonlit cumulus cloud and green in colour. The base and top elevations are measured as $25°$ and $30°$ respectively, and the azimuths of the two extremities are measured as $330°$ and $020°$ (true).

Observations in code, together with sketches, may be set out in standard format on a report sheet (Figure 20), together with date and time (UT); since observations may be prolonged past midnight, it is common to give a dual date, such as September 28/29, to avoid possible ambiguity. Geographic or geomagnetic latitude and longitude must also be given (Figure 21). Completed observations should be submitted to the B.A.A. Aurora Group, or to the Marine Department of the British Meteorological Office at Bracknell, Berks.

14.4 The auroral spectrum

Since aurorae are caused by the energy released in gaseous discharges when electrified particles emitted by the Sun make impact on atmospheric atoms and molecules, the principal emissions are at discrete wavelengths, as follows:

Wavelength (nm)	Relative intensity	Origin	
656.3 (red)	—	H	Hydrogen
636.3 ⎫ 630.0 ⎬	3.3–200	O I	Oxygen
557.7 (green)	10–600	O I	Oxygen
486.1 (double)	—	H OII	Hydrogen ⎫ Oxygen ⎬
470.8 (blue)	8.0	N_2+	Nitrogen
427.8 (band)	24	N_2+	Nitrogen
391.4 (band)	47	N_2+	Nitrogen

From a height of 400 km down to 150 km above the Earth's surface, red aurorae emitting at 630 nm are found, due to the collision of electrons with oxygen atoms. From 150 km down to 100 km or thereabouts the aurora is principally green (558 nm), due to denser layers of oxygen. Between 90 km and 85 km, the lower border may again be tinged with red, due to molecular oxygen emissions; while in aurorae of average

FIGURE 20

1977	OCTOBER			27/28			J SMITH			PAISLEY	80	59	BRITISH ASTRONOMICAL ASSOCIATION
27 2135	a_3	cF	R_2	B	2	C	90° 340°-010°	15°					
27 2200	a		H	N	2	C	5° 330°-040°		5°				
27 2210	a_2		H	A	2	C	9° 340°-020°	5°					
27 2220	P_2		R_2	B	4	C	50° 320°-040°	15°					
28 0050	a_4	F		P	3	d	40° 020°-040°	20°					

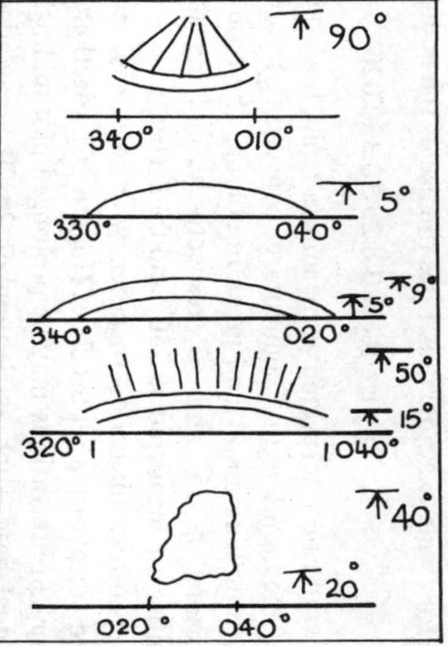

164

strength the greens and reds of oxygen may be mixed with the violet of singly ionised molecular nitrogen at 391 nm, to produce greyish yellow hues. In order of increasing rarity, aurorae are coloured green, white, red, blue and yellow.

Type A aurorae consist of high-level red emissions caused by low-energy electrons moving down the direct connections between the interplanetary and geomagnetic field lines. Type B aurorae are the result of high-energy electrons moving along enclosed field lines from the tail of the magneto-sphere and penetrating deep into the atmosphere to where molecular particles exist, thereby causing the lower red border seen in a strong auroral storm.

It has already been mentioned that the brighter spectral lines may be seen visually, using a spectroscope. Colour film can act as a crude spectro-graph; for example, the red fringe at the base of an aurora may be visually imperceptible and yet appear on a photograph. True spectrography of the aurora involves fast cameras and long exposures.

14.5 Photography

Most amateurs photograph the aurora for its artistic effect, but a good photograph, upon which the stellar background is recorded, may be used to determine the object's location; two observers, situated up to 60 km apart and preferably separated along a line of geomagnetic latitude, may arrange to take simultaneous photographs in agreed directions. The stereo-graphic pair may then be used for locating the forms. The problem lies not in the photography itself, but in the method and cost of communication between the two observers (see B.14.24).

Any camera faster than about $f/2.8$ may be used, set up on a rigid mount and pointed towards the aurora, using a cable-controlled time exposure. Fish-eye lens systems may be directed towards the zenith and used to take all-sky pictures. The exposure length should be as short as possible (to minimise fogging and star trailing), consistent with catching the desired features of the aurora; if the forms are changing rapidly, the optimum exposure will have to be judged from experience. Using Tri-X film (400 ASA), 10-sec exposures at $f/2.8$ can produce good photographs of bright aurorae; Ektachrome 200, although slower than Ektachrome 400, offers greater contrast and is probably the more suitable of the two. Agfa CT 21 (ASA 100) has also been used. Different films have different spectral sensitivity, some producing a more truthful visual rendering than others (Agfa CT 21, for example, is particularly red-green sensitive).

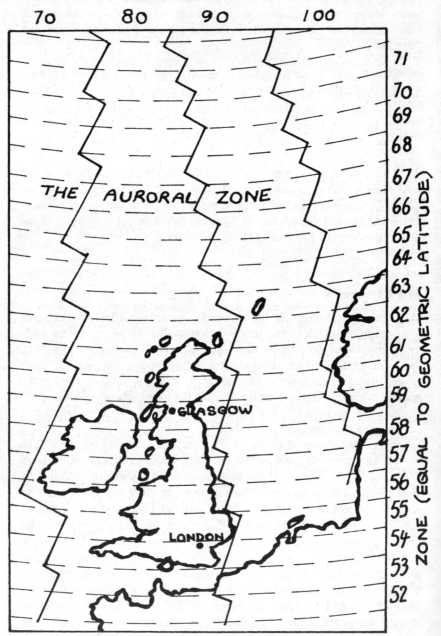

SECTOR

70   80   90   100

71
70
69
68
67
66
65
64
63
62
61
60
59
58
57
56
55
54
53
52

THE AURORAL ZONE

•GLASGOW

LONDON

ZONE (EQUAL TO GEOMETRIC LATITUDE)

FIGURE 21

14.6 Photometry

The photomultiplier offers a useful technique for the investigation of auroral phenomena. Used in conjunction with a narrow-passband optical filter, it may be tuned to a particular auroral spectral line. Two photometers, tuned individually to red and green lines, have been used to search for the centre and the lower borders of auroral forms. The photometers were so sited that as the aurora developed and migrated southwards, the green area passed out of the line of sight of the appropriate photometer, and the red fringe moved into the field of the second photometer. This technique was used to search for aurorae automatically while the observers were engaged on other work.

A very simple visual photometer, suitable for comparing the relative intensities of extended objects such as aurorae, is described in *A.A.H.*, section 21.3.10.

14.7 Analysis of observations

It is useful to adopt the technique developed by Bartels to emphasise the 27-day periodicity, where it exists. His diagram consists of a horizontal row of 27 consecutive dates, followed immediately below by the next 27 dates thereafter, and so on. The presence of aurorae and associated phenomena, such as radio and magnetic data, are plotted in the appropriate square, so that activity associated with a particular solar longitude will tend to assume a vertical alignment. Such a diagram may be used for predicting the possibility of further active periods in ensuing rotations.

If the observer also measures sunspot activity, the number of aurorae seen may be related to the mean daily frequency of active solar areas. Another useful exercise is to plot from a given date, the integrated number of auroral nights and mean daily frequencies. If the scales of the two curves are suitable, the relationship between solar and auroral activity may be examined.

If the altitude of the base of an auroral arc is assumed to be H, and its angular elevation above the true horizon is h, the distance D, from the observer to the sub-auroral point on the Earth's surface, can be derived from the diagram in Figure 22.

14.8 The airglow

The airglow is a permanent, though somewhat variable, phenomenon that affects the entire night sky, caused by photochemical processes that derive their energy from direct sunlight. Part of the energy goes to dissociating oxygen molecules into atomic oxygen (hence ozone), while the remainder is energy of ionisation. As the night progresses, molecular recombinations proceed and this energy is released as airglow. The principal emissions are the nitrogen doublet at 589.6 nm and 589.0 nm, the oxygen doublet at 636.3 nm and 630.0 nm, and the oxygen line at 557.7 nm. Although normally too faint to be distinguished visually, the total luminosity of the airglow exceeds that of integrated starlight. Abnormal airglow may be observed from time to time in all latitudes, although more frequently from higher latitudes. The following are the more likely forms:

1. Increased horizon brightness in certain directions, more likely between NW and E. The upper edge is diffuse. May be confused with the aurora.

2. As above, but with a sharp upper limit, giving the impression of a bank of luminosity.

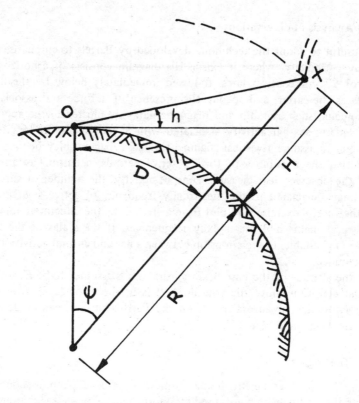

FIGURE 22 (a)

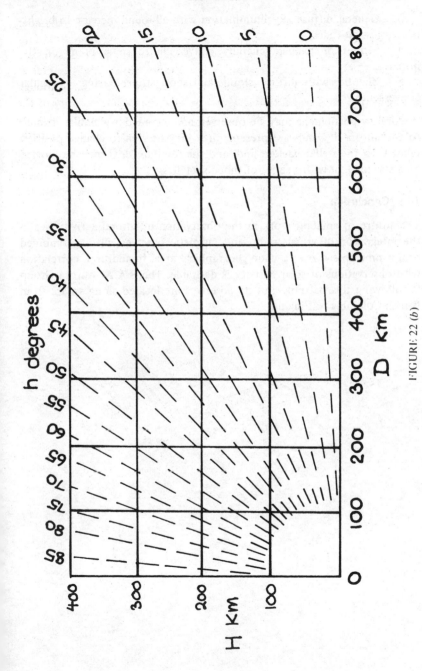

FIGURE 22 (b)

3. General diffuse sky illumination with all-round increase in brightness at low altitudes.

4. Stripes or bands, generally converging to some point on the horizon.

5. Chaotic, with diffuse clouds and short stripes having no regular convergence.

Airglow phenomena may be observed and recorded in a similar manner to the aurora. If bands are present, they may be seen to change position relative to the stellar background, and the resultant drift may be plotted on a star map, or with a series of timed sketches.

14.9 Conclusion

The future of amateur work in the observation of auroral activity lies in the integration of visual, radio and magnetic observers into a combined programme spread over a wide geographical area. In addition, correlation with observations of solar activity is desirable. The B.A.A. Aurora Group is following this pattern, and its observers are located along an arc from British Columbia to Finland.

SECTION 15

METEORS

15.1 Amateur work

Because of their evanescent nature, the observation of meteors presents special problems. First, the observer must train himself to react swiftly and with precision over watches of several hours' duration. Second, the information being sought can take various forms, each of which may demand different observing techniques and discipline.

Probably no branch of observation, apart from lunar work, has been more affected by professional inroads than has meteor work. Until the Second World War, meteor observation was almost entirely an amateur pursuit for the simple reason that the great majority of meteors could be recorded only by visual methods. The immediate post-war years saw two advances: the development of the super-Schmidt camera, which could record with great accuracy a fair percentage of naked-eye meteors; and the introduction of radar techniques, with which the ionised trails left in the upper atmosphere could be measured. By means of the latter method, meteors are observed which would have been missed by the visual or photographic observers on account of cloud; sporadic meteors and whole streams which could never have been discovered by the previous techniques (since they occur in daylight) have been detected—e.g. the unique May—August Piscids, discovered by Lovell and his colleagues at Jodrell Bank in 1947; visual systems have been studied, and their radiants determined, by an independent method; meteor velocities, a vitally important datum, have been determined; the distribution of mass within a given meteor stream has been effectively investigated.

Faced with this revolution (no other word is strong enough), meteor astronomy has tended to move away from the classical positional approach, involving the establishment of radiant positions and true paths through the atmosphere, towards (*a*) simpler documentation of magnitudes and rates of known showers; (*b*) the search for new showers and the

171

documenting of sporadic meteors; and (c) the photography of fireballs, to derive true paths and possibly to lead to the discovery of a freshly fallen meteorite.

On the other hand, it is not sufficient to say that visual observation of meteor paths is prohibitively inaccurate compared with professional techniques and leave it at that. The two-station work by Alcock and Prentice in the 1930s stands as a monument to what can be achieved in this way. It is certainly true that visual observation of such accuracy requires a rare combination of talent and enthusiasm; but it would be a sad loss to amateur astronomy if such skills as were once developed, and painfully refined, over many years, were allowed to waste away. The specialised character of high-class meteor observation is not something that can be left to chance: it must be nurtured and encouraged. While recognising that the major showers have been very well documented, there are numerous minor showers and, more particularly, evenescent showers, that may miss attention altogether, and it could be that there is scope here for the 'classical' approach to produce interesting and valuable results.

For the visual observation of meteors, the qualities most prized in the observer are accuracy, pertinacity, and a good knowledge of the constellations. These qualities in combination will produce what is most needed: a body of observational data which is accurate enough and extensive enough to merit statistical discussion. As regards the third qualification, familiarity with the night sky, a knowledge of all stars to at least mag 4.0 or 4.5 is essential. The experienced observer will learn to go lower than this, and must develop his own technique for describing stars down to the naked-eye limit without reference to a map.

15.2 Individual paths: methods of observation

Although this work is not now accorded the same weight as during the period 1833–1930, it is likely that it will again become valuable when the correct principles for the determination of group radiants have been worked out (see section 15.13). Long watches are required, and generally a watch of from 3½ to 4 hours may be considered a minimum for effective work on the minor streams, whilst watches of from 6 to 8 hours are preferable. The observer should pay careful attention to *all* the elements of the observation discussed in the subsections of 15.2 below, and not merely content himself with the determination of flight directions. It is probable that the accurate determination of magnitude, duration, and length of path will be found to be just as necessary in the determination of group radiants as the flight directions.

The isolated observer can also do valuable work in the determination of the distribution of magnitude within a major shower, discussed in section 15.5.

The normal path-plotting routine on a night when no major shower is in progress is to secure three primary data for each meteor observed:

> Direction of flight.
> Beginning and end points of its luminous path.
> Duration of visibility.

When the hourly rate is higher than about 10, it will be found that it is impossible to obtain all these data fully.

15.2.1 Memorisation of the meteor's path (both orientation and extent), and its immediate transference to a map: This method is not recommended, for two reasons:

(*i*) Inaccuracy. It is difficult to determine the direction and end points of the path with the degree of precision obtainable by the method described in section 15.2.2 below.

(*ii*) Inconvenience. The path of a meteor being an arc of a great circle passing through its radiant, special maps in the gnomonic projection are required, such that the paths are represented by straight lines tangent to the celestial sphere at the radiant point.

If this method is used, each recorded path should be labelled with a serial number (referring to additional data in the Observing Book), and the flight direction indicated by an arrowhead.

15.2.2 Separate determination of the meteor's flight direction and of its beginning and end points: These data are recorded in descriptive, not graphical form. The method—which was developed by Prentice, and has been proved by experience to be capable of great accuracy — was employed by the Meteor Section of the B.A.A.

(*i*) *Orientation of the path, or flight direction*:

Immediately a meteor is seen, hold up a straight-edge or stretched string at arm's length against the sky overlying its path and note whether this projection of the luminous path in both directions passes through or near any stars. The flight direction should be defined by three points in order to eliminate (or at least reveal) casual errors of identification or recording. It can be done by either of two methods, which are often combined in the same observation:

(*a*) Fractional method (for which the reference stars should invariably be within $3°$ of the flight direction): e.g. (Figure 23)

$$\tfrac{1}{2}\alpha\beta \quad | \quad 35 \quad | \quad \tfrac{1}{4}\gamma\delta$$

indicates that the point midway between the stars α and β, the star 35, and the point one-quarter of the way from γ to δ all lie on the flight direction.

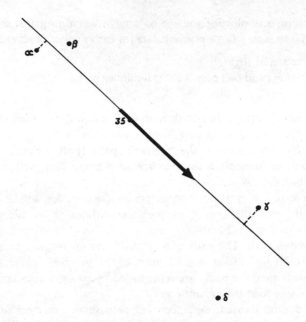

FIGURE 23

(b) Angular estimate method (for which the reference stars should be not more than 1° to 1½° from the flight direction): e.g. (Figure 23)

$$½° \, N p \, \alpha \quad | \quad 35 \quad | \quad 1° \, S f \gamma$$

indicates that the flight direction passes through a point ½° North preceding the known reference star α, through the star 35, and through a point 1° South following γ.

(ii) *Determination of the luminous path:*

Two methods are used, (a) being appropriate for the longer paths, and (b) for path lengths of less than 5° or 6°:

(a) Separate determinations of the beginning and end points:
   (A) Beginning and end points defined by perpendiculars dropped to the flight direction from individual stars.
   (B) Beginning and end points defined by cross bearings between stars lying on opposite sides of the flight direction.

In recording these observations, use is made of the following conventions:

The direction of flight is counted positive; the reverse direction negative.

The point at which the perpendicular from star α cuts the flight direction is also denoted by α.

A cross-bearing between two stars is denoted by the prefix ∠.

e.g. (Figure 24)

$$\alpha + 2° \quad | \quad ∠\beta\gamma - 2°$$

indicates that the beginning point was estimated to lie 2° along the flight direction from the point at which it was intersected by the perpendicular

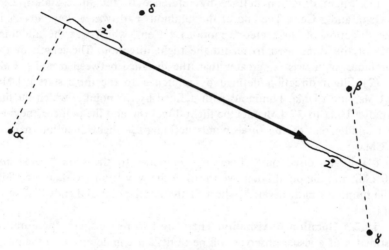

FIGURE 24

from star α, measured in the direction of flight; and that the end point was estimated to lie 2° along the flight direction from the point at which it is intersected by the cross-bearing between stars β and γ, measured in the opposite direction to that of the meteor's flight.

(b) Determination of one point (beginning or end) and the length of the luminous path: e.g. (Figure 24)

	Beginning	End	Length of path
	α + 2°		βδ
or		∠βγ − 2°	βδ

indicating that the length of the luminous path was equal to the distance separating the stars β and δ. In practice, the end point is often preferable to the beginning point (assuming equally favourably placed reference stars), as tending to be the more accurately observed of the two. The description of the estimated length of the luminous path by comparison with the separation of a pair of stars is generally preferable to its quotation in degrees.

Three further examples of observations made and recorded in this way are to be found in the sample Observing Book layout in section 15.7. It is recommended that in order to become familiar with the method one should plot the flight directions and luminous paths on a celestial globe or gnomonic star chart (or even an ordinary star atlas) with the aid of the following notes:

(1) Flight direction defined by reference to the three stars η Leo, κ Gem, and ϵ Gem. The end of the luminous path was $\frac{1}{2}°$ beyond (i.e. in the direction of the meteor's motion) κ Gem, which was the midpoint (M) of the three used to record the flight direction. The length of the luminous path was $\frac{3}{4}°$ greater than the distance between α and β CMi.

(2) Flight direction defined by reference to the three stars ν UMa, μ UMa, and ι UMa. Luminous path defined by the point at which the line from ψ UMa to 37 LMi cuts the flight direction, and the point $\frac{1}{2}°$ beyond the middle one of the three points defining the flight direction, namely μ UMa.

(3) Flight direction defined by reference to the stars ζ UMa and 30 LMi, and the point lying two-thirds of the way from χ UMa to 65 UMa. The luminous path began $4°$ short of the latter point and ended $\frac{3}{4}°$ short of it.

15.2.3 Duration of visibility: There are two reasons why the duration estimates of a visual observer will be of little use in determining the linear velocities of meteors: (*a*) most estimates of this kind are distorted by large subjective errors; (*b*) for critical work an accuracy of better than 0.1 sec is necessary, and this is beyond the ability of a visual observer. The estimation of visible duration is nevertheless a valuable part of the total observation—partly for statistical reasons and partly for purposes of control in the determination of group radiants.

The normal meteor within the visual observer's range of perception has a duration of from 0.1 to 0.8 sec: it is clear, therefore, that nothing can be done whilst the meteor is actually in flight if its position, magnitude, etc, are also to be estimated with accuracy. For the average meteor the determination of duration must be an act of pure estimation. For the exceptional meteor (lasting from 1.5 to about 15 secs), it is true, a process of simultaneous counting is both possible and more accurate.

Three methods therefore suggest themselves, to cover the variety of cases actually encountered in observation:

(a) Pure estimation. The observer should endeavour to establish a subjective, uniform, and linear time scale. This can best be done by continuous practice against short time intervals of known duration; in its way the process is not dissimilar from the establishment of the 'step' in variable star work. When established, the scale error should be determined (a factor of ×2 need not be inconsistent with good observation, provided it is known and fixed), but the scale should not be adjusted to correct the error.

(b) A semi-automatic device such as a hand-operated chronograph has been used with success. No attempt is made to operate this during the apparition of the meteor: after it has disappeared, the observer reproduces to the best of his ability a number of equivalent 'durations' on the chronograph, the weighted mean being taken. This method gives good results, but in practice is usually found to add too heavy a burden to an observer already overburdened with detail.

(c) Counting: the best way of dealing with meteors of long duration. The observer should learn to count at a uniform rate (8, 9 or 10 per 2.0 secs is usual).

Whether determining meteor durations by pure estimation or by counting, the observer will probably find that his time scale against the clock is different from his time scale in actual observation. The reason for this, certainly psychological, is obscure; but the scale error must nevertheless be determined. This can be done in either of two ways: (i) laboratory experiment, upon which few data have been published (but see B.15.4); (ii) duplicate observations of those showers for which accurate velocities are known, such as

Leonids	72	km/sec
Perseids	61	,,
Geminids	35	,,
Taurids	28	,,
Giacobinids	23	,,

By taking the observed length of path and the known velocity of the shower, the scale error in the duration estimate may be evaluated within the limits of error of the observations.

15.2.4 The full procedure, and additional data: The complete procedure for logging the flight of a normal meteor may therefore be summarised as follows. Immediately it is seen, the observer forms an intense mental image of its flight direction, its position relative to the star field, and its stellar magnitude, and he simultaneously operates the subconscious timing

177

process. With a controlled movement (and, in particular, without moving his head) he holds up the yardstick against the sky. The meteor, by this time, may have disappeared or it may still be in visible flight; but even in the former case the mental image will persist long enough to allow the observer to determine its position accurately. He notes and memorises, in order, the flight direction (three points) and the beginning and end of the luminous path. Only now can the yardstick be lowered. The observer then notes the time of observation (allowing for time lost) and extracts from his memory the stellar magnitude and the estimated duration of the meteor. All these quantities are then entered in the Observing Book.

Stellar magnitude should be determined and expressed by reference to stars at approximately the same altitude. Only for bright meteors — brighter, say, than mag 1 — need recourse be had to the numerical magnitude scale, which should be avoided wherever possible.

Its accurate determination has the following important applications: (a) the magnitude distribution within a given shower; (b) it is a necessary element in the most recent method of determining group radiants; (c) it is valuable in correcting the observed hourly rate of a major shower for the effects of moonlight (see B.15.54, 15.63).

The full observation, together with its recording in the Observation Book, will probably take at least a minute until practice in the technique has been gained; later, and assuming a detailed familiarity with the constellations, this can be cut down to about 30 secs. In addition to the basic data already mentioned, various other data are commonly obtained — as, for example, an estimate of the form of the meteor's light curve, whether a streak (or afterglow) was left, its colour, and a general description of its appearance if this is in any way unusual. Each observer must decide for himself how many such data he can include without injuring his recollection of the basic and more vital data.

Except during specific showers, the early hours and nights of moonlight or mist or high cloud are better devoted to some branch of observation other than that of meteors. Ideally, for the observation of sporadic meteors, stars down to the 5th mag should be visible at the zenith; otherwise the number of observations that will be made will hardly justify the time spent in obtaining them.

15.3 Observational cooperation

The duplicated observation of a meteor (i.e. its simultaneous observation from two fairly widely separated stations) yields information about the meteor which is of an altogether different order from that yielded by a single-station observation. The latter gives the position of the meteor against the star field, its duration, and its brightness, but nothing more. A two-station observation, on the other hand, allows the meteor's actual path through the atmosphere to be derived (i.e. the geographical position of the points on the Earth's surface vertically below its beginning and end points, and its height at each of the latter points); the linear length of its path, combined with its duration, gives its linear velocity; the intersection of the flight directions as determined at the two stations gives its absolute radiant; and from these data its orbit round the Sun can be calculated. However, the difficulties in organising this work (not the least of which are meteorological) are immense.

Simultaneous observations are best made from stations about 70–160 km apart. If they are nearer together, the parallax tends to be swamped by the observational errors. If further apart, various disadvantages follow — the loss of the fainter meteors under the inverse square law, unequal covering of the field, the increase of the linear equivalent of a given angular error, etc. If a third observation of the meteor is made, it is then possible to compute the probable error in the derived real path. Indeed, the effectiveness of the programme will be greatly increased if a network of four or more suitably placed stations can be staffed and operated. In such work the direction of the programme will naturally lie with the director of the group, and each member will do his utmost to carry out his own share of it.

The pooling of the observations of many observers, working individually and not in cooperation, is also essential: not only for the accidentally made duplicate observations that may be uncovered, but also because it provides the data required for statistical discussion. The Meteor Section Director is thus in a position to make much more fruitful use of the observations of each individual observer than that observer could himself.

See further, sections 15.14, 15.15.

15.4 Meteor storms

It is difficult for the observer to visualise beforehand how different will be the observational conditions during an exceptionally rich shower from those normally encountered. Unless the programme is carefully organised there will simply not be time to make any observations of value. Statistical work (determinations of HR, magnitude distribution, etc) must take priority over determinations of position, which are better left to the photographers or until the rate has dropped to a manageable level. For statistical work the following suggestions are made:

179

(*i*) Estimates are useless. All data should be based on careful counting.

(*ii*) Decide beforehand the exact limits of the sky area to be observed, and assess the geometrical factors involved in such choice. The area may be limited by star groups or by an artificial grid (see section 15.5).

(*iii*) Arrange means of defining smaller alternative areas for use in case the HR becomes unmanageable.

(*iv*) Arrange means whereby the observer can immediately ascertain minutes or smaller time intervals, and see that times are carefully recorded and checked. With a high HR, of the order of 200 or more, time intervals should be short—say 1, 2, or 3 minutes.

(*v*) Keep a record of magnitudes as well as of numbers. If the rate is too high, either reduce the area or record the numbers above or below a preselected magnitude level.

(*vi*) Note the minimum stellar magnitude visible in the field of observation.

(*vii*) Supplement naked-eye work with telescopic if observing resources permit. The lower the magnitude limit, the more complete will be the final picture of the swarm. In all telescopic observing keep a careful record of the aperture, magnification, and field, especially if any of these are changed during the period of observation.

(*viii*) Be ready. Yours may be the only critical observations made, as occurred with the magnitude determinations of de Roy and the telescopic observations of Sandig and Richter at the unexpected Giacobinid storm of 1933 (B.15.63).

15.5 Meteor counts and hourly rates

The procedure for determining statistically homogeneous meteor rates has been worked out by Öpik in his Double Count Method (B.15.47, 15.48), and wherever possible it is desirable that a group of observers should employ this method. In practice, however, quite consistent results can be obtained by an isolated observer, and hourly rates are usually determined in this way.

The following investigations are suggested:

(*i*) The general level of meteoric activity should be determined once in every decade, and the routine statistical work of Gravier and of Schmidt in the nineteenth century and the more recent work of Hoffmeister should therefore be repeated.

(*ii*) The intensity of the annual meteor showers should be systematically investigated every year.

(*iii*) The distribution of meteor magnitudes within each of the annual showers should be similarly investigated.

(*iv*) Most of the statistical work on meteors has hitherto been done between geographical latitudes +40° and +55°, and similar work in other latitudes would be of special importance.

In planning the programme of meteor counts the observer should follow, with the appropriate modifications, what has already been said in section 15.4, particularly paragraphs (*i*), (*ii*), (*v*) and (*vi*).

Three points are worth special mention in connexion with meteor counts:

(*i*) Both the linear area of sky that can be covered and the limiting zenithal magnitude observable will depend on the altitude of the observing field. Hence an observer watching near the horizon will see a different sample of the meteor population from an observer watching near the zenith. If it is assumed that the average height of appearance of meteors is a constant, then their distance from the observer depends on their ZD. At greater ZDs meteors will be intrinsically brighter (since more distant), and observable over a larger area for a given solid angle, than at smaller ZDs. The change in the observed frequency with changing ZD will be due to the balance of (*a*) decrease of absolute frequency with increasing luminosity, (*b*) increase of absolute area observed. This has been investigated by Öpik (B.15.42). Taking the average height of appearance as 100 km, and an arbitrary unit of solid angle, the frequency of meteors observable by eye per unit angle at six ZDs from 0° to 82° is given below (Δm = difference between apparent and absolute magnitude):

		ZD					
		0°	45°	60°	75°	80°	82°
Δm {	distance	0.0	0.8	1.5	2.7	3.4	3.7
	absorption	0.0	0.1	0.3	0.6	1.0	1.2
Δm total		0.0	0.9	1.8	3.3	4.4	4.9
Limiting (zenithal) mag assuming 4.5 = effective apparent mag limit		4.5	3.6	2.7	1.2	0.1	−0.4
Number		252	110	42	5.3	1.15	0.57
Area		1.00	2.81	7.34	41.0	9.35	138
Observable frequency		252	309	308	217	108	79

Thus it can be seen that the maximum observable frequency may be expected between ZDs of approximately 45° and 60°, and the figures in the final row may be used as correction factors dependent upon the altitude of the area of observation.

(*ii*) Counts of hourly numbers are by themselves of only secondary importance: the magnitude distribution is also required. The observer should therefore make a practice of recording the magnitudes of the meteors which he counts, and in order to enable these to be converted to zenithal magnitudes he should also record some nearby reference star from which the altitude may be derived.

(*iii*) Meteor counts can be combined very well with the routine recording of meteor paths up to an hourly rate of 40 at least, and probably 60 — i.e. for all normal returns of the annual major showers. If the standard B.A.A. practice of recording in complete darkness (entailing a guide to keep the written record on a single line, and some ingenuity in describing stars whose names are unknown) is adopted, no allowance need be made for the time spent in recording; if otherwise, this will be necessary.

The number of meteors visible per moonless hour varies to some extent from one observer to another, as also from one latitude to another. Figure 25 (*a*) gives some idea of the diurnal rates, averaged over the whole year. Observation before about 8^h GMAT is less productive than observation between midnight and dawn, the average HR at 16^h GMAT being almost double that at 8^h. The evening (overtaking) meteors nevertheless require investigation as much as the more attractive morning ones.

The hourly number for a given hour (say 12^h–13^h) at different times of the year will be subject to considerable fluctuations due to the incidence of the major showers; and this is superimposed upon a well-marked season variation in the mean hourly rate dependent on the mean altitude of the anti-solar point and of the Apex of the Earth's Way. Thus during the latter half of the year, in the latitude of England, the HR between given limits of time is about twice the corresponding rate during the first six months (Figure 25(*b*)).

Observations should be made under standard conditions, to facilitate their comparison. Observation periods should be continuous, and the direction of observation and sky state always mentioned in the record. To facilitate subsequent discussion, the observation of an equal area of sky each night can be assured by the use of a grid. This is a wire circle one metre in diameter, on whose central axis is mounted a smaller wire ring ('eyepiece'), from which eye position the large ring subtends the required angle. It will be found that a circular area of sky, diameter about 50°, is as much as can be watched effectively. The distance of the eye

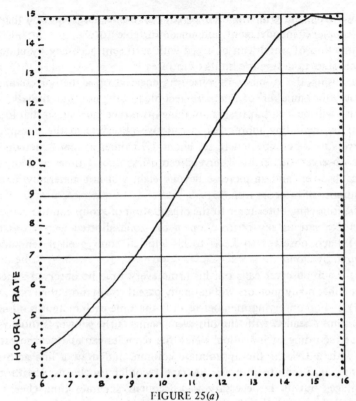

FIGURE 25(a)

Diurnal variation of HR of meteors, averaged over the year (Olivier)

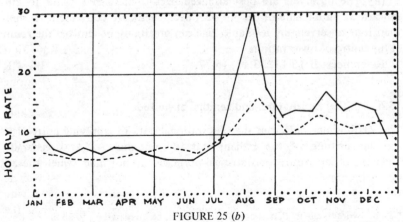

FIGURE 25 (b)

Annual variation of HR of meteors: ————— fortnightly averages (Denning
- - - - - - - monthly averages (Olivier)

position from the grid will then be 107 cm. Some observers find the wire ring 'eyepiece' inconvenient and uncomfortable to use, and in this case the direction of watch can be fixed with sufficient accuracy by choosing suitable stars to define the limits of the area.

Increasing the number of observers engaged on a meteor count will increase the number of meteors recorded, but not indefinitely. The increase will be rapid at first, smoothing off rather quickly, so that little is gained by including more than about five observers in the group: two observers may be expected to see about 1.75 times as many meteors as a single observer, but eight observers record less than 4 times as many as a single observer, and an increase beyond eight will not materially increase the number of meteors seen.*

The following notes refer to the organisation of group counts:

(*a*) No serious sky obstruction in the zone allotted to any observer.

(*b*) Each observer to keep to his allotted zone, even if temporarily obscured by cloud.

(*c*) Each observer calls out his name every time he observed a meteor in his zone; many meteors will naturally pass through more than one zone.

(*d*) A non-observing member records the time each meteor is observed in a column headed with the observer's name. If the shower is too prolific for the recording of individual times, the recorder can mark each observation with a tick in the appropriate column, and draw a line across all columns at 5-minute intervals. A tape recorder can be used with great advantage, but it is necessary that frequent accurate time checks are recorded along with the observational material.

(*e*) The observers are best arranged back-to-back in a circle, looking upward at an altitude of 30° to 45°; deck-chairs are convenient. If more than four observers are available, one can profitably be allotted the zenith as his centre of observation.

References: B.15.11, 15.12, 15.77.

15.6 Zenithal hourly rate, and zenithal magnitude

The ZHR is derived from the observational data by applying corrections for the altitude of the radiant and for moonlight, if any; generally, however, observations for statistical purposes are not made when there is strong moonlight.

---

* Some workers claim that the numbers seen can be increased up to about 7.5 times before the curve of numbers seen against number of observers falls off appreciably (B.15.76). The figures quoted in the text are probably nearer the mark.

If R is the observed hourly rate, then R_c (the HR corrected for radiant altitude a) is given by

$$R_c = R \operatorname{cosec}(a + 6°)$$

The correction for moonlight depends upon the lunar phase and the incidence of diffusing matter in the atmosphere, and must be determined empirically by comparing the numbers per magnitude with similar figures relating to a previous, moonless, return of the swarm. (See also section 15.18.1.)

The zenithal magnitude is the stellar magnitude a meteor would have if it were at the observer's zenith and a standard height of 100 km. The correction, in units of the first decimal, to be applied to the observed magnitude in order to obtain the zenithal magnitude is given in the following Table (B.15.51) for a range of values of the meteor's distance, r, and height, H (km).

r \ H	60	80	100	120	140	160	180	200
60	+11							
80	4	+5						
100	−1	−1						
120	6	5	−4	−4				
140	10	9	8	7	−7			
160	14	12	11	11	10	−10		
180	17	16	15	14	14	13	−13	
200	−20	−18	−17	−16	−16	−16	−15	−15
220	23	21	20	19	18	18	17	17
240	25	23	22	21	21	20	20	19
260	28	26	24	23	23	22	22	21
280	30	28	26	25	24	24	23	23
300	−32	−30	−28	−27	−26	−26	−25	−25
320	34	32	30	29	28	27	27	26
340	37	34	32	31	30	29	29	28
360		35	34	32	31	31	30	30
380		37	35	34	33	32	31	31
400		−39	−36	−35	−34	−33	−33	−32

15.7 Errors in meteor observation

The various sources of error in the observation of meteors have been rather fully investigated and are summarised below. Considering the shortness of the apparition of the average meteor (less than 1 sec) and the number of separate facts that have to be noted and memorised, before being recorded in the Observing Book, it is a matter for surprise that errors are not larger than, with experienced observers, they are. Forgetfulness, rather than carelessness or misobservation, is probably the most important single cause. The effects of fatigue appear to be less serious than might have been expected.

15.7.1 Hourly rates: It had for many years been assumed that fatigue causes an under-estimation of the HR towards the end of long spells of observation. A thorough investigation by Prentice (B.15.55), however, suggests that the error from this cause is negligible.

15.7.2 Flight direction: The errors here depend very much on the experience of the observer, the technique employed, and the type of meteor. With inexperienced observers the errors are usually very large: thus Watson and Cook (B.15.66), basing their conclusions upon the direct plotting of about 1500 Leonids (1933) on a gnomonic projection centred on the Leonid radiant, found that for inexperienced observers the probable error in an individual plot averaged $\pm 12°$ ($\pm 11°$ to $\pm 19°$), compared with only $\pm 4°$ for a bright meteor recorded by an experienced observer.

Similarly with technique. Öpik (B.15.3, 15.43-46, 15.58) quotes $\pm 8°$ as the average probable error of all the members of the Arizona expedition; and the special technique used by this expedition was apparently unreliable where determinations of position were concerned. Porter, at any rate, from a discussion of 70 triple and multiple observations made by B.A.A. observers using the technique described in section 15.2.2, found a median error in flight direction of only $\pm 0°.7$.

Experiments made at the Hague Planetarium in 1948 are not directly comparable with the figures quoted above, but are nevertheless of interest. It was found that when a string or yardstick was used to define the flight directions, these passed through a circle $4°$ in diameter, centred on the radiant; without a string or straight-edge, through a circle $5°$ in diameter.

15.7.3 Beginning and end points: Displacement of either end point along the flight direction by one or both observers has no effect upon the derived radiant, but introduces an error in the derived heights.

The Hague Planetarium tests gave $2°.5$ as the mean error in the beginning and end points. As an example of the sort of accuracy obtainable by experienced observers, the mean deviations of the beginning and end points of one set of 309 meteors were 3.2 and 2.7 km respectively. Porter's 'sliding errors' are in close agreement with the Hague figure (average errors $\pm 2°.5$; median, $\pm 1°.7$, $\pm 1°.7$).

15.7.4 Time of flight: Boyd (B.15.4), investigating the duration estimations of a group of 11 students (method employed is not mentioned, but was presumably counting or direct estimation), found that in 50% of the cases the scatter was random, while 50% showed a bias towards systematic under- or over-estimation. Observers should therefore test themselves with artificial 'meteors' to determine the nature of their personal error.

With durations between 0.38 and 1.80 secs, the errors ranged from 14% to 49%. Boyd found that although the error increased with increasing duration, it did so rather more slowly: hence the percentage errors decreased slightly as the duration to be estimated increased. Roughly speaking, an error of the order of 25% may be expected in the estimation of durations exceeding 0.5 secs; and a considerably larger percentage error for shorter durations.

15.7.5 Angular length of luminous path: There appears to be a pretty universal tendency among untrained observers to over-estimate the length of a meteor path, as of other angles and distances in the sky (cf. size of Sun or Moon when near the horizon). Thus members of the general public commonly over-estimate path lengths by as much as a factor of 5. Yet in the Hague Planetarium experiments the estimated paths were, surprisingly, 13% too short on the average.

The existance of a serious systematic error in angular length as a function of the angle of foreshortening has been known for some years, but at the time of writing no critical examination of this error has been published.

15.7.6 Magnitude: Trained observers—especially those used to the observation of variables—are surprisingly precise in their estimates of stellar magnitude. The magnitudes of 100 meteors, when reduced to zenithal magnitude (section 15.6), were found to be accurate to within 0.56 mag. Though this compares unfavourably with the accuracy obtained by variable star observers, the difference is hardly surprising considering the dissimilarity between the two types of observation.

15.7.7 Miscellaneous illusions and errors: Commonly encountered in the reports of fireballs made by inexperienced observers are: (*a*) 180°-error in the direction of flight; this strange error may be allied to the 'mirror failure' which makes it difficult for some people to distinguish quickly between b and d, 69 and 96, etc; (*b*) 'hearing a light', i.e. imagining that a synchronous 'swish' is audible, even though the meteor may be so distant that its noise, if any and if loud enough, could not reach the observer for many seconds; (*c*) inexperienced observers also tend to over-estimate the altitude of any celestial object; the majority of people, if asked to point to a spot at an altitude of 45°, will indicate an altitude in the neighbourhood of 30°. See also B.15.41, 15.66.

15.8 Photography

15.8.1 Direct: Photography is a powerful technique for the study of meteors, owing primarily to the precision of its data; both their apparent paths across the sky, and their angular velocity at any instant along that path, can be measured (the former, assuming that the camera is driven to follow the stars; the latter, that it is equipped with a rotating shutter in front of the lens to occult the view at a known frequency per second). Apparent path can only be estimated visually with an appreciable margin of error, even by a trained observer (see section 15.2.2), and the latter not at all. It is the most direct—or only—method for obtaining accurate data concerning stationary radiants, variations of brightness and speed during the meteor's visible flight, spectra, and the data necessary for the calculation of real paths and velocities.

It is, moreover, a field in which the sort of equipment found in amateur hands can be put to very good use and is an important complement to visual observations generally. It does, however, demand considerable patience from the observer, though perseverance and a rationally planned programme are all that are required to ensure eventual results.

The efficiency of a lens for meteor photography is determined by the intensity of its images and the size of the field over which they are tolerably defined. If a point image crosses a plate with an angular velocity of ω, the intensity of the photographic image is, per unit area, a function of $D^2/\omega F$, or, as in the case of a trailed asteroid (section 12.2),

$$b_p = k \cdot \frac{D^2}{F}$$

for a given value of ω. It is clear that, ω always being large, F must be made as small as possible for a given value of D. That is, considerations of image intensity alone indicate the use of small focal ratios.

On the other hand, the meteor-recording efficiency of a lens is directly proportional to the usable area of its field. Combining this factor with the above expression for the photographic brightness, and putting d = angular field diameter, the efficiency of the lens is given by

$$E = k \cdot \frac{(D \cdot d)^2}{F}$$

providing that, for an anastigmat, $F \geqslant 60$ mm approximately, otherwise the image of an average meteor is in danger of being narrower than the grain of the emulsion. Hence the linear dimensions of the lens are irrelevant to E; furthermore, since d is inversely proportional to D/F, a lens of small focal ratio, though photographically fast, is not necessarily the most efficient for 'catching' meteors. In other words, a compromise must be struck between the conflicting demands of d and D/F. This, as a very rough guide, is represented by the lens of smallest focal ratio that is capable of giving a well-defined field of about 30° or 40° diameter; what its focal ratio will in fact be when this condition is satisfied will of course, depend on both the type and quality of the lens, though it will probably not be less than about $f/3$. The simplest way of testing the relative suitability of two lenses in practice is to expose trails of the same stellar field, for the same length of time and under identical atmospheric and other relevant conditions (such as altitude), with each lens in turn, and then to count the number of trails on the two negatives.

Results obtained experimentally by Waters and Prentice with an artificial meteor show clearly that the essential factor is focal ratio, or a function of focal ratio, and that the linear values of D and F as such are irrelevant—though, of course, for a given value of the focal ratio the image intensity varies directly with D:

D	F	D/F	D^2/F	Image intensity
1.3	6.0	0.22	0.28	1
2.3	10.5	0.22	0.50	3
2.0	6.0	0.33	0.67	6
4.25	8.5	0.50	2.17	9
2.75	6.0	0.46	1.26	10

The penultimate result appears to be 'wild', but the general drift is clear.

Unless the amateur can afford cut film or Polaroid, in which case he could consider using an ultra wide-angle lens such as the ex-government Ross Xpres 125 mm $f/4$ (designed for aerial mapping purposes) in a special camera made for the purpose, he will have to join the great majority of observers in using the 35 mm format. Within this constraint, instrumental requirements are not elaborate; any lens working at $f/2$ or $f/2.8$ and of about 50 mm F will serve, although the field of view (theoretically about 42° × 28°, about 1,200 square degrees) will be limited by off-axis vignetting to an effective full-aperture coverage of only 900 square degrees or so. The degree of vignetting varies according to the quality of the lens, but in many types the effective aperture at the extreme corners of the frame can be $f/5.6$ or even less. It is also possible to buy lenses of wider coverage (shorter F), up to the extreme fish-eye type capable of photo-graphing a field 180° across; but, with the exception of the fireball work described in section 15.11, such lenses are progressively less efficient because of the very small image scale. Speaking very generally, an $f/2$ lens will record most meteors of mag 1 and brighter, unless their angular velocities are abnormal, mag 2 meteors providing that they are fairly slow-moving and that other conditions are favourable, and meteors fainter than mag 2 only if they are abnormally slow-moving or if they possess a train. Lenses of focal ratios from about $f/3$ to $f/4.5$ will normally record only the brightest meteors or ones of abnormally low angular velocity.

To cover a really large area of sky, then, will require a battery of cameras arranged so that their fields of view overlap slightly; five 35 mm cameras could between them cover an approximate circle $75°-90°$ across. Schmidt camera systems, although attractive in some ways, are costly and awkward, besides offering no decisive advantage in focal ratio and field of view over ordinary lenses. But, whatever approach is adopted, it is essential that the camera does function properly: in particular, the exposures must be terminated before fogging becomes excessive (experience will determine the optimum exposure time, under various conditions), and dewing must be eliminated. Long lens hoods are out of the question when a wide field is being photographed; the only sure solution is to use a length of suitable resistance wire, or several torch lamps wrapped in opaque foil, arranged around the front element of the lens and operated either by mains transformer or, preferably, from a 12-volt car battery (see *A.A.H.*, section 20.15).

It is essential that the camera, or battery of cameras, be used in a way calculated to increase the naturally rather poor chances of a meteor bright enough to be recorded crossing the field within a reasonable aggregate exposure time. Exposing at random for sporadic meteors is a waste of time. Be guided by the hourly rates quoted in sections 15.5, 15.16, concentrating particularly on the maxima of showers occurring when there is no Moon. From an analysis of some 16,000 hours of exposures made with 35 mm cameras, Hindley has suggested that the optimum direction at which to aim a camera during shower nights is between $35°$ and $55°$ from the radiant, and at an altitude of between $40°$ and $60°$, the zenith being less productive because a smaller area of the actual meteor-producing layer in the atmosphere is being covered. Far from the zenith, of course, numbers are again reduced by the increasing distance of the meteors and absorption in the intervening air.

The choice of emulsion for meteor work presents no particular difficulty: Tri-X or HP5, the fastest available, offer the best chances of success, since graininess is of little account; and their strong reciprocity failure, normally disadvantageous in stellar photography, is now of some benefit in inhibiting the formation of sky fog without detracting from the film's sensitivity to the virtually instantaneous impression of a meteor. Various speed-increasing techniques can be tried, such as development for longer than normal time in very dilute developer: this and other methods are referred to in *A.A.H.*, section 20.

The meteor photographer, despite the benefits of fast lenses and hypersensitive emulsions, must nevertheless anticipate a high proportion of fruitless exposures; even on nights of strong shower activity, disappointingly few trails may be recorded. However, those that do appear, if measured accurately and reduced properly, will carry a weight far higher than most visual flight determinations. It must also be borne in mind that the continuous accumulation of negatives will carry useful records of things other than meteors: variable stars certainly, asteroids possibly, and even a comet or nova, may be picked up. On the very rare occasions of storm activity, as happened with the Leonids in 1966, negatives may be dense with trails—one exposure on Tri-X, developed in D-19, using a 50 mm *f*/2 lens, showed a meteor for each second of the 43-second exposure! (See B.15.38.)

A meteor camera may be either stationary or equatorially mounted and clock-driven. The latter is preferable, for the following reasons:

(*a*) With a camera recording star trails the only way of relating a meteor image to the positions of the stars is to terminate the exposure immediately a meteor is seen which crossed the field of the camera and which might have been bright enough to be recorded. Hence a visual watch must be kept throughout the photographic session (conveniently with the aid of a wire grid which, from the observer's chair, outlines the limits of the camera field against the sky), the film changed, and the time noted, each time a possibly recorded meteor appears.

(*b*) With a camera recording point stellar images, on the other hand, it is not necessary to know the time at which a meteor appeared in order that the negative may be used to determine the radiant. Hence a night-long visual watch is not necessary—for work of this type guiding is not essential, so long as there is no systematic gain or loss in the drive—film exposure being dictated by fogging only, and being operated automatically at regular intervals. If fogging does not necessitate termination of exposure throughout the night, the final closing of the shutter before dawn, or when the radiant becomes unobservable, can easily be operated by means of an alarm clock.

(*c*) The importance of training the camera on shower radiants has already been stressed, and implies a clock-driven equatorial.

(*d*) Meteor trails are much more easily overlooked when examining a negative of star trails than one of point images. This becomes increasingly so in the case of faint meteor images and/or images which are approximately parallel to the star trails.

A rather strange device was constructed at Harvard many years ago which was in effect a half-way house between a stationary and a clock-driven meteor camera. The camera was mounted on a polar axis whose driving circle was a 144-toothed cog. This wheel was advanced mechanically one tooth every 10 minutes, the stars having been trailing during that period. The polar axis was slightly displaced from the NCP, with the result that each 10^m exposure left a separate trail of every star in the field. The time of appearance of every meteor had to be noted, so that its image could be related to the correct set of star trails.

Photography is particularly valuable in cooperative work by two or more observers designed to obtain parallactic observations of identical meteors from separated stations. From such accordances, true paths, heights, and mean geocentric velocities can be calculated. A baseline of from about 40 to 80 km is suitable; if three observers are cooperating, their stations should lie as nearly as possible at the corners of an equilateral triangle. If the separation of the stations is too small, accuracy suffers through the measured parallax likewise being too small; if too great, there is a danger that a meteor included wholly in the field of one observer will be partially or even wholly outside the field of the other. The procedure is to synchronise exposures at the two stations, both cameras being set to the same RA and Dec and clock-driven.

An interesting account of the Harvard 2-station programme employing specially designed Schmidts fitted with rotating sectors (see below) is given in B.15.67.

In connexion with the reduction of meteor photographs, reference may be made to B.15.61.

If some recognisable modification of the meteor trail on the plate is made at equal known intervals of time, their linear separation will give the meteor's angular velocity at all points along its flight path. If an accordance has been obtained from another station, not only the angular velocity but also the linear geocentric velocity at all points, and hence the deceleration (the latter giving valuable information to meteorologists regarding atmospheric pressure at high altitudes) may be calculated.

Three main methods have been employed to introduce the modifications of the recorded trail:

(a) Vibrating the camera itself, a device suggested by the accidental discovery at Harvard many years ago that meteor trails taken with a certain camera were shown under microscopical examination to be regularly sinuous. The cause was traced to vibration imparted by the driving mechanism. The amplitude of the vibration in this particular case was less than 2 mm, but even so was accurately measurable with a microscope.

(*b*) Mounting an oscillating plane mirror in front of the camera objective in such a way that the reflected stellar point images are drawn out into circles; the image of a meteor then describes a symmetrical sinuous line.

(*c*) The most commonly employed device is a rotating shutter which passes between the objective and the sky at regular and known intervals. The optimum speed of rotation for a given meteor with a given equipment is determined by the meteor's angular velocity and the plate-scale (hence F) of the camera. No precise rule can be given, therefore, but interruptions of the order of 15–20 per sec will be found satisfactory in most cases. The

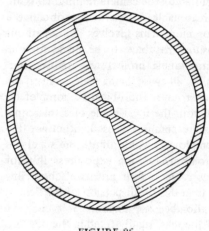

two opposite dangers to be avoided are the difficulty of detecting the trail interruptions, even under high magnification, if the sectors are revolving too fast, and insufficient accuracy resulting from too few and two widely spaced interruptions if they are revolving too slowly. A 3-ply or aluminium disc with two or three open sectors cut in it is probably the most satisfactory arrangement, allowing lightness to be combined with strength. The disc should be as small as is compatible with cutting the whole field of the camera, or possibly battery of cameras; it should therefore be located close to the dewcaps. A synchronous motor or a mechanically governed non-synchronous electric motor would be suitable for the drive, and the speed of revolution must be accurately determined by a rev-counter, strobo-scope, or similar device. A 2-vane home-made rotating sector of this type (Figure 26) is described in B.15.61 (see also B.15.69); in this instance a sector speed of 480 r.p.m. (giving 16 interruptions per sec) was found to be generally satisfactory with focal lengths of from 150 to 210 mm.

FIGURE 26

Negatives should be laid on a ground-glass screen, illuminated from below by a 150 watt lamp, and examined with a magnifying-glass for trails.

As a matter of routine, an eye should always be kept open for comets and novae on non-trailed (stellar) negatives. Alternatively, the negative can be projected on a screen.

Loss of meteor trails that should theoretically have been recorded is commonly due to: stars trailed (stationary camera); bad focusing; over-development of fog; dewing-up of lens during long exposures.

As regards the latter point, long dewcaps are essential, and it is also wise to install some form of heater—such as a coil of resistance-wire mounted on an insulating ring surrounding the lens cell.

15.8.2 Spectrographic: The photographic recording of meteor spectra has been undertaken by very few amateurs—understandably enough, in view of the difficulties. The only reasonable prospect of success is offered by an objective prism or grating—most amateur work has been done with prisms, but large replica gratings of the transmission kind are now available, and would appear to be more suitable. The camera requirements are wide angle, small focal ratio, and reasonable F in order to obtain the necessary resolution: say, 100 mm or more. This involves the use of cut film, and since several dozen exposures may have to be made for one successful spectrum, a serious spectrographic project will involve considerable financial backing.

The objective prism or grating must cover the objective completely. For stellar work with the film exposed in the focal plane of a telescope, small-angle (3° to 5°) crown prisms are commonly used, otherwise the dispersion (at the relatively long F employed) would dilute the spectrum to an extent that would entail prohibitively long exposures. But for meteor spectrograms with short-focus cameras flint prisms of about 25° to 30° angle are more suitable; if less than this the dispersion is insufficient, and if much more, light loss by absorption becomes excessive.

The prism is mounted in front of the lens, rigidly fixed in the position of minimum deviation relative to the optical axis, BC (Figure 27). For an average flint prism of angle $A = 25°$, the angle of minimum deviation, ω, will be in the region of 15°, and γ, the angle between the forward face of the prism and the normal to the optical axis, about 5°.

Focusing should be carried out by exposing trial negatives of star trails, using the absorption lines in the spectrum of a mag 1 star which is allowed to trail in the direction perpendicular to the direction of dispersion.

In operation, the camera is oriented so that the direction of dispersion is perpendicular to the line joining the field centre and the radiant (since the spectrum of a meteor travelling parallel to the direction of dispersion will not be recorded). If exposing for sporadic meteors (by comparison, a waste of time and film) the camera and/or prism are best oriented so

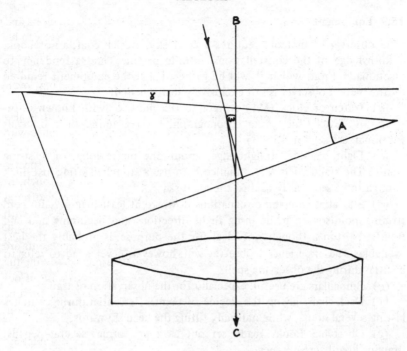

FIGURE 27

that the direction of dispersion is perpendicular to the hour circle through the centre of the field.

As instances of what can be achieved, Ridley, in the years 1954–57, exposed 325 plates, an aggregate exposure time of 90 hours, for 9 spectra; the Lloyd Evans brothers, in the period 1958–65, using plates and cut film, made 1,016 exposures (aggregate time 555 hours) for 17 spectra. Ridley used a Kodak Aero-Ektar lens, while the Lloyd-Evans team employed a battery of three Ross Xpres lenses of the type mentioned above.

The three main types of meteor spectra, referred to by Millman's classification, are as follows:

x Predominant feature, lines of magnesium or sodium
y Predominant feature, lines of ionised calcium
z Predominant feature, lines of iron or chromium

A separate class, w, includes those objects not covered by the main categories.

15.9 Equipment

'The observer of meteors requires a clear sky, a thick coat, a notebook, a knowledge of the constellations, infinite patience, and a tendency to insomnia.'* From which it will be gathered that the equipment required by the meteor observer is not exclusively instrumental.

(*a*) Watch or clock (MT) correct to 0.5 mins, or with known error.

(*b*) A device, such as a chronograph, for registering time intervals (optional: see section 15.2.3).

(*c*) Light wooden straight-edge, about one metre long, or a 'string wand'. The standard B.A.A. practice is to use a stretched string just thick enough to be seen easily against the night sky.

(*d*) Star atlas showing designations down to at least mag 5.7, for confirming points used in defining flight directions and beginning and end points; Norton's, though not ideal for the purpose, is probably the best available. The experienced observer will, however, never need to refer to an atlas during the observing spell.

(*e*) Binoculars are useful, especially for the observation of trains.

(*f*) Deck-chair, giving the degree of comfort essential during watches lasting several hours, while naturally tilting the head skywards.

(*g*) Observing Book, ready set out as a pro-forma; several pencils; drawing-board across knees.

(*h*) Dim, red-screened light clipped to back of chair and shining over the observer's shoulder, the switch being mounted in a convenient position such as the chair arm (optional).

(*i*) A thorough knowledge of the constellations, so as to eliminate time spent (and memory strained) poring over the atlas.

(*j*) Several thick rugs, since keeping warm is most important. An electric tubular heater for the hands is used by some observers.

(*k*) Vigilance, a good memory (at any rate over short periods), pertinacity, and some of the characteristics of a polar bear.

15.10 Meteor trains

Only a very small percentage of meteors leave trains visible to the naked eye, and of these the vastly greater number are ephemeral. On any occasion when an enduring train is seen, its observation should be given priority after the normal record of the meteor, and all other work temporarily suspended in order to record its appearance (colour, continuity, width, brightness, etc) and particularly its movement. It is

---

* J. B. S. Haldane.

absolutely essential that the latter should be recorded with accurate timing. The trail's linear velocity will probably be of the order of 100 to 300 km/hr, and therefore observations should be timed to the nearest 0.25 min, and the timepiece checked against a signal as soon afterwards as possible. The inadequacy of the time measurement is where nearly all previous observations have come to grief.

When nearing the limit of naked-eye vision, observation may be continued for a further period by binoculars or by a telescope using the lowest magnification available. Many bright meteors which appear to have no train will be found to have had one, if a telescope is immediately directed to the area; these, however, are usually too ephemeral to permit of more than the initial observation.

Numerous observations in the past have suggested that trains are tubular in structure; the physical and mechanical conditions of their formation are not completely understood. Duplicate observations of meteor trains give valuable information about the meteorology of the upper atmospheric levels.

Observations of an enduring train may be recorded in either of two ways: (a) drawing its position on a star chart at convenient intervals (one per minute should be possible); each position is labelled with the time, the clock error being known or subsequently corrected; (b) determining the orientation and end points of the train by the method already described for meteors in section 15.2.2; if the train becomes distorted it will be necessary to determine several points along it.

The aim should be to compile a record giving a detailed, accurate, and continuous picture of the train's shape, size, brightness, colour, position, and movements—a tall order, but not an impossible one.

15.11 Fireballs

Data which should be included in any report of a fireball or exceptionally bright meteor are:

(a) Date and time. If a casual timepiece is used, check it against a time signal as soon as possible. Summer Time, even if in force, must never be used in reporting fireballs (or, indeed, any other astronomical observation).

(b) Coordinates of beginning and end points:

Night: as for an ordinary meteor (section 15.2.2).

Day: note the horizon features immediately below the beginning and end points, the altitude of both points, and the exact position from which the observation was made. Later, measure the bearings (state whether true or magnetic are given) of these horizon marks by compass. The altitude will necessarily be less precisely determined than the azimuth, since a point in the featureless day sky cannot be memorised accurately enough to justify subsequent measurement. (See *A.A.H.*, section 36, for methods of making angular estimates.)

(*c*) Duration of visibility (see section 15.2.3).

(*d*) Details of appearance and behaviour:

Colour.
Brightness.
Explosions, if any. If possible, time the interval between seeing and hearing any explosion.
Curvature or other irregularities in the path.
Any variations of angular velocity along the path.

(*e*) Train, if any:

Night: see section 15.12.
Day: 1-min or ½-min drawings, each including a distinctive feature on the horizon, and an estimate of the altitude of the train.
If too ephemeral for drawings to be made, estimate its duration in secs.

Subsequently the following data or amplifications should be added:

(*f*) Longitude and latitude of observing station (from 1:50,000 O.S.).

(*g*) Any sounds: detonations, thunder-like echoes, etc. If possible, relate these temporally to the visible phenomena.

(*h*) Whether one or more bodies were seen.

(*i*) Condition of sky at the time.

(*j*) Go back over the above items, indicating where necessary the estimated degree of accuracy of each.

It should not be, but is, necessary to stress the fact that unless a fireball report is accurate it is not worth making, and certainly not worth submitting.

In recent years there has been considerable interest in the tracking of fireballs to see whether the freshly fallen meteorite (if any) can be recovered. Professional research, using networks of cameras in Canada and Czechoslovakia, has led to disappointingly few recoveries, and it seems that exceedingly few fireballs do, in fact, reach the Earth's surface intact. The Meteor Section of the B.A.A. has set up a network of over 50 amateur-operated fireball cameras in the hope of duplicating this work.

Since fireballs are, by their nature, bright objects, large apertures are not needed, and part of the success of the scheme depends upon this fact: that at relatively large focal ratios ($f/11$, say) sky fogging is so slow that the camera shutter can be left open for long periods without needing attention. Despite the lack of concrete results, the work has led to the accurate determination of a number of fireball orbits (see B.15.24).

15.12 Telescopic meteors

Telescopic meteors are exceedingly numerous; rather untrustworthy extrapolation suggests that although about 75 million naked-eye meteors enter the atmosphere each day, the total down to mag 10 is about 7,500 million, their numbers increasing by a factor of about 2.5 per magnitude step. Relatively little serious work has been done on them, however, most reports coming from observers who happen to note them during low-power work on other objects. The field offers scope for both individual and cooperative observation by anyone with the patience to tackle it.

The human eye, with its enormous real field of view, is more efficient at detecting meteors when used alone than in conjunction with any telescope, unless conditions are excellent and the observer is skilful and concentrating particularly hard. Meteor rates recorded by, for example, Alcock during 241 hours of comet and nova sweeping with 45 mm binoculars were less than one meteor every two hours; intensive work by Flain, using a 125 mm refractor with a magnification of ×20, produced 83 meteors in 40 hours. During comet-sweeping work by a team of observers at Skalnate Pleso, Czechoslovakia, using 100 mm binoculars, 3,925 meteors were noted during 1,117 hours of observing; on the other hand Denning, during 727 hours' comet sweeping with a 25 cm reflector, ×40, recorded 635 meteors. Denning's rates, considering his small field of only about 1°, suggest that large field of view does not necessarily compensate for smaller aperture, and that with his ability to see much fainter meteors he was able to match the results of others using less powerful equipment. Hindley's results with a short-focus 125 mm refractor have suggested that telescopic meteor rates can match naked-eye rates, but only over the brief periods during which intense concentration can be maintained. Machholz (California), using a short-focus 25 cm reflector, ×42, under excellent conditions, recorded 1,312 telescopic meteors during 553 hours of comet-sweeping in 1976, an average hourly rate of 2.4; and, as would be expected in the case of sporadic meteors, the morning rate (2.8) was higher than the evening one (1.5). Houston (Kansas) has claimed rates of between 12 and 25 meteors per hour, during deliberate watches with a 15 cm reflector, ×20, field of view 3½°. (See B.15.79 for an interesting discussion of the subject.)

Failing a specially-constructed wide-field telescope,† binoculars probably offer the best chances of success; they should have an aperture of at least 6 cms or so, giving a magnitude threshold of 8 or 9. One set of observations with 6-cm and 8-cm binoculars (limiting mag 8.5, field about 8°) produced 14 meteors in 7½ hours. With this equipment, 47 meteors showed durations from 0.05 to 1.0 sec (peak at 0.3 sec) and a magnitude range of 0–8.5 (peak at 6.0). With apertures smaller than this, numbers are likely to be unimpressive; Muirden recorded only 61 during 250 hours of nova sweeping with 8 × 30 binoculars. See also B.15.3, 15.49 and 15.70.

The threshold of visibility of meteors is from 1.0 to 1.5 mags higher than that of stars, with given equipment; the faintest visible to the naked eye, for example, are of approximately mag 5.5.

The stellar magnitude of a meteor observed telescopically can be estimated in one of two ways. A telescope of aperture D increases the naked-eye brightness of a star* by

$$\Delta m' = 5 \log D - 5 \log \delta$$

The true apparent magnitude of a 'line' source, such as a meteor, whose apparent magnitude in a telescope is threshold (say, 5.0), is given by

$$m = 5 - 5 \log \delta + 5 \log D - 2.5 \log M \qquad . \quad . \quad . \quad (1)$$

where M is the magnification. The telescope of aperture D increases the naked-eye brightness of the meteor by

$$\Delta m'' = 5 \log D - 5 \log \delta - 2.5 \log M$$

Therefore,

$$\Delta m' - \Delta m'' = 2.5 \log M$$

Hence:

(a) Estimate the true apparent magnitude (m_s) of a star in the field, such that it and the meteor are equally bright. Then the true apparent magnitude of the meteor is given by

$$m = m_s + 2.5 \log M$$

† See *A.A.H.*, section 3.10, for an analysis of the difficulty and means of overcoming it.

* See, further, *A.A.H.*, section 1.1.

Or:

(*b*) Estimate the brightness of the meteor by comparing it with that of a star observed with the naked eye, and use equation (1) above.

The anomalous fact that many telescopic meteors appear to cross the field with angular velocities comparable with those of naked-eye meteors is now known to be physiological in cause. Part of the explanation may lie in the eye's persistence of vision. Certainly, some telescopic meteors appear to move so slowly that details of the head and train (if any) can be perceived, but others, presumably far from the radiant, move so quickly that the observer's impression is of no more than an instantaneous line of light in the field. Very possibly, many meteors that would otherwise come above the visual threshold are lost because of the rate at which they are carried across the retina. Variations of brightness frequently occur, and the observer should attempt to record, however inadequately, a 'light-curve' of the meteor during its passage.

The manner in which the meteor intersects the field boundaries can be recorded using the following B.A.A. Meteor Section convention:

00 Entered and left field.
0a Entered and ended inside field.
a0 Began inside field and left field.
aa Began and ended inside field (give path length).

In addition to recording magnitude and appearance, the observer should also note the PA of the track, and, of course, the RA and Dec of the field centre. This information, when added to an accumulation of reports, will allow statistical analysis of the numbers and flights of these objects.

The observations of a single observer can be of value in improving our sparse knowledge of (*a*) the numbers of meteors of each telescopic magnitude in the major showers, and any variation in these data from year to year; (*b*) the determination of radiants by the group radiant method; this will be possible if the shower has a sharp radiant, but not if it is diffuse; (*c*) their appearance: colour, whether stellar or non-stellar, occurrence of streaks, etc; (*d*) statistics relating numbers seen with magnitudes and apertures.

What is really needed is a mass of two-station observations, allowing the real paths to be derived. The most hopeful way of overcoming the practical difficulty of the two observers obtaining observations of the same meteor within a reasonable number of observing hours would be to direct the telescope at the radiant of a major shower, using a low magnification and a large-scale map (e.g. the *Atlas Coeli* chart of the radiant area, enlarged several times by hand). Even so, the observers would probably have to be in continuous telephone contact to ensure the correct identification of the meteors. Any other way of organising a 2-station search for telescopic meteors would be virtually a waste of time.* Fatigue should be avoided, since it both reduces the accuracy of the observations and tends to promote the imagined observation of non-existent meteors; it is recommended that the observers should rest their eyes for a specified period at 5-min intervals (their watches being synchronised if they are not in telephone communication).

15.13 The determination of group radiants

In principle this is a simple matter. It is assumed that the meteors of a given shower enter the Earth's atmosphere in parallel paths, whence it follows that by an effect of perspective they appear to emanate from an infinitely distant point, known as the radiant. We therefore only have to observe the great-circle tracks of the meteors and plot them on a map of suitable projection in order to determine the radiant either by inspection or (if greater accuracy is justified) by a least square solution.

In the case of a dominant major shower with a sharp radiant this procedure can be carried out successfully. It can also be used to determine an approximate radiant centre for those more numerous major showers which have a diffuse radiant, such as the Quadrantids, Perseids, and Geminids. But the method does not give significant results where there are three, four, or more radiants in simultaneous activity, as occurs in the diffuse major showers and (at a different intensity level) in all observations of minor streams. Here, if reliance is placed on the intersection of flight directions alone, a great number of spurious intersections are obtained, arising from real differences in the radiants; and among this tangle of intersections the correct radiants or radiant subcentres cannot be found.

Some authorities have secured results from such data by arbitrarily rejecting those meteors which did not conform to their view of where the radiant should be. Such a process is obviously worthless. The only method of treatment which has given some successful results has been that of Hoffmeister, in which reliable radiants have in certain cases been derived by a process of successive elimination. The observations of many years were considered together, and group radiants were determined in the usual way. Then those which did not recur in n or more years were rejected, and their constituent meteors were combined to form other groups, the

* Öpik, observing the Perseid radiant with an aperture of 12.5 cm, field diameter 2½°, on 1921 August 10 and 12, mapped 15 telescopic meteors ranging from mag 5.5 to 9.5.

process being continued through five such stages of elimination. The method has numerous disadvantages. Even if the basic assumption is correct that a given radiant should be expected to recur two or more times in n years (and this may be doubted), the method is extremely laborious and at the same time wasteful of a large proportion of the observational material. It does, however, represent the only work on group radiants which can be considered to give even partially reliable results.

Certainly the simple process described earlier in this section—by which group radiants were determined during the period 1833 to about 1830—yielded results that were quite unreliable. Historically, the inherent weakness of the method was obscured by errors of technique whose correction took many years. A brief summary of these is given below as a guide to the observer in the avoidance of known pitfalls:

(*i*) It is erroneous to combine, for group radiants, sets of observations made on different nights, when the Sun's longitude differs by, say, more than 2° or 3°. This fundamental mistake has vitiated most of the work of the older observers, who frequently based group radiants on observations gathered over periods of up to 30 days. Such radiants are quite worthless, and it was to correct these that the American Meteor Society's rules for group radiants were formulated by Olivier in 1916. These are enumerated in paragraph (*iii*) below, and are a reliable guide for the combination of observations.

(*ii*) It is quite legitimate, on the other hand, to combine sets of observations made in different years, but when the Sun's longitude was the same.

(*iii*) The A.M.S. rules, so far as they concern the combination of observations, are as follows:

(*a*) A radiant shall be determined by not less than four meteors whose projected paths all intersect within a circle of 2° diameter and which are all observed within a period of at most four hours on one night by one observer.

(*b*) Or, by three meteors on one night and at least two on the next night, seen during the same approximate hours of GMT, and all five intersecting as described in (*a*).

(*c*) Or, by one stationary meteor.

(*d*) Under no circumstances shall a meteor be used to determine a radiant point whose projected path passes more than 3½° from the adopted point, and it is recommended that 2½° should be adopted as the usual limit.

(*e*) Three meteors which fulfil condition (*a*) shall be considered enough to give a radiant in confirmation of one determined on the same date of a previous year, i.e. where L, the meteoric apex, differs by less than $2°$.

These rules are considered to be valuable as a guide to the combination of observations, though they are not themselves sufficient for the determination of group radiants.

(*iv*) The correction for zenithal attraction must often be applied. The correction is greatest where ϵ (the elongation from the apex) is large and h (the radiant altitude) is small; for most purposes the correction may be neglected when it amounts to less than $1°.0$ to $1°.3$. Where the Z.A. correction is applied to the individual meteors, that part of rule (*a*) of paragraph (*iii*) which limits the period of observation to at most four hours may be ignored. The correction for diurnal aberration is less important, and may be neglected for the purpose of classifying meteors for group radiants.

But although the observer may be careful to avoid the errors committed by his predecessors, he is still faced with the inherent weakness of the method of group radiants discussed at the beginning of this section. Evidently additional criteria must be found to enable us to distinguish between true radiants and chance points (or centres) of intersection. It is probable that these criteria will be found in the hypothetical lengths of paths of meteors of varying magnitudes and radiant altitudes, from which the approximate angle of foreshortening of the path may be deduced.

15.14 The determination of real paths

From a good double observation of a meteor its real path in the atmosphere (and hence its height at every point along this path, and its individual radiant) may be computed. The normal heights of meteors are now well recognised, but collections of data on these heights are still of considerable value.

Ideal separation of observing stations is about 80–160 km (see also section 15.3). If less than 30 km the observational errors are too great; if more than about 200 km, not only is identification more difficult but the errors of observation become progressively more harmful.

The paths of the observed meteors may be described in either of two ways:

(*a*) With reference to the stars, as already described in section 15.2.2; this is the better method.

(*b*) With reference to an artificial coordinate system, stationary with respect to the Earth, which is duplicated at the two stations. This is provided by a wire grid at each station, identical in construction and orientation. Similar orientation of the two grids can be obtained either by compass and level or by reference to the stars at a predecided moment, the observers' watches being synchronised. A convenient design of grid consists of five concentric wire circles, of radii 8.75, 17.63, 26.79, 39.40, and 46.63 units. Six sets of cross-wires divide the circles into twelve 30° zones, and provide rigidity. The eye position is defined by a small wire ring located on the axis of the grid at a distance of 100 units from its centre; it should be just large enough for the outer-most ring to be visible through it when the eye is placed as close to it as is comfortable. From this position the diameters of the rings will subtend angles of 10°, 20°, 30°, 40°, and 50° respectively. Beginning and end points of any meteor observed through the grid are described in terms of position angle and distance from the centre of the grid, estimated to the nearest 0.1 division; time is given to the nearest 0.5 min.

The advantages of this method, compared with (*a*), are more continuous observation of the sky (since less time spent making the records), and the more convenient form for reduction of the observations. Method (*a*), on the other hand, has been proved to be of far superior accuracy.

The most persistent source of error in such work is the lack of simultaneity between the first seeing of the meteor by the two observers, so that their estimated beginning points correspond to no single position of the meteor. This error, however, is taken care of in Davidson's method of reduction (see below).

Reductions can be made by a variety of methods:

(*a*) A 'beginner's method', employing only simple trigonometry, is described below:

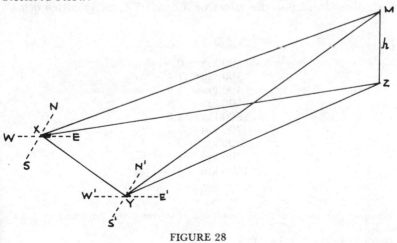

FIGURE 28

The observed RAs and Decs, or grid coordinates, from the two stations are first converted to altitude and azimuth.* If X, Y are the observing stations, and Z the point in the plane containing X and Y which is vertically below M, the beginning or end point of the path, then:

$$
\begin{aligned}
\text{distance between the stations} &= XY \\
\text{azimuth of } X \text{ from } Y &= S'YX \quad \text{or } \alpha \\
\text{azimuth of } Y \text{ from } X &= \text{ext } SXY \text{ or } \beta \\
\text{azimuth of meteor from } Y &= \text{ext } S'YZ \text{ or } \gamma \\
\text{azimuth of meteor from } X &= \text{ext } SXZ \text{ or } \delta
\end{aligned}
$$

all of which are known. Now

$$
\begin{aligned}
ZYX &= \text{ext } S'YZ - S'YX = \gamma - \alpha \\
ZXY &= \text{ext } SXY - \text{ext } SXZ = \beta - \delta \\
XZY &= 180° - (ZYX + ZXY) = \epsilon
\end{aligned}
$$

and

$$
XZ = \frac{XY \sin (\gamma - \alpha)}{\sin \epsilon}
$$

$$
YZ = \frac{XY \sin (\beta - \delta)}{\sin \epsilon}
$$

Errors in h due to neglecting the curvature of the Earth† can be brought within the range of ordinary and unavoidable observational error by adding to the observed altitudes MXZ and MYZ a quantity c whose values, dependent upon the values of XZ and YZ, are tabulated below:

XZ or YZ	c
200 km	$0°.9$
300 km	1.4
400 km	1.9
500 km	2.3
600 km	2.8
700 km	3.3
800 km	3.7
900 km	4.2
1000 km	4.7

---

* Formulae for the commoner coordinate conversions will be found in *A.A.H.*, section 31.

† Dip of 12.4 cm/km, or 1 km in 89.

Then the height of the meteor at point M can be independently derived from

$$h = XZ \tan (MXZ + c)$$
$$h = YZ \tan (MYZ + c)$$

and the mean taken.

M' and Z', respectively the other end point of the observed path and the ground point vertically beneath it, are derived in the same way, whence the length of the path is directly calculable. The weakness of this method is that it assumes the observations were simultaneous (i.e. both observers saw and described the same extent of the luminous path); nine times out of ten this condition is not satisfied, and the results are accordingly misleading.

(b) The neat method due to Davidson (B.15.10) was designed to overcome this difficulty. The RA and Dec of the beginning and end points as observed from each of the two stations are first converted to azimuth and polar distance. These four positions are then referred to a system of rectangular coordinates (two horizontal, one vertical) set up on each station as origin. Since an observer often misses part of the luminous path immediately following the true beginning point, the larger of the two stations' z-coordinates (vertical) is taken as correct; since, on the other hand, an observer is unlikely to imagine he continues to see a meteor after its true end point, the smaller of the two z values is in this case taken.

Porter's simplification of this method, replacing the rectangular by equatorial coordinates, is described in B.15.51 and was the method used by the B.A.A. computers. See also B.15.15, 15.52, 15.68.

(c) The best semi-graphical solution, employing a celestial globe and squared paper, is the so-called Newton-Denning or Tupman-King method; it is described in detail in B.15.17 and 15.40. The essential stages are: first, the determination of the azimuth of the radiant (point of intersection on the star sphere of the two flight directions produced backward) and of the Earth point (i.e. the azimuth in which the altitude of the flight direction is zero) as seen from each station; these azimuths are then laid off from their respective stations plotted on squared paper, their point of intersection being the geographical position of the Earth point; secondly, the ground line of the meteor is laid off from the Earth point in the azimuth of the radiant; finally, the azimuths of the beginning and end points are drawn to intersect the ground line, and the two heights worked out by elementary trigonometry. The reduction of meteor observations is a good introduction to astronomical computation generally, for those with a mathematical turn of mind.

15.15 The determination of meteor orbits

The requisite data are the apparent velocity and the radiant. Thence the geocentric velocity and radiant, and finally the heliocentric velocity and radiant, are calculated. From these the elements of the orbit are derived.

A graphical method described by Davidson (B.15.14) involves the assumptions that the Earth's orbit is circular and that the meteor's velocity is parabolic; nevertheless, in most cases it is sufficiently accurate for the type of observation. See also B.15.72–74 for a concise and useful account; also B.15.13, 15.41, 15.71.

15.16 Meteor diary

15.16.1 Major streams: The data contained in the Table opposite refer to the major night-time meteor streams:

CLASSIFICATION OF STREAMS:
Meteor streams are classified under the following types (column 4 of Table opposite):

A. Streams giving regular annual meteor showers of good strength (ZHR 20–60) or intermediate strength (ZHR 10–20).
   Examples: Quadrantids, Lyrids, Perseids, Orionids, Geminids, and probably also the December Ursids.
B. Streams which have occasionally given exceptional meteor storms (ZHR 2000–30,000) but which are normally very weak.
   Examples: Leonids (period 33 years), Giacobinids (period 6.6 years).
C. Streams which have given occasional or periodic storms or rich returns in the past, but which are now extinct.
   Examples: Bielids, June Draconids (Pons-Winnecke meteors).
D. Major streams which owing to their latitude give very weak displays in north temperate latitudes.
   Examples: η Aquarids, δ Aquarids.
E. Streams only observable in daylight by radar techniques (see section 15.16.2).
F. Minor streams (see section 15.16.3).

Date UT	⊙	Stream		Radiant			ZHR *	LMT of Transit (UT)	Normal Limits
		Name	Type	α	δ	Type			
Jan. 4	282°.8	Quadrantids	A	232	+50	Diffuse	110	8ʰ5	Jan. 1–6
Apr. 21	31.4	Lyrids	A	272	+32	Multiple centres	8 to 15	4.2	Apr. 19–24
May 6	44	η Aquarids	D	336	00	Multiple centres?		7.6	May 1–8
July 28	124	δ Aquarids	D	339	00	Double		2.4	Jul. 15
				339	–17				Aug. 15
Aug. 12	139.3	Perseids	A	46	+58	Diffuse	68	5.6	Jul. 25– Aug. 18
Oct. 10	196.3	Giacobinids	B	262	+54	Sharp?		16.2	<6 hrs
Oct. 21	207	Orionids	A	96	+15	Multiple centres?	30	4.4	Oct. 15–25
Nov. 7	225	Taurids	A	56	+14	Double	12	0.6	Oct. 20– Nov. 30
				56	+22				
Nov. 14	232	Bielids	C	23	+43	Diffuse		22.0	?
Nov. 17	235	Leonids	B	152	+22	Sharp	10?	6.5	Nov. 15–20 (?)
Dec. 14	262	Geminids	A	112	+32	Multiple centres	60	2.0	Dec. 7–15
Dec. 22	270	December Ursids	A?	217	+78	Sharp	1?	8.4	Dec. 17–24

* The ZHR is given for type A showers only. For types B, C, and D refer to text below.

HOURLY RATES:

The hourly rate given in column 7 of the Table is the zenithal hourly rate at maximum. In order to forecast the HR at a given time it is necessary to know (*i*) the correction for radiant altitude, (*ii*) the rate of change of ZHR for the interval between the time of observation and the time of maximum, (*iii*) the annual variation in the intensity of the stream. Of these, (*ii*) and (*iii*) can only be stated with an approach to accuracy for the streams having sharp maxima, such as the Quadrantids and Giacobinids. In any case considerable statistical fluctuation in the tabulated values should be expected—particularly when the radiant is at a low altitude, and casual variations will have a correspondingly greater effect.

(*i*) Correction for radiant altitude:

The radiant altitude of the principal streams for an observer in latitude +52° is given in the following Table at intervals of $t = \pm 0, 1, 2, \ldots$, etc hours from the time of transit, and is based upon the radiant data in the Table above.

Altitude of Shower Radiants ($\phi = +52°$)

	Time from Transit in hours											
	0	1	2	3	4	5	6	7	8	9	10	11	12
Quadrantids	—	—	71	62	53	45	37	30	24	19	15	13	12
Lyrids	71	68	61	52	43	34	25	17	10	3	—	—	—
η Aquarids	—	—	—	—	—	10	1	—	—	—	—	—	—
δ Aquarids	22	21	17	12	4	—	—	—	—	—	—	—	—
Perseids	—	—	72	64	56	48	42	36	30	—	—	—	—
Giacobinids	—	—	72	63	55	47	39	33	27	22	19	17	16
Orionids	53	51	46	38	30	21	12	3	—	—	—	—	—
Taurids	56	54	47	41	32	24	14	6	—	—	—	—	—
Leonids	60	58	52	44	35	26	17	8	1	—	—	—	—
Geminids	70	67	60	52	43	33	25	16	9	3	—	—	—
Dec. Ursids	—	67	65	61	57	53	49	45	42	39	37	36	35

The correction factor to be applied to the ZHR in order to correct for radiant altitude is given in the following critical Table, and is based on the expression

$$OHR = ZHR \sin (h + 6°)$$

Altitude	Factor	Altitude	Factor
	0.0		0.5
0.0		27.4	
	0.1		0.6
2.6		34.5	
	0.2		0.7
8.6		42.5	
	0.3		0.8
14.5		52.2	
	0.4		0.9
20.7		65.8	
	0.5		1.0
27.4		90.0	
	0.6		

In critical cases ascend*

This correction applies to visual work only—not to radar, where wholly different factors are involved.

* I.e. when the argument is one of the tabular values themselves, the functional value to take is the one immediately above, e.g. the factor for altitude $20°.7$ is 0.4, not 0.5.

(ii) *Correction for* $\Delta \odot$:

This correction is more uncertain. The correction curves for the Quadrantids and Giacobinids are quite well known, these being of short duration; the December Ursids are probably similar. They are less well established for the streams with long maxima, such as the Perseids and Geminids. The Table below gives the interval from the time of maximum in which the ZHR reaches a stated fraction of the ZHR at maximum:

	Units	$\Delta t -$		Max.	$\Delta t +$
		0.1	0.5	0.5	0.1
Quadrantids	hours	12	4	4	12
Lyrids	days	3 ?	1	1	3
Perseids	days	8	1	1	3
Giacobinids	hours	2.4	0.5	0.5	2.4
Orionids	days	8	5	5	8
Leonids	days	—	—	—	—
Geminids	days	3	1.0	0.8	1.2
Dec. Ursids	hours	12	4 ?	4 ?	12

(iii) *Correction for annual variation:*

(a) This does not apply to streams of types A and D, which are of fairly uniform annual richness. Occasional variations by a factor of ×2 or ×3 appear to be mainly of a random nature.

(b) The date for the recurrent streams are:

June Draconids: Associated with Pons-Winnecke's comet, period 6.1 years. The shower was discovered on 1916 June 28, ZHR c. 60. It made only one (perhaps three) known appearances, and is now extinct owing to the increase in q through perturbations.

October Draconids (Giacobinids): Associated with Giacobini's comet, period 6.6 years. The shower was discovered on 1926 October 9, ZHR c. 20. Rich returns (i.e. meteor storms) in 1933 and 1946; noteworthy return at sub-storm level (c. 200 per hour) in 1952.

Bielids: Associated with Biela's comet, period 6.6 years. Gave rich returns at major shower level frequently during the nineteenth century, with meteor storms on 1872 November 27 and 1885 November 27. Last known return in intermediate strength, 1899 November 24. Now extinct.

Leonids: Known from at least 1799. Associated with Tempel's comet, period 33.2 years. Major storms on 1833 November 13 and 1866 November 15. Owing to perturbations the 1899 return failed, but the shower returned at major shower intensity (ZHR c. 60) in 1932 and

1933, though far below storm intensity. Another storm occurred on 1966 November 17.

December Ursids: Possibly associated with comet Méchain-Tuttle. Discovered 1945 December 21 at major shower level (ZHR *c*. 100). In subsequent years has returned at 10–20 per hour, although recently rates of only 5 per hour have been reported.

15.16.2 Daylight streams: The following Table contains data concerning the principal daylight streams, i.e. those discovered and exclusively observable by radar techniques:

Maximum	$\odot$	Stream	α	δ	Transit HR	Normal limits
May 13	52°	υ Psc	17°	+26°	16	May 12–13
15	54	o Cet	26	− 3	25	May 14–23
June 8	77	ϵ Ari	46	+21	60	May 29–June 18
8	77	ζ Per	63	+21	40	June 1–16
27	95	β Tau	88	+17	25	June 24–July 5

The approximate date of maximum is given in column 1; usually it extends over many days. In column 5 is given the HR at transit for a receiver sensitivity adjusted to give approximately the HR of a visual observer under normal conditions.

15.16.3 Minor streams: The following Table includes the best of the minor streams found by duplicate observation mainly during the period 1946–1949; it is in no way a complete survey. These streams are all very weak, giving a ZHR at maximum of one meteor every one or two hours.

Maximum	$\odot$	Stream	Radiant α	δ	Normal limits
Apr. 9	19°	θ Vir	*196°	−2°	Apr. 3–11
15	25	τ Her	*248	+46	Apr. 11–19
Jul. 26	123	θ Aql	*299	− 3	Jul. 22–27
29	126	α Cap	307	− 16	Jul. 23–31
Aug. 14	141	ζ Dra	261	+63	Aug. 14–16
14	141	θ Cyg	290	+54	Aug. 14–17
Sep. 8	165	δ Psc	8	+12	Sep. 5–11
30	187	ξ Psc	23	+2	Sep. 27–Oct. 3
30	187	ρ Cyg	325	+45	Sep. 27–Oct. 2
Oct. 11	198	ξ Ari	*36	+12	Oct. 2–16
21	207	ϵ Gem	*99	+26	Oct. 14–26

* These radiants exhibit the motion in longitude of approximately 1° per day that theory requires. The time interval in most cases is too short for the effect to be demonstrable.

SECTION 16

COMETS

16.1 Amateur work

Comets provide a fruitful field for amateur observers. The work includes:

(a) Observation of the appearance and structure of comets.
(b) Measurements of position, the fundamental data for the computers.
(c) Searching for new comets, and for returning periodic comets.
(d) Photometry.
(e) Spectroscopy (though with normal amateur equipment this is restricted to the brighter comets).
(f) Computation of orbits and ephemerides.

The observations may be made either visually or (see section 16.3) photographically.

The regular cometary observer will need to subscribe to a service such as the B.A.A. Circulars, which give early notification of cometary discoveries, reappearances, and other news.

Would-be members of the Comet Section of the B.A.A. are asked to supply the Director with the following information:

Details of equipment, including charts and catalogues available.
Types of work in which particularly interested.
Local observing conditions (horizon, artificial lights, etc).
Observational experience, and mathematical standard.
Approximate amount of time available.
Telephone number.

16.2 Visual observation

The continuous visual observation of comets is a necessary accompaniment of photography, since finer details and rapid changes cannot generally be

213

so well recorded photographically. Observations of sudden developments should be reported immediately, and the wider the longitude distribution of observers the less likely it is that such developments will be missed.

As with the Moon and the planets, visual observation with a given aperture can show detail invisible in photographs taken with telescopes of many times larger aperture. Only with the smallest instruments is visual observation less useful than photographic, and even here this is far from being invariably the case.

Accurate timing of observations of position, and of sudden or rapid changes, is essential; an accuracy of one-tenth of a minute of time is usually sufficient. Other observations may be timed to the nearest minute and reported to one-tenth of a day.

For glimpsing the outermost regions of the tail very low magnifications, approaching M',* must be used. It is no uncommon experience to find that a comet (as also some very diffuse nebulae) can be clearly seen in the finder but is invisible in the main instrument. The general rule is to employ the smallest aperture that will show the comet clearly. Only when the nucleus and brighter central regions are being examined must the full aperture (and the highest magnification permitted by the atmospheric conditions) be used.

The visual observation of comets falls primarily into three categories:

(*a*) Structure (for which the head generally demands fairly high magnification—say about ×150—with progressively lower magnifications for the fainter outlying regions, so as to gain maximum contrast with the background):

Nucleus (not to be confused with the central condensation; the nucleus is typically either a small—few " arc—uniform disc with a sharp boundary, or a star-like point): single or multiple; if multiple, the relative positions of the components; disc or stellar; size of disc; sharp or diffuse.

Central condensation (if any): size and shape; if elliptical, p.a. of major axis.

Coma: size and shape (variations in these); hoods or haloes (if any): diameters, development, p.a. of major axes (if elliptical), whether separated by dark interspaces; jets (if any): length, p.a. (of tangent to the jet at the nucleus, if curved), p.a. and distance of several points on a curved jet (if possible).

Anomalous tails (i.e. directed towards the Sun): length p.a. or coordinates, curvature.

---

* Minimum useful magnification: see *A.A.H.*, section 3.3.

Tail: length and position (p.a. and distance, or coordinates, of points along its axis); any distortions or divisions; condensations: position measures as often as possible; any fluctuations of light (points or areas) moving along the tail—probably physiological.

(b) Stellar magnitude of nucleus, and integrated magnitude of the whole comet (section 16.6); variations in these.

(c) Position (section 16.4): the point to measure is the nucleus, if present; failing that, the central condensation of the coma. If the comet is of sufficient angular size, its outline may be transferred at intervals to a tracing from a suitable star chart. This, however, is best done photographically.

16.3 Photographic observation

Comets should be photographed at every possible opportunity (pictures being taken in duplicate), so as to get as complete a record as possible both of position and of brightness variations. Structural detail is not likely to be shown effectively by amateur equipment except in the case of the largest comets, but given a reasonable plate-scale and adequate exposure, a comet's position, shape, and brightness can be deduced at leisure from a photographic plate with an accuracy at least comparable with that of visual observations. With careful guiding, and access to a plate-measuring machine, positions accurate to $1''$ arc or better can be obtained, even using an ordinary Newtonian reflector and 35 mm film, if the comet shows a well-defined nucleus, if there are convenient reference stars, and if the focal length is adequate. In the case of a newly discovered comet, whose position is uncertain, even a measure to within about $30''$ arc may be useful, and this can be achieved with nothing more elaborate than a millimetre scale. In all positional measures of this type, the source from which the comparison stars' positions were obtained must be quoted, and the section in *A.A.H.* on reduction of measures to the equinox of date should be studied.

Newly announced discoveries may often be picked up photographically before they are even visible in the guide telescope. If the announcement includes information regarding the comet's motion, allow for this in the guiding; otherwise the image will be, at best, blurred, and at worst will not register at all. If d is the comet's diameter, and x mins the minimum exposure needed to record it, then no increase of exposure will show it if its motion is greater than d in x mins unless the camera follows it. Therefore displace the camera in the direction, and at the velocity, of its announced motion.

When searching for a returning comet with a battery of cameras, displace them in the direction, and by the amount, of its anticipated motion. Even partial elimination of its image's drift across the plate may make all the difference between visibility and invisibility. Inevitably, any image recorded thus will be so blurred as to be useless for any purpose other than the establishment of its mere existence; but this is all that is required in the case of rediscoveries. It is helpful to know the magnitude threshold of each camera of the battery for various exposures.

Fogging* is especially relevant to cometary photography, since comets are often near the Sun. The relation between exposure and visual magnitude varies considerably from comet to comet owing to different relative visible and actinic brightness; often comets are remarkably rich in actinic wavelengths, particularly if they contain a preponderance of gas rather than dust. Long dewcaps help to reduce the speed of fogging to some extent. Experience with a given instrument, familiar films, past successes and failures—these are the only reliable guide to estimating exposures in relation to fogging.

It is particularly important in the case of comets to record the times of beginning and end of each exposure.

As regards equipment, the great majority of comets have overall dimensions of less than $1°$; hence, pictorially, a lens of between 300 mm and 500 mm F (frame coverage on 35 mm film $7° \times 4\frac{1}{2}°$ and $4° \times 2\frac{3}{4}°$ respectively) would be a good choice for general work; for positional measurements, where the object is to record no more than the central condensation, the longest possible F that will still include satisfactory reference stars is preferable.

For short exposures on remote comets, it may not be necessary to adjust for the comet's own orbital motion during the exposure, even with large F; similarly, a bright comet being photographed with a standard 35 mm lens will show no drift unless it is exceptionally near the Earth, the normal exposure limit at $f/2$ being about 10 minutes, even under rural conditions, before fogging becomes objectionable. If, however, a very faint comet is being sought with an instrument of long F and requiring an exposure of up to 30 minutes or so, some form of special guiding will be required. The image of a tailed comet should be offset in the direction of the tail so that when the head is at the cross-webs of the guide telescope the maximum extent of tail lies across the film area of best definition, i.e. the image is central in the camera field when the head is central in the guide field (Figure 29 (a)).

---

* The whole subject of sky fog, and magnitude and exposure limits, is discussed in greater detail in A.A.H., section 20.5.

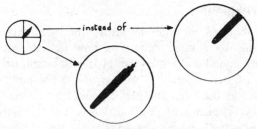

FIGURE 29 (a)

Where a nucleus or small central condensation is visible it is best to dispense with cross-webs—which are necessarily thick, and therefore hide the nucleus—and to guide by keeping the nucleus central on the tip of a tapered pointer, the so-called 'toffee-apple' method (Figure 29 (b)).

There are a variety of methods of guiding on a comet lacking a nucleus or central condensation (and a nucleus is usually absent just when most needed—with faint comets):

FIGURE 29 (b)

(a) Treat the coma like a star, and attempt to keep the intersection of the webs superimposed upon the same point in the image—the least satisfactory method.

(b) Compute the direction of motion and arrange the webs (which can afford to be rather thicker than usual) so that the coma is tangent to them and its direction of motion bisects the diametrically opposite quadrant (Figure 29 (c)).

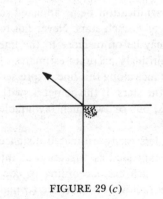

FIGURE 29 (c)

(c) The best method of all is to compute the comet's angular motion; set the cross-webs of a filar micrometer on a guide star; orient the micrometer to the p.a. of the comet's motion; at pre-calculated intervals (depending on the comet's angular velocity) displace the movable web by the appropriate amount opposite to this motion, and bring the star back to the intersection of the webs. These intervals must be sufficiently short to prevent any perceptible elongation of the comet's image from appearing on the negative.

Anyone intending seriously to observe comets will find subscription to the Smithsonian Astrophysical Observatory *Announcement Cards*, or to one of the various 'early-warning' systems operated by larger societies, practically essential.

16.4 Measures of position

The accuracy required in order to justify the measuring of a comet's position depends to a large extent on how well it has already been observed. Where a good preliminary orbit has been established, it is a waste of time – that of the observer as well as of the computer, whose chief grouse is always at the large number of poor observations – to make and report rough eye-estimates of position. Where a preliminary orbit is not available, on the other hand, or where a check is required on an ephemeris, approximate measures of position can be valuable. Even an eye-estimate of the comet's position relative to known stars can be useful if it should happen to fall in a gap in the series of available micrometer measures, or before the first or after the last of these (e.g. when the comet is so near the Sun as to be barely visible). At all other times an accuracy of a few " arc must be achieved if the measures are to be of any value to the computer, and even so he will ignore such observations in favour of any likely to be more accurate. All observations claiming precision must be timed to the nearest 0.1 min.

Unless circle positions can be relied upon, good charts are essential to ensure the correct identification of the field stars. Given accurate circles, a diagram drawn at the telescope can be checked subsequently with a suitable catalogue (even without charts), and provides useful confirmation; or the comet's position can be indicated on a tracing from *Atlas Coeli* or similar atlas.

(a) By eye:

Made by reference to field stars, the magnification being adjusted so that there are, if possible, not less than 5 or 6 such stars. Never fail to take advantage of the fact that the comet may lie on or close to the line joining two stars: the eye is capable of surprisingly accurate estimates of such a configuration and of the comet's distance along this line, expressed as a fraction of the total distance between the stars. If the comet is swift-moving in a region of the sky rich in stars, one or two such favourable configurations may be observable per night.

Otherwise recourse must be had to the less easily estimated distance of the comet from each of the field stars, expressed as a fraction of the separation of any two of them, chosen in each case according to convenience. In the case of the field illustrated below, for instance (comet at x), it may be judged that

$$xa = \tfrac{3}{4}ca$$
$$xc = 3\,bd$$
$$xb = \tfrac{1}{2}ab = \tfrac{3}{5}cb$$

218

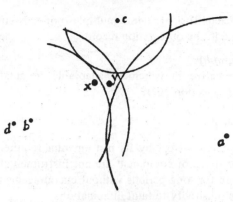

FIGURE 30

The field having been identified, the reference stars are plotted for any convenient (e.g. Catalogue) epoch,* the position of the comet plotted (y in Figure 30), its RA and Dec determined, and the precession correction applied to bring the position up to the mean equinox of the year of observation.

An accuracy of the order of 1′ can be achieved by this bow-and-arrow method; this may be superior to direct readings of the position circles unless these are finely divided and the mounting is very precisely adjusted.

As an alternative to the field sketch, $\Delta\delta$ between the comet and the comparison stars may be estimated in terms of the known field diameter, and Δx measured by the method of transits across a vertical (N–S) wire.

(b) By micrometer:
Any form of micrometer, even the simplest, can be used to advantage. Considering the ease with which a ring or cross-bar micrometer can be home-made, and the simplicity of the equipment with which it can be used,† no amateur intending to observe comets seriously need make use of the rather cumbersome and doubtfully accurate methods described above.

With a home-made cross-bar micrometer the position of a comet with a well-defined nucleus can be determined to within 2″ or 3″; if there is no nucleus, and especially if the coma is strongly elliptical, the inaccuracy may be as large as 10″, or one-sixth that to be expected from the eye method.

Owing to the dependence of the accuracy of the measures upon the

---

* Generally speaking, comparison stars brighter than mag 9 should be used wherever possible.
† See, e.g., *A.A.H.*, sections 18.4, 18.5.

nature of the nucleus (bright, faint, multiple, non-existent), this should always be specified in the observation record.

(c) *By photography:*
Methods of negative measurement adaptable to amateur needs are discussed in *A.A.H.*, section 20.3.

16.5 Comet seeking

Well suited to amateur effort by its instrumental requirements. Indeed, the most valuable items of equipment are not instrumental at all, but the ability to persevere for long periods without encouragement, and a retina of at least average sensitivity to faint illuminations.

Since 1600, some 700 comets have been discovered and observed well enough for their orbits to be determined; and, between 1800 and 1940, 105 comets were visible to the naked eye—approximately 20% of the total. Since very few periodic comets ever attain naked-eye brightness—Halley's being an obvious exception—we find that new naked-eye comets occur very roughly at the rate of one every 18 months (subject, naturally, to large variations); and this rate still applies to the comets discovered in the past 40 years, for the obvious reason that such bright objects are fairly readily detected by routine sweeping.

Where the picture has changed is in the discovery of much fainter comets, picked up on professional photographs taken with large apertures and usually, if not always, exposed for other purposes (e.g., detection of new asteroids). This work has produced a crop of discoveries far below the possible visual range of detection, say between magnitudes 12 and 18, and these can be divided into two classes: (a) comets which would probably have been detected later in any case, since they attain quite a bright magnitude; and (b) comets which always remain faint, and would otherwise pass undetected. Of the nine comets discovered in this way in 1973, for example, seven appear never to have exceeded mag 12, and so belonged to class (b), while of the two class (a) objects, one attained mag 10 at maximum brightness and the other, comet Kohoutek, became a conspicuous naked-eye object. It would therefore seem that the amateur's worries about professional 'poaching' of his objects is groundless, the scope of cometary discovery being almost as wide as it ever has been. In the past decade, of 82 new comets discovered, exactly half were found by amateurs, and most of the 41 professional accidental discoveries were excessively faint objects with large perihelion distances. On average, then, about five comets per annum attain sufficient brightness to be discoverable with amateur equipment.

The pattern of comet discoveries, 1962—79, is shown in Figure 31, while Figure 32 indicates the magnitudes, at discovery, of both (a) amateur and (b) professional discoveries for the same period. These analyses do not include deliberately sought recoveries of known objects, for which the magnitude is very much fainter, typically 18—20. It is worth emphasising that photographic magnitudes frequently refer only to the brightness of the central condensation (denoted by m_2), whereas visually the total magnitude (m_1) is usually quoted; hence, magnitudes measured from photographs may considerably under-estimate the comet's visual brightness.

The comet seeker's chief challenges, then, are (probably in order of significance) the weather; moonlight; and other observers; and there is little that he can do about any of them, except to take his chances when they occur. Instrumentally, the demands are not exacting; few successful observers have used apertures greater than 30 cm, and, of the comets discovered since 1962, the great majority were found using apertures of 10—20 cm, both refractors and reflectors. The team of observers at Skalnate Pleso, Czechoslovakia, who discovered so many comets during the period 1945—59, used 25 x 100 binoculars, and Alcock has used 25 x 105 binoculars for his discoveries. Bennett, who discovered the brilliant comet of 1970, worked with a 'Moonscope' refractor of 12.5 cm aperture, originally intended for satellite tracking; the Australian observer Bradfield, the most successful amateur comet-hunter of the century, uses a home-made refractor built around an old photographic portrait lens of 15 cm aperture. The Japanese amateurs, who were particularly active in the 1960s, used Newtonian reflectors, none larger than 20 cm in aperture. Magnifications were in all cases x40 or lower, the object of any comet-sweeping telescope being to cover as much sky as is commensurate with reasonably detailed examination. Too low a magnification, below M' certainly,* wastes light; too high a power reduces the field of view so severely that the work of covering a given region of sky is unduly prolonged, and, if excessive, may cause a diffuse object to become invisible through lack of contrast.

Focal length is not relevant to light grasp, although it is sometimes supposed that the large relative aperture of 'comet-seekers' contributes in some way analogous to camera 'speed' to increased brightness of extended images. On the contrary, F does not appear in the expression for the telescopic brightness of an extended image, but only in that for the

* See *A.A.H.*, section 3.3.

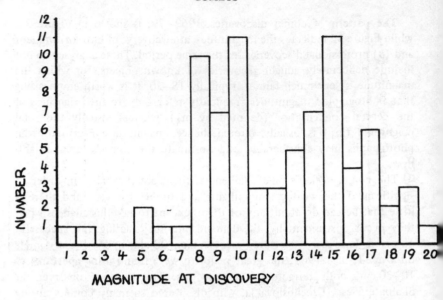

FIGURE 31
Discoveries of comets 1962–1979

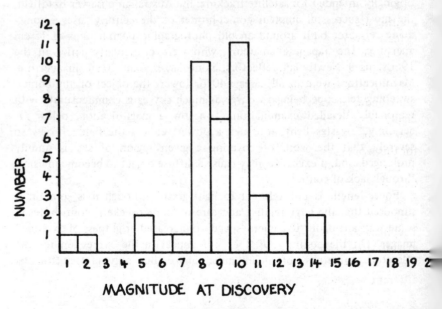

FIGURE 32 (a)

222

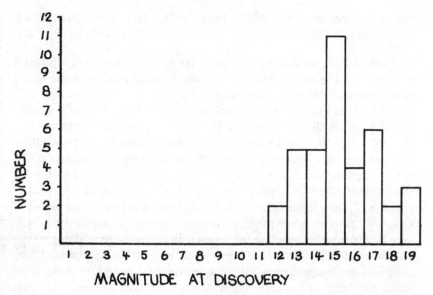

MAGNITUDE AT DISCOVERY

FIGURE 32 (b)

photographic intensity.* Hence the brightness of an extended image is not increased by reducing F (the claims of some manufacturers of 'comet-seekers' notwithstanding), and the advantages of a small f/ratio for comet hunting are to be sought elsewhere: in the possibility of utilising the wide fields offered by LP oculars without increasing the diameter of the field lens to an unmanageable degree. Chromatic aberration, which would be intolerable with high powers when an ordinary Huyghenian is used with, say, a f/9 comet-seeker, is in fact not objectionable with the magnifications under about ×40 which are commonly used for sweeping. With a highly corrected ocular, even smaller f/ratios could of course be used.† Oculars providing flat fields, well defined to the edges, are most suitable for this work.

Large aperture (with low limiting magnitude) and short focal length (allowing wide fields) are therefore the instrumental characteristics to go for. A typical 75 mm, f/9 comet-seeker, magnifying about ×20 (which is well over the safety margin, being about $2M'$), gives a field of nearly 3°. The magnitude threshold should be at least as low as 9, and sufficient magnification should be used to show a nebulous object 2′ in diameter. The following objects (all of which resemble telescopic comets with different apertures) should be clearly shown by any instrument intended

---

* See, respectively, *A.A.H.*, section 1.3, equation (*a*), and section 1.6.
† See *A.A.H.*, section 8.2.1.

for comet seeking: NGC 1068, 1501, 1514, 2022, 2392, 2440, 3242, 3379, 3587, 4472, 4736, 5024, 5272, 6093, 6341, 6402, 6779, 7089, 7099.‡

From *A.A.H.*, section 1.3, equation (*a*), it is apparent not only that a $f/3$ and a $f/15$ telescope will yield equally bright cometary images with a given magnification, providing that their apertures are the same, but also that two telescopes of unequal apertures D_1 and D_2 will yield equally bright images with magnifications M_1 and M_2 respectively, provided that $D_1 : M_1 :: D_2 : M_2$. The absolute magnification, rather than the magnification per cm of aperture, is therefore the important consideration. To reduce the magnification until it approaches M' is to run the risk of losing small comets through insufficient angular size to catch the eye; under about ×20 there is also the wastage of time caused by mistaking small groups of faint stars for nebulous objects meriting HP examination. ×20 may be taken as the absolute minimum with apertures up to about 10 cm, after which a magnification of from $1\frac{1}{2}M'$ to $2\frac{1}{2}M'$ should be used. The danger of over-magnification (quite apart from the reduction of the field) is the possibility of losing the comet owing to reduced contrast between it and the background.* Hence with any comet there is, depending upon its size and brightness, an optimum magnification.

Nevertheless, for general sweeping the following figures represent an effective compromise:

Aperture	Magnification
75–100 mm	× 20
12.5 cm	22–37
15 cm	27–45
17.5 cm	32–52
20 cm	36–60
et cetera	

In all cases the field diameter should not be less than about $1°.5$, and if it is nearer $3°$ so much the better, since four times the area of sky can be covered in the same time. The problem of arranging for wide-field eyepieces with all but the smallest instruments of normal f/ratio is referred to in *A.A.H.*, section 3.10.

‡ The following nebulae and clusters were originally described as 'cometic' by William Herschel (NGC numbers):

57	2592	3342	3872
524	3070	3423	4032
596	3078	3491	4578
1055	3166	3640	5248
1579	3169	3822/3825	

* See also *A.A.H.*, section 1.3.

Regular sweeping, carefully and systematically carried out, is the mode of working most likely to bring success. Even one hour's sweeping every clear night (atmospheric transparency is the *sine qua non*) of the moonless half of each month is better than spasmodic sessions of several hours' duration; a single constellation swept over with such care that the observer may be confident that it contains no comet within reach of his telescope is preferable to a much wider area skimpily covered. The beginning of each moonless period should not be missed, since any comet present may have brightened considerably since the area was last observed.

All comets must at some time pass near (angularly) to the Sun, and the majority of new comets are inside the orbit of Mars before they are spotted; furthermore they increase in brightness as they approach perihelion. Hence the western or NW sky should be swept as soon as it is quite dark, and the eastern or NE sky before sunrise (in the S hemisphere, W and SW, and E and SE, respectively); sweeping should be started in the evening, and finished in the morning, well outside the period of twilight. The low northern sky during the summer months is similarly a good region to sweep. Regions about 90° from the Sun should be given second priority, and both the plane and the poles of the Milky Way avoided; in such regions as Coma, Ursa Major, and Virgo the concentration of faint nebulae is so great that an undue proportion of the available observing time will be spent with one's nose in the catalogues. As sweeping moves further from the region of the Sun it should be carried out more and more carefully, since any comet encountered is likely to be faint and angularly small.

Despite such general indications, however, the truth remains that a new comet may appear in any part of the sky whatever.

Sweeps should be made slowly enough to permit the scrutiny of all parts of each successive field by indirect vision. Sweep, say, 20° in RA or azimuth; then alter the Dec or altitude setting of the telescope by not more than two-thirds of the diameter of the field, and sweep back again; this back and forth sweeping of overlapping strips is continued throughout the period of observation.

The sky contains innumerable objects (nebulae, clusters, and small star groups*) which under low magnification resemble telescopic comets; it is essential, in the interests of time-saving, to work out a quick method of checking these as they are swept up:

(*a*) An ocular adaptor, which allows the HP scrutiny of each suspicious

---

* Ghosts of bright stars are not likely to give any trouble, since the telescope is moving during sweeping, and the consequent motion of the ghost relative to the image of the star producing it is immediately noticeable.

object without incessant changing of oculars and re-focusing, is invaluable. Failing a rotating adaptor (e.g. B.16.52), in which the alternative eyepieces are ready mounted, a great deal of time can be saved by scrapping the screw fitting of the oculars, mounting them instead in sections of brass tube which slip into the draw-tube; refocusing can be avoided by fixing flanges or stops on the eyepiece extensions in the required positions for a given setting of the drawtube adjustment.†

(b) If the instrument is fitted with circles, it is probably quickest to refer straight to a catalogue of nebulae and clusters; the *Revised N.G.C.* (B.27.12) is probably the most suitable.

(c) If the instrument is not fitted with circles, the identification of the field on a star chart, the determination of the approximate RA and Dec of the object, and finally reference to the catalogue, is a long process. Reference to a chart being quicker than to a catalogue, it is a good plan to keep a set of charts showing stars to mag 9 or thereabouts, and to insert in ink beforehand those NGC objects in the area of the night's sweeping which might be mistaken for a comet; any other object which is suspicious in the LP sweeping ocular is also inserted, in pencil, at the telescope. Before the next observing session the pencilled additions (it being assumed that their non-cometary nature has been established) are also inked in. *Atlas Coeli* (B.20.2) is recommended for this purpose. As experience is gained, reference to charts and catalogues will become less and less frequent.

An occurrence with which the comet seeker soon becomes familiar is the passage across the moving field of a faint though perfectly visible nebulous object which vanishes as soon as the motion of the telescope is arrested for it to be more carefully inspected. This curious effect of a faint enough object being visible only while its image is moving relatively to the field has never been satisfactorily explained. It is quite independent of the indirect vision effect, since such an object is often perfectly visible to direct vision while in motion across the field. By the same token, it has nothing to do with motion of its image across the retina. A possible analogous example of apparently increased contrast between an object and its surroundings by the introduction of movement is the varying visibility of a page of print through a sheet of plain paper laid upon it: with a sheet of the right thickness it will be found that the print beneath is quite invisible when the sheet is lying motionless on it, and perfectly visible, even legible, when the sheet is moved about.

When an object not clearly non-cometary under HP is discovered, an

† See also *A.A.H.*, section 22.9.

accurate map of the field must be made; the field must be readily identifiable (RA and Dec if fitted with circles, or otherwise tied to some easily
recognisable star) and the position of the object relative to the nearest
three field stars put down with the greatest precision possible; a couple of
micrometer measures of $\Delta\alpha$ and $\Delta\delta$ are better still. The field is then
reobserved after the lapse of an hour or so; if no trace of motion on the
part of the object is detectable with the HP ocular, it is almost certainly
not a comet; in most cases ¼ hr is a long enough interval for change of
position to become perceptible.* As a matter of routine, it is well to start
each night's session with a check-up on any unidentified objects,
apparently motionless, that were found during the previous night. An
object showing no change of position relative to the stars after 24 hrs, in
a HP ocular, is certainly not a comet.

Once proper motion has been confirmed, the discovery should be
reported with the minimum of delay, so that other observers may be
notified and observations in all terrestrial longitudes be initiated as early as
possible (also in the interests of establishing priority). The announcement
should be made by telegram or phone to the nearest observatory (in the
British Isles, to the Royal Greenwich Observatory, Herstmonceux, Sussex,
or to the Director of the Comet Section of the B.A.A.). See section 16.7.

16.6 Photometric observations

The laws governing the variations in the brightness of comets are still
imperfectly understood, a fact not unconnected with the practical
difficulties involved in obtaining accurate observational data.

Presumably the apparent brightness of a comet varies inversely as the
square of its distance both from the observer and from the Sun, in so far as
it shines by reflected light. That is to say, its brightness at any moment
will be proportional to

$$\frac{1}{r^2 \Delta^2}$$

where r is its heliocentric distance and Δ its geocentric. If it is further
assumed that the proportion of its total radiation which consists of
emission is produced through the agency of solar radiation acting upon the
material in the nucleus, we might expect its brightness due to this factor
to vary with $1/r^2$. Hence the preliminary expression for the brightness of
a comet is

---

* A daily motion of only $1°$ is equivalent to a displacement of $2''.6$ per minute. This
would be clearly perceptible in 12 minutes $(0'.5)$ providing there are reasonably
nearby stars for comparison.

$$b = \frac{b_0}{r^4 \Delta^2} \quad \cdots \cdots \cdots \quad (1)$$

where b is its observed brightness, and b_0 a constant.

However, no comet behaves accurately in accordance with this expression, which has to be made more general:

$$b = \frac{b_0}{r^n \Delta^2} \quad \cdots \cdots \cdots \quad (2)$$

where the value of the exponent n is different for different comets, and even for a given comet may vary (e.g. before and after perihelion); and the fractions of the total brightness due to reflected light and to solar action are respectively proportional to

$$\frac{1}{r^2 \Delta^2} \quad \text{and} \quad \frac{1}{r^{n-2}}$$

Expression (2), converted to terms of stellar magnitudes and adopting the more convenient logarithmic form, becomes

$$m = m_0 + 2.5n \log r + 5 \log \Delta$$

where m = observed magnitude,

$$m_0 = \text{magnitude when } r = \Delta = 1, \text{ or } b = b_0.$$

Values of m for different values of r and Δ are given by observation, and m_0 and n are derived by the method of least squares.

The brightness of the majority of comets is at any rate roughly represented by the so-called 'r^4 law' or 'r^6 law'; i.e. by putting $n = 4$ or $n = 6$ in equation (2) above, we obtain, respectively,

$$m = m_0 + 10 \log r + 5 \log \Delta$$
$$m = m_0 + 15 \log r + 5 \log \Delta$$

Generally the former is likely to be more correct for new comets (parabolic), and the latter for short-period comets.

The different values of m_0 for different comets presumably reflect differences in their physical and chemical composition, but other factors must be looked for to explain, *inter alia*:

(*a*) Fluctuations unrelated to Δ occurring in comets whose eccentricity is so small that r is also insufficiently variable to account for them. For example, Comet Schwassmann-Wachmann, 1925*ii*, $e = 0.142$, r varying by only 1.8 A.U., whose average magnitude of 18 is liable to sudden decreases of several mags—in one case a brightening of 5 mags within 4 days. 1925*ii* is visible at aphelion, and has been continuously observed since its discovery.

(b) Fluctuations occurring when r is so large that the temperature is too low for the expulsion of gaseous material from the nucleus, as a result of solar heating, to be feasible. For example, at mean solar distance Comet Whipple-Fedtke, 1942g, has a temperature in the neighbourhood of $-160°C$; yet the presence of gaseous material in comets is proved spectroscopically.

(c) Sudden decreases in brightness, not susceptible of explanation by the hypothesis that seeks to account for some fluctuations in terms of collision with meteor swarms, or of that which suggests that ultraviolet radiation or streams of ionised particles from the Sun produce an excitation effect in the gaseous material of the coma and tail. As regards the latter hypothesis, a possible correlation was established between the sudden brightening of Comet Whipple by 3 mags with solar activity in February 1943, at a time when both r and Δ were increasing.

(d) The fading of many comets before they reach perihelion.

(e) The rough periodicity in the light variation of many comets, investigated by Bobrovnikoff (B.16.7). Comet Whipple-Fedtke, for example, brightened during the 1943 return five times, these exhibiting a periodicity of roughly 30 days (B.16.10).

The photometric behaviour of a comet is most conveniently studied in terms of the difference of its observed brightness from that predicted, that is, of $O - C$, where C is the computed value of its magnitude at suitable intervals, and O is the corresponding observed value. A plot of the values of $O - C$ will show clearly the nature and extent of any anomalous variations of brightness. The calculated values are derived as follows. Suppose that on a given date

$$\left. \begin{array}{l} \Delta_1 = 1.126 \\ r_1 = 1.626 \end{array} \right\} \text{ from the ephemeris}$$
$$m_1 = 6.7 \qquad \text{from observation}$$

It is required to find m_2, the comet's magnitude on some subsequent date. From the ephemeris, for this latter date,

$$\Delta_2 = 2.542$$
$$r_2 = 2.321$$

Then* $\log r_2^4 \Delta_2^2 - \log r_1^4 \Delta_1^2 = 0.4\,(m_2 - m_1)$
hence $2.2730 - 0.9474 = 0.4\,(m_2 - 6.7)$
$$m_2 = 10.01$$

The r^6 law would in this case yield $m_2 = 10.76$.

* From equation (1), p. 228. and $A.A.H.$, section 1.7, equation (e).

The 'Light' of a comet is given in ephemerides that have been tele-graphed. Since the telegraphic code has to cover minor planet discoveries the 'Light' is derived from the proportion

$$\frac{r^2 \Delta^2 \text{ at discovery}}{r^2 \Delta^2 \text{ at ephemeris date}}$$

Although not really satisfactory for comets, this quantity does indicate if the brightness is likely to be increasing or decreasing.

When a nucleus is present, its magnitude can be estimated by any of the visual methods employed by variable star observers. Except when the comet is faint, and showing little coma, the photographic method tends to be unreliable, since the photographic image of the nucleus is probably several times larger than the true image and therefore includes light from the surrounding central condensation.

The brightness of a comet is, however, normally taken to mean the integrated magnitude of the coma, and such estimates are very much more difficult to make. Comparisons of the comet with nebulae or clusters of known magnitude (e.g., B.27.12) can at times yield good estimates of the integrated magnitude, but this method is necessarily limited in its scope. The usual, and most satisfactory, method of making visual estimations of the integrated magnitude of a comet is to compare it with extrafocal stellar images, expanded till their angular size is similar to that of the comet's focused image. Two applications of this method are available:

(*a*) (*i*) Observe the comet, memorising its brightness and angular size.

(*ii*) Rack out the drawtube till the images of the field stars are of the same size as the focused image of the comet.

(*iii*) Attempt to find one or more nearby stars, as nearly as possible at the same altitude as the comet, the brightness of whose expan-ded images matches that of the comet. If their altitude is much different from that of the comet, a correction for atmospheric absorption must be made.

(*iv*) Take the position of each comparison star (circles or chart) and identify it later. If brighter than mag 9 it is most likely to be in the Smithsonian Astrophysical Observatory *Star Catalogue*, but with faint comets the difficulty of obtaining the magnitudes of the comparison stars is a perennial source of trouble, and method (*b*) will probably have to be used.

(*b*) (*i*) and (*ii*) as before.

(*iii*) Still carrying the memory of the comet's brightness in mind, swing the telescope on to a suitable nearby variable-star field and match it with a star of known magnitude. Allowance must in this case be made for the difference in altitude.

Several independent estimations should be made during the course of each night's work.

Factors affecting accuracy include:

(*a*) Nearness of suitable comparison stars, permitting the comparisons to be made quickly.

(*b*) Brightness of the comet: the brighter it is, the further one will probably have to look for comparison stars; on the other hand, if it is fainter than about mag 9 the mags of the comparison stars may have to be established by comparison with a suitable variable-star field, or B.23.2.

(*c*) Number and brightness of the comparison stars: ideally not less than three should be used—one slightly brigher than the comet, one slightly fainter, and one matching.

(*d*) Altitude of the comparison stars must be approximately the same as the comet's—or the necessary correction made.

(*e*) Comparison stars should be of roughly the same colour as the comet, i.e. in most cases of types B, A, or F.

(*f*) The derived magnitude will depend heavily on the angular expanse of comet visible, and this in turn depends upon the instrument used. The diameter of Comet Schaumasse (1951), for example, was given as 8' in ordinary telescopes and eyepieces, and at the same time was seen as 20' diameter in a finder, while a photoelectric photometer recorded it over a diameter of 160'. Under such circumstances, more or less true of all comets, the term 'integrated magnitude' loses its meaning.

It was in an endeavour to avoid this last source of error that Merton once proposed a method based on a standard angular area of comet, suggested as $1'/\Delta$ (Δ = geocentric distance), which would roughly correspond with a fixed volume of comet at different times. 1' arc was chosen since it gives an area which most observers can observe in most comets, and therefore eliminates the uncertain and often large peripheral area which may or may not be seen and recorded. The technique is to expand the image of a nearby star, not until it is the same *size* as the image of the comet, but until its surface *brightness* matches that of the comet's central condensation. The diameter of the expanded image is measured, and the light for the standard $1'/\Delta$ area computed and translated into magnitudes.*

The consistency of the results is as important as their absolute accuracy, and with experience a consistency of about one-third of a magnitude may be hoped for. The absolute values of the derived mags

---

* *Trans. I.A.U.*, 8, 1952; see Commission 15.

will be much wider of the mark—by as much, in difficult cases, as 2 or even 3 mags—since differences of brightness in extended images are much harder to perceive and estimate accurately than those between point images. Factors affecting the consistency of the results include:

(*a*) Nature of the background against which the comet, with its extended image and indefinite boundary, is observed. A bright ground will appear to dim a comet out of proportion to the dimming of star images, though this effect is reduced by placing the latter out of focus. Moonlight, haze, high clouds, and the nature of the stellar background all introduce variable factors into a set of measures made at different times, and must therefore be specified in the observational record.

(*b*) Magnification, for the same reason, affects the results, since increased magnification dilutes the extended image more than the point images; it likewise must always be recorded. It is usually wise to use the lowest magnification that permits easy observation of the comet.

(*c*) Aperture and type of instrument must be specified. If more than one instrument is used during the observations of the same comet (e.g. naked eye and binoculars, binoculars and telescope, or telescope with reduced and full aperture) a series of overlap estimates should be made to determine the systematic difference, if any, between them. Speaking very generally (since the angular size of the coma is also relevant), the limiting magnitude for comets, with a given aperture, is about 2 mags higher than for stars. The normal tendency is to under-estimate the brightness of a comet, partly, no doubt, owing to neglect of this higher threshold in the case of extended objects; announcements of discoveries not infrequently under-quote the brightness by as much as 2 mags.

(*d*) The observational record should contain sufficient indication of the photometric method used, and the identity of the comparison stars.

Though the use of extrafocal images for comparison represents an improvement upon that of focused images, a telescopic comet (or the head of a brighter one) is nevertheless very unlike the comparison objects, both as regards the intensity gradient across its diameter and the definiteness of its boundary. A system of instrumental photometry, giving more consistent results than the stellar comparison method, might be based upon the injection into the field of a comparison image more closely resembling the comet.

The whole subject of cometary brightness is still in a rudimentary state, and more observational data are urgently required for discussion. There is great scope here for amateur work, both in the accumulation of reliable data and in the investigation of possible correlations between cometary fluctuations and, for example, spectrohelioscopic solar events.

16.7 Communication of observations

If, for the announcement of a cometary discovery, a telegraphic code is not used, the telegram should include the following data:

(*a*) Nature of object.

(*b*) Position: For a preliminary announcement an accuracy of 2' or 3' is good enough to allow the object to be identified. This order of accuracy can be obtained for α by timing the interval between the transits of the object and of a known star at a vertical (N–S) web in the eyepiece, and for δ by eye estimation of $\Delta\delta$ from a known star based upon the known field diameter. If the telegram is followed up with a chart of the comet's position, ensure that (*i*) the field is made readily identifiable by linking it to some nearby bright star or stars, (*ii*) the relative positions of comet and field stars are put down as accurately as possible, (*iii*) the scale and orientation of the chart are given.

(*c*) Direction (p.a.) and estimated daily rate of motion.

(*d*) Estimated integrated magnitude.

(*e*) Whether there is a nucleus; if so, estimated stellar magnitude.

(*f*) Whether there is a tail; if so, direction (p.a.) and length.

(*g*) Name and address of discoverer.

(*e*) and (*f*) may be omitted, but the remainder are essential.

It is recommended, however, that the standardised telegram code adopted by the I.A.U. be used. This is adaptable for a number of purposes besides the announcement of new comets, and consists of a number of groups from which the appropriate ones must be selected. The code is divided into three sections, as follows:

First section: Indicates the type of object being reported, the observer or discoverer, and the equinox.

Second section: Gives either the *position* of the object, its *orbital elements*, or an *ephemeris* (in the case of a discovery telegram, of course, only the position can be given).

Third section: Includes check-sums, to ensure that the figures of the code have been transmitted correctly, together with any additional remarks, and the name of the communicator.

The full schedule is set out on page 234.

ASTRONOMICAL TELEGRAM CODE

First groups for all telegrams

DISCOVERER OBJECT OBSERVER AAAAB

1. The name of the discoverer(s) and/or other designation, such as year and letter for a comet, constellation for a nova, galaxy for a supernova (N = NGC, I = IC, M = Messier), etc.
2. The type of object (COMET, OBJECT, NOVA, SUPERNOVA, VARIABLE STAR = VSTAR, etc.).
3. The name of the observer(s) and/or computer(s).
4. AAAA = equinox (assumed to be the mean equinox of the beginning of the year specified).

B = 1 for an approximate position;
  = 2 for an accurate position;
  = 3 for orbital elements of an object moving around the Sun;
  = 4 for an ephemeris.

Note: When an ephemeris follows the elements, group 4 is replaced by the word EPHEMERIS, and the epoch of the ephemeris is assumed to be the same as that of the elements.

Middle groups for a position

Approximate: CDDEE FFFGH IIJJJ LMMNN PQRRS TUUUU VWWXX
Accurate: CDDEE FFFGH IIJJK KKKLM MNNPP PQRRS TUUUU VWWXX

5. Date of observation:
C = final digit of year
DD = month (01 = January . . . . . 12 = December);
EE = day (U.T.).

6. FFFGH = time of observation in decimals of a day (U.T.). This group may be omitted entirely in the case of observations of stationary objects.

7. *Approximate position*:
Right Ascension II^h JJ^m . J
Declination   L MM^o NN'
$P = 0$

  Accurate position:
Right Ascension II^h JJ^m KK^s. KK
Declination   L MM^o NN' PP". P

L = sign: 2 = positive, 1 = negative.

Q = 1 for "total" magnitude (m_1) for a comet;

   = 2 for "nuclear" magnitude (m_2) for a comet;

   = 3 for visual (m_v) $\left.\begin{array}{l}\\\\\\\end{array}\right\}$ magnitude for objects

   = 4 for photographic (m_{pg}) of stellar appearance

   = 5 for photovisual (m_{pv}) (including minor planets).

RR = magnitude; if the magnitude is negative, add 100.

S = appearance according to the table below; or, *if the object is not a comet*, S = tenth of a magnitude.

	Nothing reported about tail	Tail $< 1°$	Tail $> 1°$
Stellar appearance	0		
Nothing reported about appearance of object	1	2	3
Object diffuse, without central condensation	4	5	6
Object diffuse, with central condensation	7	8	9

8. *For comets and minor planets*, the daily motion:

In Right Ascension T UU^m. UU

In Declination V WW$°$ XX$'$

For supernovae, the offset from the nucleus of the parent galaxy:

In Right Ascension T UUUU$''$

In Declination V WWXX$''$

T, V = sign: 2 = positive (east or north), 1 = negative (west or south).

Group 8 may be omitted if unknown or irrelevant.

Middle groups for orbital elements

CDDEE FFFGH IIIII JJJJJ KKKKK TTTTT UUUUU

5. Date of perihelion passage:

C = final digit of year;

DD = month (01 = January . . . . 12 = December);

EE = day

6. FFF = time of perihelion passage in decimals of a day (E.T.).

G = time interval in days (rounded to the nearest integer) between the first and last positions used for the computation; 0 = 10 days or more.

H = the number and quality of the observations on which the computation is based, or by which the elements have been checked, according to the following table:

	Maximum residuals		
	$5''$	$1'' - 5''$	$< 1''$
Fewer than 3 accurate positions*	1	2	3
3 accurate positions	4	5	6
More than 3 accurate positions	7	8	9

* Or other factors decreasing the reliability of the orbit, such as approximate, doubtful or unsatisfactorily distributed positions.

7. $III^o . II$ $= \omega$
   $JJJ^o . JJ$ $= \Omega$
   $KKK^o . KK = \iota$
8. $T . TTTT$ $= q$ (in A.U.).
   $U . UUUU = e$

The group for e may be omitted if the orbit is parabolic (e = 1).

Middle groups for ephemerides

CDDEE (IIJJJ LMMNN 9TTTT 8UUUU) cddee

5. Date of the first position in the ephemeris:
C   = final digit of year
DD = month (01 = January . . . . . 12 = December)
EE = day
6. It is understood that ephemerides are given for 0^h E.T.
7. Right Ascension $II^h \ JJ^m . J$
Declination $L \ MM^o \ NN'$
L = sign: 2 = positive, 1 = negative.
8. $T . TTT$ = geocentric distance in A.U., preceded by a 9
$U . UUU$ = heliocentric distance in A.U., preceded by an 8
Groups 7 and 8 are repeated as often as is necessary to complete the ephemeris. Positions should be given at uniform time intervals (10, 5, 2 or 1 days). Group 8 need not be given for every line in the ephemeris.

cddee is the date of the last position in the ephemeris, in the same form as CDDEE.

Final groups for all telegrams

YYYYY ZZZZZ REMARKS COMMUNICATOR

9. YYYYY = the last five figures of the sum of all the groups of digits including and following group 4 (containing the equinox).

ZZZZZ = the last five figures of the sum of the groups giving the right ascension, declination and magnitude only (or ω, Ω, and ι in the case of orbital elements). In practice this second check-sum includes only the items under group 7 (or involving the letters I through S).

Note: Any digit that is unknown, not significant, or otherwise to be withheld should be replaced by a slash (/); this is to count as zero in the summations. A group of five slashes (/////) should be avoided. If two or more observations are being telegraphed in the same message, check-sums should follow each one. A check-sum of all numerical information is useful even if the telegram is not sent in code.

10. Any additional remarks in qualification of the observation or the computation. In the case of an accurate position the location of the observing station could be specified.

11. The name of the communicator(s).

EXAMPLE

CLARK COMET CLARK 19501 30610 66 /// 20540 13130

01135 2015/ 10002 81068 34805 GILMORE

Gilmore reports that Clark has discovered and observed a comet as follows:

1973 U.T.   R.A. (1950.0) Dec. m_1
             h   m   o
June 10.66   20   54.0   −31   13
Object diffuse without central condensation, tail $< 1°$.
Daily motion: in R.A. $+1^m$. 5 in Dec $−2'$.

Announcements of discoveries should be sent to the nearest large observatory (see section 16.5).

When communicating routine observations to the B.A.A. Comet Section, the following data are required, a separate sheet being devoted to the observations of each comet:

(a) Date and time (UT), or date and first decimal.

(b) Place of observation.

(c) Approximate RA and Dec.

(d) Estimated magnitude.

(e) Diameter of coma (' arc) and other descriptive data, including degree of condensation of coma:

    0 no condensation,

    9 sharply condensed at centre.

(f) Aperture and magnification.

(g) Observational conditions: sky, Moon, etc. Scale adopted for describing atmospheric transparency:

0 hopeless (mist, fog, etc),
5 average,
6 Milky Way pretty clear,
9 perfect: clear down to horizon.

(*h*) If field sketch is added: location, orientation, scale.

All observations of sudden changes or developments should be reported immediately.

Below is printed a facsimile of the observing card now used in the Section for all observation reports, exemplifying a convenient form of layout:

OBSERVATION OF COMET..

Date (and decimal U.T.)   Place
Approx. R.A.   Dec.
Obs. conditions
Magnitude estimates (and comp. stars)
Description (Coma diam.; degree of condensation; nucleus; tails, length and p.a.) and
　　remarks
Instrument (and x)   Observer
　　　　　　　　　　　　　　　　　(Continue over, e.g. field sketch)

16.8 Orbits

From a series of observations of the position of a comet, its orbit may be computed. The more widely spaced in time and the more numerous the observations, the more precisely may the orbit be derived. Crommelin's 3-observation method (B.16.20) is a good introduction to the subject, but yields less reliable elements than the more complicated methods involving long series of observations (see, e.g., B.16.3, 16.16, 16.71). B.16.37 is a more modern treatment of the subject.

Methods of orbit computation fall into three main groups:

(*a*) *Gauss's method* of which Merton's is a modern version (B.16.54, 16.55; B.16.44 gives a detailed account of the method, with applications). The observational data are converted into the comet's heliocentric coordinates at two different times, the elements then being derived from these two completely defined positions.

(*b*) *Laplace's method*. In its formulation by Leuschner this method has been much used in America (e.g. B.16.9, 16.49), but generally elsewhere the Gaussian method has been preferred. From the observational data are derived the three heliocentric coordinates of the comet at a single time, and the three components of its velocity at the same time. The elements are then deduced from this complete description of its position and velocity at a single moment. B.16.69 describes Väisälä's Laplacian modification, which is interesting and practical for the general orbit.

238

(*c*) *Olbers' method*, for parabolic orbits.

(*d*) *Graphical*. The main advantage offered by graphical methods is that they allow the worker who is new to the subject to apprehend more clearly what is involved and the direction that successive steps are taking, and to form a clearer mental picture of the whole problem than the analytical methods provide. It is, however, prohibitively inaccurate unless the observations are very widely spaced; when the comet's angular motion is large an observational period of several weeks may provide data from which an orbit can be derived with an accuracy comparable with that given by computation, but in most cases several months are required. See B.16.60 of which B.16.36 is a detailed account, with applications; and B.16.12.

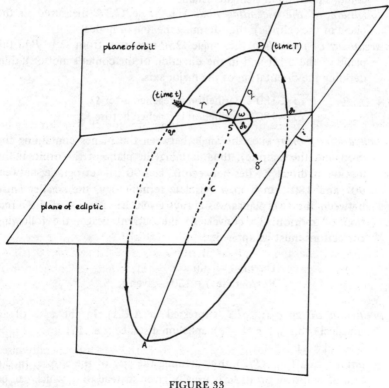

FIGURE 33

Comet direct : therefore *i* = smaller angle between the two planes
Comet past perihelion: therefore *t* is positive, (*t* − *T*), and *v* is positive
Perihelion reached before Ω: therefore ω < 180°
Orbit elliptical: therefore *e* = *CS/CP* < 1

239

The following are recommended to the newcomer to this type of work: B.16. 25–27, 16.30–32, 16.43.

An elliptical orbit, and the position of the comet in it at a given time, can be completely defined by the six elements T, Ω, ω, i, a, and e; a parabolic orbit by the five elements T, Ω, ω, i, and q (see Figure 33). i and Ω determine the plane, passing through the Sun's centre, in which the orbit lies; q or a and e describe the size and shape of the conic section; ω defines the conic's orientation in the plane of the orbit. Additional data are usually included in ephemerides.

In detail, the elements and associated quantities are:

T: *time of perihelion passage* (i.e. comet at P); thus, with ω, defining the comet's position at a single instant.

Ω: *longitude of the ascending node* (i.e. angle $\Omega S \Upsilon$, measured in the plane of the ecliptic); thus defining the line of nodes.

ω: *argument of perihelion* (i.e. angle ΩSP measured from Ω to P in the plane of the orbit and in the direction of the comet's motion); thus defining the orientation of the major axis.

If $\omega < 180°$: perihelion is reached after Ω,
$\omega > 180°$: perihelion is reached before Ω.

i: *inclination of orbit* (i.e. the angle between the planes containing the orbit and the ecliptic); thus defining the plane of the orbit. If the motion is direct,* i lies between $0°$ and $90°$; if retrograde, between $90°$ and $180°$. In a now obsolete terminology, the smaller angle between the two planes was always given, irrespective of the direction of motion. To convert to the current usage, the following corrections must be applied:

Direct: none,
Retrograde: $i = 180°$ −given i,

q: *perihelion distance* (i.e. PS, expressed in A.U.s). In the case of an elliptical orbit, $q = a\,(1 - e)$; aphelion distance (i.e. AS): $Q = a\,(1 + e)$.

a: *semimajor axis* (i.e. $AP/2$); thus defining the size of the ellipse. In the case of elliptical orbits, a is usually given instead of q, which can be derived from it by means of $q = a\,(1 - e)$.

e: *eccentricity* (i.e. CS/CP in the case of an elliptical orbit, C being the midpoint of AP); thus defining the shape of the conic.

* Anticlockwise as seen from the North Pole of the ecliptic.

If $e < 1$: ellipse,
$e = 1$: parabola,
$e > 1$: hyperbola.

ϕ (or φ): may be used instead of e, ϕ being the angle such that

$$\sin \phi = e = \sqrt{\frac{a^2 - b^2}{a}}$$

ϕ is related to the semimajor axis by $b = a \cos \phi$.

P: *period* in years, a function of a. In the case of a comet (mass assumed zero) moving in an elliptical orbit:

$$P = a^{3/2} \text{ years (sidereal)},$$
$$= 365.256898a^{3/2} \text{ mean solar days},$$
$$= 360/n \text{ mean solar days},$$
$$\text{or, } \log a = \tfrac{2}{3} \log P \text{ (from Kepler III)}.$$

Epoch: The date for which the elements are considered correct; given only when perturbations have been applied.

n (obsolescent: μ): *mean daily motion*. Previously expressed in $''$ arc, now usually in decimals of a degree:

$$n^\circ = \frac{0.9856077}{a^{3/2}}$$

M_O: *mean anomaly at the Epoch*. The value of the mean anomaly, M, varies at a uniform rate of n from 0° at time T.

t: time interval between T and the present position:

after perihelion: t is $+$,
before perihelion: t is $-$.

v: *true anomaly* (i.e. angle at S from P to position at time t, measured in the plane of the orbit):

after perihelion: v is $+$,
before perihelion: v is $-$.

r: *radius vector* measured in A.U.s from the Sun's centre at time t.
Δ (or ρ): distance from the centre of the Earth, measured in A.U.s at time t.

16.9 Ephemerides

The full procedure for obtaining a search ephemeris for a periodic comet is to apply to the best available orbit from the comet's last return the perturbations of all the planets and to compute the ephemeris from this corrected orbit. If the perturbations to which the comet has been subjected during its last revolution are neglected, the resulting ephemeris will naturally be less accurate and not good enough for search purposes with large reflectors of restricted field.

The first stage in the computation of an ephemeris for a comet moving in either a parabolic or an elliptical orbit is to derive the six vector elements Px, Py, Pz, Qx, Qy, Qz. These are functions of ω, Ω, ι, and ϵ (obliquity of the ecliptic) and are independent of time.

$$Px = \cos \omega . \cos \Omega - \sin \omega . \cos \iota . \sin \Omega$$
$$Py = (\cos \omega . \sin \Omega + \sin \omega . \cos \iota . \cos \Omega) \cos \epsilon - \sin \omega . \sin \iota . \sin \epsilon$$
$$Pz = (\cos \omega . \sin \Omega + \sin \omega . \cos \iota . \cos \Omega) \sin \epsilon + \sin \omega . \sin \iota . \cos \epsilon$$
$$Qx = -\sin \omega . \cos \Omega - \cos \omega . \cos \iota . \sin \Omega$$
$$Qy = (-\sin \omega . \sin \Omega + \cos \omega . \cos \iota . \cos \Omega) \cos \epsilon - \cos \omega . \sin \iota . \sin \epsilon$$
$$Qz = (-\sin \omega . \sin \Omega + \cos \omega . \cos \iota . \cos \Omega) \sin \epsilon + \cos \omega . \sin \iota . \cos \epsilon$$

For those having a calculator with a polar/rectangular facility, the following will afford a quick means of obtaining the vector elements: Notation: P/R = polar/rectangular, INV = inverse, STO = store, RCL = recall, EXC = exchange, x = the memory used in conjunction with the P/R key, y = one other memory.

For Px, Py, Pz	For Qx, Qy, Qz	Subroutine
(1)	(2)	STO y
Enter 1	Enter 1	Enter ι
STO x	STO x	EXC y
Enter ω	Enter $-\omega$	EXC x
P/R	P/R	INV P/R
EXC x	Use subroutine	$+ \Omega$
Use subroutine		=
		P/R Display
		EXC x (1) P x (2) Qx
		RCL y
	ϵ 1950.0 = 23°.44579	INV P/R
		$+ \epsilon$
		=
		P/R (1) Pz (2) Qz
		EXC x (1) Py (2) Qy

The following numerical checks should always be calculated:

$$Px^2 + Py^2 + Pz^2 = 1$$
$$Qx^2 + Qy^2 + Qz^2 = 1$$
$$Px.Qx + Py.Qy + Pz.Qz = 0$$

In the case of an elliptical orbit it is convenient to employ E, the eccentric anomaly, as a parameter for calculating the heliocentric rectangular coordinates.

E is a function of the interval in days from perihelion $(t - T)$, the mean daily motion $(n°)$ and the eccentricity $e = \sin \phi$ and must be found from Kepler's equation:

$$E - e°.\sin E = n(t - T) = M$$

where $e° = 180e/\pi = 57.295780e$, and all are in degrees. This equation has to be solved by trial, which may be done quickly with a calculator, and an approximate value E_0 found, which may then be corrected by calculating the equivalent M_0:

$$M_0 = E_0 - e°.\sin E_0$$

and the error

$$\Delta M = M - M_0$$

Then the correction to E_0, usually with sufficient accuracy, is

$$\Delta E = \frac{M}{1 - e.\cos E_0} \quad \text{(using } e, \text{ not } e°)$$

Users of modern calculators may prefer to use M as the first approximation to E and successively substitute the value of E found until the solution converges.

The heliocentric rectangular coordinates x, y and z are then derived from:

$$x = aPx (\cos E - e) + bQx.\sin E$$
$$y = aPy (\cos E - e) + bQy.\sin E$$
$$z = aPz (\cos E - e) + bQz.\sin E$$

where a and b are respectively the semi-major axis and the semi-minor axis, a $\cos \phi$, $(e = \sin \phi)$.

For those using a calculator with a polar/rectangular facility, aPx, bQx etc., may be obtained directly by substituting a for 1 in (1) and a.$\cos \phi$ = b for 1 in (2).

243

The following numerical check should be applied:

$$aPx.bQx + aPy.bQy + aPz.bQz = 0$$

The radius vector r is obtained from $r = a(1 - e.\cos E)$.

In the case of a parabolic orbit the true anomaly v is derived by substituting the required value of $(t - T)$ in the equation

$$\tan \frac{v}{2} + \frac{1}{3} \tan^3 \frac{v}{2} = \frac{3.14159\,(t-T)\,2}{365.257\,q^{3/2}} = \frac{0.0121637\,(t-T)}{q^{3/2}}$$

and solving the cubic for v by means of tables of $V + \frac{1}{3} V^3 \sim v$ (where $V = \tan v/2$) which are to be found in, e.g. B.16.60; or a first approximation may be obtained by means of Figures 34 (a) and (b). Alternatively the value of $\frac{1}{2}\tan v = V$ may be solved directly by the following means:

Multiply the right-hand side of the equation by 1.5 to give
$$\frac{0.0182456\,(t-T)}{q^{3/2}} \qquad \text{Call this quantity } x$$

Then $y = \sqrt{x^2 + 1}$ and $\tan\frac{1}{2}v = V = 3\sqrt{x+y} - 3\sqrt{y-x}$

The heliocentric rectangular coordinates are then derived from:

$$x = qPx(1 - V^2) + 2qQx.V$$
$$y = qPy(1 - V^2) + 2qQy.V$$
$$z = qPz(1 - V^2) + 2qQz.V$$

where q is the perihelion distance.

For those using a calculator with a polar/rectangular facility, qPx, 2qQx etc., may be obtained directly by substituting q for 1 in (1) and 2q for 1 in (2).

The radius vector r is obtained from

$$r = q \sec^2 \frac{v}{2} = q\left(1 + \tan^2 \frac{v}{2}\right)$$

For both types of orbit, it remains to convert the heliocentric rectangular coordinates to geocentric rectangular coordinates (ξ, η, ζ) and thence to the equatorial coordinates α and δ.

Using the values of the Sun's equatorial rectangular coordinates X, Y, Z given in the $A.A.$, ξ, η, and ζ are given by

$$\xi = x + X = \Delta \cos \delta \cos \alpha$$
$$\eta = y + Y = \Delta \cos \delta \sin \alpha$$
$$\zeta = z + Z = \Delta \sin \delta$$

244

The RA and Dec of the comet at time t are thus derived from

$$\tan \alpha = \frac{\eta}{\xi} \quad \text{(observing the sign rule of Figure 35 and converting from arc to time)}$$

$\Delta \cos \delta = \xi \sec \alpha$ (if $\cos \alpha > \sin \alpha$ numerically)
$ = \eta \operatorname{cosec} \alpha$ (if $\sin \alpha > \cos \alpha$ numerically)

$$\tan \delta = \frac{\zeta}{\Delta \cos \delta}$$

$$\Delta = (\Delta \cos \delta) \sec \delta = \sqrt{\xi^2 + \eta^2 + \zeta^2}$$

Once again a user of a calculator with a polar/rectangular facility may obtain a quick solution. Using the notation under vector elements:

Enter ξ STO x
Enter η INV P/R   RA displayed
Enter ζ INV P/R   Dec displayed
RCL x   displayed

At the present time all cometary elements are given for the standard equinox 1950.0 and it is important to use the Solar Coordinates for the same equinox.

Note: Recently the I.A.U. has decreed that all ephemerides should be astrometric (i.e. corrected for light time) and not geometric as hitherto. In practice, once Δ has been obtained, the correction $0.005776\,\Delta$ is subtracted from the ephemeris date in question and fresh heliocentric coordinates calculated from the revised time. The Solar Coordinates are not affected.

16.10 B.A.A.H. data

Elements and ephemerides of periodic comets whose returns are expected during the year. Ephemerides contain the following data:

Date.
RA $\big\}$ for the mean equinox of the beginning of the year, or for a
Dec $\big\}$ standard equinox, e.g. 1950.0.
r (solar distance, in A.U.s).
Δ (distance from Earth, in A.U.s).
Magnitude.

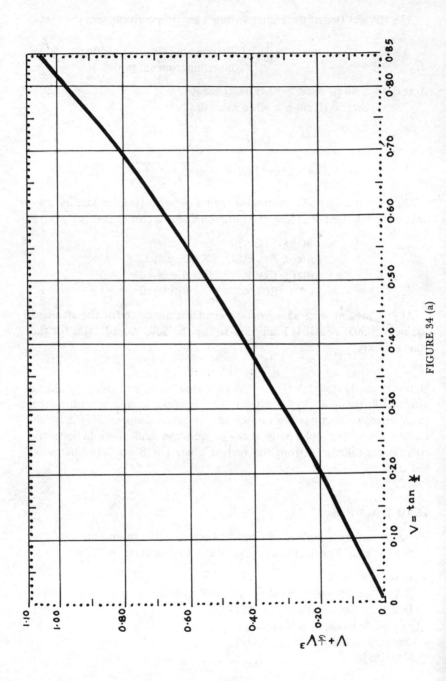

FIGURE 34 (a)

$V = \tan \frac{v}{2}$

$\frac{1}{3}V^3 + V$

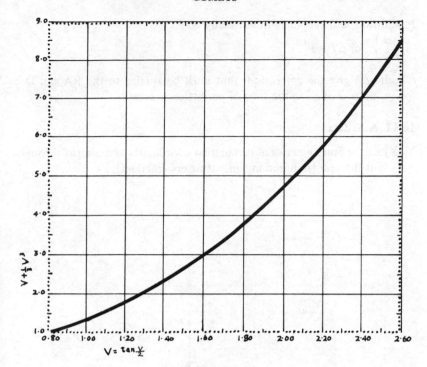

FIGURE 34 (b)

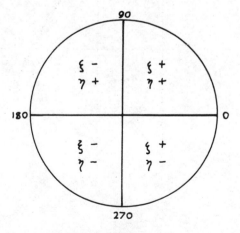

FIGURE 35

In addition, search ephemerides usually give:

$$\left.\begin{array}{c}\Delta\alpha\\ \Delta\delta\end{array}\right\} \text{ for } \Delta T = 1^{\text{d}}.$$

$\Delta\alpha$ and $\Delta\delta$ give the corrections that must be applied to the RA and Dec for an increase of 1^{d} in the time of perihelion passage.

16.11 A.A. data

X, Y, Z: the Sun's equatorial rectangular coordinates at midnight throughout the year (required in computing ephemerides).

SECTION 17

VARIABLES

17.1 Amateur work

The majority of the tens of thousands of variables that have been studied at all have still been very inadequately observed. There are, in addition, the thousands of stars whose suspected variability has never been confirmed, and the equally numerous cases where variation has been established but which have never been classified owing to insufficient observation.

Variable stars provide one of the most noteworthy examples of a field in which the amateur can render valuable assistance to the professional astronomer. The technique of observation is within the accomplishment of anyone, and even the most modest types of amateur equipment can be turned to valuable work.

Instrumental photometry, as distinguished from visual methods, is demanded by a small minority of variables whose insignificant amplitudes render them unobservable visually or photographically, and for the detection of secondary variations in stars whose general characteristics are already well known. An obvious field for photoelectric investigation is the red giant stars, the vastly greater number of which are slightly variable, the amplitude in most cases being below the limit of effective eye-estimation.

Photography (e.g. section 17.11) in amateur hands is of limited though real usefulness in the observation of variables.

Observations—however accurate and however badly needed—are wasted if they never get further than the pages of the Observing Book. The value of pooling observations of certain types of variable*—since it allows smoothed interpolation curves to be constructed from a very much greater number of observations (and with corresponding weight) than any single observer could make—indicates the desirability of joining an observ-

---

* E.G. LPVs, and some rapidly varying stars like R CrB and U Gem.

249

ing group, such as that provided in this country by the Variable Star Section (V.S.S.) of the B.A.A., or in America by the American Association of Variable Star Observers (A.A.V.S.O.). The V.S.S. keeps several hundred variables of different type under observation. Anyone wishing to strike out on his own will find many interesting objects in the *General Catalogue of Variable Stars* (B.25.9) and the accompanying *Catalogue of Suspected Variable Stars*.

The V.S.S. issues general instructions, lists of comparison stars, and the necessary charts—the latter often based on A.A.V.S.O. material, considerably modified to suit B.A.A. methods and procedures. The Director and his helpers also perform such tasks as the construction of the standard light curves, the determination of periods, and the determination of maxima and minima. Observations should be submitted annually; they are discussed in Reports in the *Journal* and in the occasional *Memoirs (Mem. B.A.A.)*.

The frequency with which variables should be observed depends upon their period. The following instructions are intended as no more than rough guides:

Eruptive variables:† Nightly, or (e.g. I.1a and Is) at the beginning and end of each night's work. During periods of rapid change, observations can profitably be made at almost hourly intervals.

Long-period and semiregular variables: Twice a week, or even less frequently, is adequate in most cases.

Eclipsing binaries and other short-period variables: Hourly, nightly, or still less frequently, according to the lapse of time required for the brightness of the star to change appreciably.

It is important to reduce, as much as possible, each variable's annual season of invisibility when it is near to the Sun. Observations in twilight (both evening and morning), though made under difficult circumstances, and therefore of reduced accuracy, are nevertheless particularly valuable. An approach to continuity of observation is especially desirable in the case of eruptive variables.

17.2 Visual estimation of stellar magnitudes

The methods which permit the magnitude of a star to be accurately derived by direct eye-estimations fall into two fundamentally dissimilar

† See section 17.13.

categories: Step methods and Fractional methods. These are described in detail in the subsequent sections, but some preliminary remarks of a general and ground-clearing nature may be helpful.

All branches of observation are to some extent an art as well as a science—in so far as proficiency involves mental processes of which the observer is largely unconscious—and this is truer of the visual estimation of stellar magnitudes than of most. Although there are a favoured few who appear to be able to divine the brightness of a star to one-tenth of a magnitude by some aptitude akin to clairvoyance, such feats are not really inspired shots in the dark, but rather the result of unconsciously applied experience accumulated through years of observation according to the 'rules'. There are no short cuts in astronomical observation: proficiency can only come from experience in the use of those techniques which trial has proved to be the most reliable. Such techniques in the case of variable-star observation have been worked out in some detail: but it is worth emphasising that although the 'rules' may be more numerous than in some other branches of observation, they have not been formulated for the fun of it, but simply because they represent the most efficient and reliable means of attaining the desired goal—accurate results.

It is precisely because the visual estimation of stellar magnitudes is so beset with pitfalls that special techniques have been evolved; and if he is to avoid these pitfalls, the newcomer to variable-star observation cannot afford to disregard the 'rules'. In comparing by eye the brightness of a variable and a comparison star the observer is performing an exercise in point photometry. He must allow the image of each star in turn to fall upon the same spot on his retina, and by means of the techniques described in the following sections make an estimate of the stars' relative brightness. The purpose of these techniques, as was said above, is simply and solely to enable the observer to make the estimate with reasonable accuracy: hit-or-miss methods just will not produce accurate results, and the apparent complexity of the recognised methods of observation is the inescapable result of the necessity for circumventing the various errors (mostly physiological) that afflict visual point photometry.

It is also essential to be quite clear in one's mind regarding the precise nature of the Step and Fractional methods, and the manner in which they differ from one another. In making an observation by a Step method, the elements involved are 2 stars and the light interval (or brightness difference) between them; the variable may afterwards be compared with other comparison stars, but each such comparison is an entirely independent operation, each involves the same three elements—2 stars and 1 interval—and each consists, first, in forming a correct impression of the brightness of the two stars, and then in assessing the light interval between them in terms of a scale of 'steps'.

With the Fractional method the unit observation involves a greater number of elements—3 stars and the 2 light intervals between them (the variable, and a brighter and a fainter comparison star)—and the assessment consists in weighing these two intervals against one another, not against a known scale of 'steps'.

The making of an estimate by the Fractional method is to this extent a longer and more complicated process than making a Step estimate. For this reason the latter is generally to be preferred when comparison stars at small enough light intervals are available. Both the Fractional and the Pogson Step methods are employed by the V.S.S.

For a fuller discussion of these methods, and their relative advantages and weaknesses, see B.17.27.

17.2.1 Pogson's Step method: Like the Herschel-Argelander method (section 17.2.4), this is a true Step method: it can be used with a single comparison star, and in ignorance of any actual magnitudes. The only difference is that whereas the former is based upon a scale of subjective units, Pogson's step is a consciously memorised difference of 0.1 mag. (That the Herschel-Argelander physiological step also happens in most cases to be *about* 0.1 mag is purely fortuitous.)

The Pogson method thus involves a preliminary training of the eye to recognise intervals of 0.1 mag, 0.2 mag, etc, when it sees them. Comparison stars are the most convenient objects to use when memorising the appearance of the Pogson steps, since they are easily found and cover a wide range of accurately determined magnitudes.

To avoid confusion with observations made by other methods, Pogson step observations are recorded as follows: the comparison star is named first, and is followed by + and the number of steps that the variable is brighter than it, or − and the number of steps that the variable is fainter than it. Thus $a-2$ means that the variable is 2 steps (i.e. 0.2 mag)* fainter than star a, $b+4$ that it is 4 steps brighter than star b, etc.

The derivation of the variable's magnitude from the observational data (columns 6 and 7 of the Permanent Record: see section 17.6.2) is performed as follows. At the end of each observational session the mags of the comparison stars used are inserted against the observations in the

---

* Note that in the Argelander method, $a(2)V$ would also mean that the variable is 2 steps fainter than a, but in this case we could not go on to say that the difference was 0.2 mag.

'Light Estimate' column. A typical entry in this column, with the corresponding entry in the 'Deduced Magnitude' column, might then read:

Light Estimate	Deduced Magnitude	
$e-2$ (7.0)	7.2 ⎫	
$a-1$ (6.9)	7.0 ⎬ 7.1	
$b+4$ (7.5)	7.1 ⎭	

It is important to realise that—this being a Step and not a Fractional method—these three observations are absolutely independent of one another. Although the second and third observations quoted above do in fact place the variable between a and b at a point one-fifth of the total a-to-b interval fainter than a, and four-fifths of the interval brighter than b, these two observations were made quite independently of one another in two separate operations, each of which involved 2 stars and 1 interval— and not, as with the Fractional method (section 17.2.4), in a single operation involving 3 stars and 2 intervals.

Also it should be clearly realised that the observations are, at the time they are made, absolutely independent of any known magnitudes. It is, indeed, a thoroughly undesirable practice to go to the telescope armed with the magnitude values of the comparison stars that one intends using.

17.2.2 Pogson's Mixed method: Pogson himself proposed a development of his true Step method, which is a sort of half-way house between the Step and the Fractional methods. It remains a Step method, since the estimation is made in terms of steps, while resembling the Fractional method in that each single observation involves 3 stars and 2 intervals.

In the Mixed method the observer goes to the telescope knowing the mag interval between the comparison star immediately brighter than the variable and that immediately fainter than it. Suppose, for the sake of illustration, that the mags of the two comparison stars, a and b, are respectively 10.0 and 10.5. The observer then knows that the number of steps in his observation comparing the variable with a, and the number of steps in his observation comparing the variable with b, must sum to 5. He goes ahead and makes his two independent step estimations, each concerning 2 stars and the 1 interval that separates them, and if these estimates should be $a-1$, $b+4$ or $a-2$, $b+3$ or $a-3$, $b+2$ or $a-4$, $b+1$, all is well. But if his observation is, let us say, $a-2$, $b+4$ he cannot let it rest at that (as he could with Pogson's pure Step method, when he is ignorant of what the sum of the steps in the two intervals really is) since $4+2 \neq 5$, and his observation must be in error. He therefore has to recompare a and V, and b and V, and decide whether he wants to change his observation to $a-2$, $b+3$ or to $a-1$, $b+4$. But he does this by a reconsideration of the number of $0^{m}1$ steps in the two intervals, and not (as in the Fractional method) by thinking in terms of the proportion of 1 : 4 or 2 : 3.*

It is the standardised procedure in the V.S.S. to record the observations themselves, and to deduce the magnitude of the variable from them later (see section 17.6.2). In the A.A.V.S.O., on the other hand, the result is written straight down in mags, and the 'Light Estimate' column goes out of commission.

17.2.3 The Fractional method: Two comparison stars are chosen, one slightly brighter and one slightly fainter than the variable, thus immediately fixing the magnitude of the latter between definite limits. In constructing the B.A.A. sequences of comparison stars the aim has been to avoid jumps of more than 0.4 mag. Where this has not been practicable, however, the Fractional method is preferable to one of the Step methods, since the assessment of intervals up to 1 mag is a feasible undertaking by its means.

The brightness difference between stars a and b is divided mentally into any convenient number of parts, depending on the size of the interval (3, 4, 5, or 6 for normal intervals; only exceptionally more than 7), and the position of the variable within this interval is estimated in terms of the fraction of the total interval separating it from each star. For example, if the brightness of the variable is reckoned to be one-quarter of the way from a to b, and ¾ (a-to-b) brighter than b, the observation would be recorded in the form $a,1,V,3,b$†; if the variable is estimated to be midway from a to b, the observation would be recorded as $a,1,V,1,b$; if the difference between the comparison stars is great enough for one-fifth of it to be perceptible, and the variable is judged to be three-fifths of the way from a to b, the observational record would read $a,3,V,2,b$.

Interval assessments of from 1 : 1 up to about 1 : 5 (e.g. $a,1,V,1,b$ to $a,1,V,5,b$) can be made with sufficient accuracy to justify placing reasonable confidence in them. But mentally to divide a light interval into 7 parts (as is involved in the observation $a,1,V,6,b$), or more, and to place the variable correctly within this range, is too difficult for much weight to be attached to such an observation.

Pickering's decimal notation, in which the numerators of the two fractions into which the variable divides the whole interval between a and

---

* If the observation were made fractionally, it would be recorded in the form $a,1,V,4,b$ or $a,2,V,3,b$ (see section 17.2.3).
† The brighter star is always written first.

b always sum to 10 (e.g. $a,6,V,4,b$, $a,2,V,8,b$, etc), is open to the double objection that it either encourages the observer to claim a precision which in fact he cannot attain, or is capable of simplification. Thus 5 : 5 is the same as 1 : 1; 6 : 4 simplifies to 3 : 2; 7 : 3 is in practice indistinguishable from 2 : 1; 8 : 2 simplifies to 4 : 1; and a 9 : 1 observation means in practice that the difference between V and b is only *just* distinguishable, and is therefore given the minimum value of 1, the other 9 being in no sense an accurate estimate of the size of the a-to-V interval relative to V-to-b, but merely being dictated by the fact that the sum must be 10.

It can be seen that it is unnecessary to know the mags of the comparison stars at the time the observation is made. The observation is reduced subsequently as follows:

Observation: $a,3,V,2,b$.

Mags of comparison stars: a 5.27,

$\qquad\qquad\qquad\qquad b$ 6.05.

Then mag difference $(b-a) = 0.78$,

$\qquad\qquad \therefore \frac{1}{5}(b-a) = 0.156$.

Hence,

Mag of variable = $6.05 - (2 \times 0.156) = 5.738$

$\qquad$ or $= 5.27 + (3 \times 0.156) = 5.738$ $\Big\}$ = 5.7.0.

Observations should be reduced to the nearest 0.01 mag.

It is impossible to say anything about the relative accuracy of the Fractional and Step methods without specifying the conditions under which they are being used: thus a step comparison when the available comparison stars are spaced at not more than about 0.5 mags is likely to be as accurate as a fractional assessment when the interval is larger than this. It is the present practice of the V.S.S., in systematising the observations of its members, to give double the weight to a fractional observation as compared with a single step observation. The reason for this is that in the former case 2 comparison stars were employed, and in the latter only 1, i.e. unit weight is given to each comparison star. Though this may appear reasonable enough at first glance, it is by no means certain that the practice is always and necessarily valid, for it does not follow that an observation like $a,1,V,4,b$ must be twice as valuable as $a-1$: an experienced observer's estimate of 1 step is so precise that the accuracy of $a-1$ is unlikely to be doubled merely by tying V to a second comparison star b and making a fractional assessment of 1 : 4.

A small point that might be noticed in passing is that, since the magnitude scale is logarithmic, it is the ratios of the brightness of two pairs of stars which should be compared in the Fractional method, rather than the differences in brightness. In practice, however, this does not present any problem, since it is brightness ratios rather than brightness differences that the observer perceives. Thus, if the brightness of V appears exactly midway between a and b (as estimated by the Fractional method), and the mags of a and b are respectively 3.5 and 5.0, then the magnitude of the variable is not 4.01 (corresponding to the arithmetical mean of the brightnesses) but 4.25 (the arithmetical mean of the magnitudes). The estimate $a,1,V,1,b$ implies that $[a] \div [V] = [V] \div [b]$, where $[x]$ signifies the brightness of star x.

17.2.4 The Herschel-Argelander Step method: Commonly called, simply, the Argelander method. Independent comparisons are made of the variable with as many comparison stars in turn as are available. The selected comparison star is brought to the centre of the field, steadily regarded for not longer than several seconds, and an impression taken of its brightness; the same operation is then carried out with the variable; on the basis of these two impressions, two judgments are made—first, a qualitative one as to which star is the brighter, and then a quantitative one as to the size of interval between them, expressed in terms of the scale of steps described below.

The smaller the mag difference between each comparison star and the variable, and the more evenly their brightnesses are distributed on either side of the variable, the better will be the derived magnitude. Unlike the Fractional method, the Argelander and other Step methods thus permit the estimation of the brightness of a variable by comparison with a single field star, though it is desirable that more than one should be used if available.

The mags of the comparison stars do not have to be known at the time of observation—in fact it is psychologically most undesirable that they should be. The relative brightness of the variable and each usable comparison star in turn is estimated as so many 'steps'. The step is, strictly speaking, incapable of general definition, since it is a subjective unit. But broadly speaking it may be described as the smallest difference of brightness between two stars observed in fairly quick succession that is clearly distinguishable. Subjectively, the steps may be described as follows. Suppose that at a single casual glance two stars, a and b, appear to be equally bright. Suppose, further, that a is now observed intently for several seconds, its brightness memorised, star b similarly observed, a judgment made as to which is the brighter (assuming that, as will often be the case, they no longer appear to be *exactly* equal), and the whole procedure then repeated several times. Then if a was judged to be brighter than b as often as b was judged to be brighter than a, they may be taken to be as nearly equal as the eye is capable of distinguishing, and their step difference will be 0. If one star is only occasionally judged to be brighter than the other, then their difference is 1 step; if one is never brighter than the other, but occasionally equal to it, the difference is 2 steps; if one is always brighter than the other, but at times only just so, their difference is 3 steps; and if one is continuously and clearly brighter than the other, 4 steps or more. Thus the final opinion regarding the step-difference between two stars is made up by the accumulation of individual estimates. It can be seen that the Herschel-Argelander method is precisely that one which attempts to utilise the observer's eye to its most critical limit.

For the average observer with a trained eye, the value of the step is not far from 0.1 mag; and although its value varies slightly from one observer to another, it appears to remain tolerably constant for a given observer over long periods. With practice, differences of up to 3 or 4 steps can be estimated quite accurately and consistently, but if the magnitude difference exceeds about 0.4 or 0.5 it is better to use the Fractional method.

When recording observations, the brighter of the two stars is given first and the fainter last, with the step-difference in brackets between them. Thus $b(1)V$ means that the star designated b is 1 step brighter than the variable, $V(2)a$ that the variable is 2 steps brighter than star a, etc.

The symbols $>$ and $<$ should not be used for 'brighter than' and 'fainter than', the correct symbols being respectively] and [.* Thus [37 would mean that the variable was fainter than the comparison star 37, the faintest visible on that occasion (i.e. the variable was below the threshold of the telescope); this might be recorded more specifically as [37 N.S. (variable fainter than star 37, and not seen); or amplified to [37 gl. (variable fainter than star 37, but glimpsed).

On every occasion when a variable is invisible (no matter what method is being used), the faintest visible comparison star should be specified in this way.

Since the value in terms of magnitude of the observer's step will not, in general, be known, the Herschel-Argelander method more closely resembles the Fractional than the Pogson Step method in its reduction of step estimates to magnitude. If only two comparison stars have been used, the magnitude of the variable can be obtained by interpolation exactly as described above (section 17.2.3). If several comparisons are made, however, the comparisons with stars brighter and fainter than the variable must first be averaged. The following example illustrates the procedure.

Suppose that four comparison stars are used, their magnitudes being as

* Round brackets,) and (, are used in typewritten work, and sometimes in printed work also.

follows: a, 6.9; b, 7.5; c, 7.3; and d, 7.0. The estimates would then be treated as in the fictitious observation below.

Light estimate	Average
$d(2)V$	
$a(1)V$	$\}\ 6.95(1.5)V$
$V(4)b$	
$V(4)c$	$\}\ V(4)7.4$

Hence, step value $= \dfrac{7.4-6.95}{1.5+4} = 0.082$

Mag of variable $= 6.95 + (1.5 \times 0.082) = 7.073$
$\left.\begin{array}{l}\\\\\end{array}\right\} = 7.1$
or $= 7.4 - (4 \times 0.082) = 7.072$

Argelander's method is often used when comparison star magnitudes are unknown, the purpose of the observations being to obtain a general idea of the shape of the light curve or the time of maximum or minimum (particularly in the case of eclipsing binary stars). In such cases, a scale of grades may be derived from the estimates by the procedure illustrated in the following example.

GMAT	Light estimate			Deduced Grade
08.01	$a(3)V,$	$V(1)b$		3.4
08.32	$a(5)V,$	$V(0)b,$	$V(2)c$	4.7
08.58	$c(1)V,$	$V(4)d$		7.1
09.24	$c(4)V,$	$V(0)d,$	$V(3)e$	10.1
09.50	$c(3)V,$	$V(0)d,$	$V(3)e$	9.8
10.15	$b(3)V,$	$V(0)c,$	$V(3)d$	7.0
10.39	$a(5)V,$	$V(1)b$		4.4
11.00	$a(2)V,$	$V(2)b$		2.4

The intervals in steps implied by the estimates are as follows:

ab	4, 5, 6, 4	mean 4.75
bc	2, 3	2.5
cd	5, 4, 3, 3	3.75
de	3, 3	3.0

Assuming a grade of 0.0 for a, the grade values for b, c, d and e are then 4.75, 6.25, 10 and 13 respectively. Since it is known that one step is equal to an interval of one grade, the observations can be reduced by a calculation analogous to that described by Pogson's Step method (17.2.1), giving the results shown above under *Deduced grade*. The procedure for deriving the time of minimum from the deduced grades is described in section 17.16.

258

When taking up this type of work, the eye should be subjected to a thorough course of training in the estimation of relative brightnesses. This is most conveniently undertaken by making long series of observations among the comparison stars on the V.S.S. charts, their magnitudes being referred to later and the accuracy of the observations checked.

17.3 Making the comparisons

In point photometry the most important condition for securing consistency of results is that the images being compared should be received by the same spot on the observer's retina. In practice this must mean the fovea, whence each star must be brought in turn to the centre of the telescopic field, and there regarded (for not more than 2 or 3 seconds) by direct vision. The fact that a *simultaneous* comparison is not possible – the impression of the brightness of one star having to be carried in the memory while the next star is being brough to the field centre and observed – does not in practice appear to be a source of material error.

Simultaneous observation of two stars by averted vision, the eye being fixed on a point midway between them, results in large and unsystematic errors due to the enormous range of sensitivity of different regions of the retina: for whereas when the two stars are looked at directly (though necessarily in succession) the image of each is known to fall on the same retinal spot – the fovea – in the case of indirect vision there is no such guarantee. The indirect method is also subject to position-angle error* and to error arising from the fact that a star tends to appear slightly brighter at the edge of the field than at the centre. The danger to be constantly on one's guard against when using foveal vision is the unconscious use of averted vision – by allowing the fixation point to wander from the star whose brightness is being impressed upon the visual memory, or by allowing the extrafoveal image of the first star, as the telescope is being shifted to bring the second to the field centre, to influence the original impression of its brightness gained whilst it itself was at the centre of the field.

However, foveal vision is not universally adopted by variable-star observers, because the change from rod to cone vision, which occurs gradually with increasing brightness over the range in which variable-star estimates are usually made, gives rise to inconsistencies such as the Purkinje effect. A solution to this difficulty is to use averted vision at all times (making every effort to employ the same point on the retina), and

---

* *A.A.H.*, section 24.9.

to throw the field out of focus whenever the stars are sufficiently bright for colour to be perceptible, thereby dimming the intensity and allowing rod vision to be used consistently. A possible disadvantage of this method is that red stars now appear appreciably fainter than as measured in the photoelectric V system, which corresponds approximately to cone vision. Pure cone vision is not, however, realised in the observation of stars. It is important for the observer to decide on one method as quickly as possible, and thereafter to use it consistently.

Averted vision is certainly justified when no observation at all would be possible without it—i.e. when either the variable or the comparison star is so near the magnitude threshold of the instrument that even with magnification pushed to the limit (to increase the star's visibility by darkening the sky background) it is still inaccessible to foveal vision. Every effort must then be made to fix the eye on the same point in the field when each star is at the field centre.

17.4 Comparison stars

Charts are specially constructed for variable-star observers, and issued by such bodies as the A.A.V.S.O., and the V.S.S. of the British Astronomical Association. Each chart shows the area of sky containing the variable in question; those field stars suitable for use as comparison stars are indicated by letters, numbers (referring to a separate table of magnitudes), or according to Pickering's system whereby they are denoted on the chart by their magnitudes (to tenths), decimal points being omitted (e.g. 12 = mag 1.2, 101 = mag 10.1, etc).

The magnitudes of the standard comparison stars have been very carefully determined; but errors do occur, many having been corrected in the past, and among the thousands of comparison stars in use there are still wrongly ascribed magnitudes. The detection of these errors lies with the variable-star observers themselves, and any anomalous observations, suggestive of wrongly ascribed comparison-star magnitudes, should be reported to the V.S.S. Director.

V.S.S. magnitudes are based on the scale of the Harvard photometries, or are photoelectric V measures, though in most cases 'adopted magnitudes'—derived from both photometric measures and step estimates—are used. These are often the mags of the A.A.V.S.O. charts.

Each variable should ideally have a set of comparison stars whose mags are accurately known and which are distributed in not more than half-magnitude steps over the whole of the variable's amplitude plus 0.5 mag at each end. Thus at any time two comparison stars could be found whose mags are within 0.5 above and below that of the variable. They should be included in the same telescopic field which shows the variable clearly, but should be neither too close to one another nor to any much brighter star; double stars are unsuitable. Finally, they should be of the same colour as the variable.

Needless to say, all these conditions cannot in most cases be fulfilled.

17.5 The observation of coloured variables*

The kernel of the difficulty has been put succinctly by P. M. Ryves: 'In comparing a red star with a white star we are trying to equate things which are essentially different; rather like trying to decide whether one cup of tea is as strong as another is hot.'

There are several physiological bases, among them:

1. The effect of colour upon the perceived brightness of two equally bright stars.

2. The eye's non-instantaneous adaptation to red light. Thus, as a red star is stared at, it appears to grow brighter over a period of a few seconds. This is less troublesome than 1, and can be avoided by substituting quick glances for a steady scrutiny.

3. The Purkinje effect. If R and W are the intensities of a red and a white star respectively, and it is estimated that $R = W$, then $2R$ will appear to be greater than $2W$. Examples: if $R = W$ when the stars are viewed with the naked eye, R will appear up to 0.5 mag brighter than W when seen with a telescope of moderate aperture, owing to the increased intensities of the telescopic images; again, U Cyg at maximum appears, with moderate apertures, virtually equal to a bright nearby comparison star, but if the aperture is reduced till the variable is almost invisible it appears about 0.5 mag fainter than the comparison star; again, T Cas observed with a 25-cm is estimated to be about 0.5 mag brighter than with a 75-mm.

The varying effects of colour can be clearly seen when combining the observational data of different observers: whereas in the case of red stars the greatest scatter is found around maxima, in the case of blue stars it increases with decreasing brightness.

The error can be circumvented in various ways, of unequal practicality:

(a) Only compare a red variable with a red comparison star. This is the radical and ideal solution, but infrequently possible.

(b) Use a red filter to reduce the difference between the tints of the two stars. The disadvantages here are (i) the mags of all the standard

---

* This section should be read in conjunction with *A.A.H.*, section 24.4.

comparison stars would have to be redetermined, (*ii*) identical filters would have to be used by all observers whose observations are to be combined for discussion, (*iii*) the visible range of each variable would be reduced, (*iv*) a discontinuity with previous estimates would be introduced.

(*c*) Reduce the aperture by an amount (depending on the brightness of the variable at the time) that brings its intensity down to its minimum as seen with full aperture, or to a faint but clearly perceptible intensity, whichever is the greater. The justification for this procedure is that whereas the eye's sensitivity to colour decreases with decreasing intensity, it becomes more sensitive to intensity-differences at a low intensity level (providing this is not too near the threshold). In this way the Purkinje effect is harnessed to reduce the perceived colour difference between the variable and the comparison star. It should be noted (*i*) that the presence of haze or high cloud will affect the estimated brightness of a red star, (*ii*) that owing to the darker field of a HP ocular, a red star (compared with a white one) will appear brighter with a high magnification than with a low.

(*d*) Use indirect vision, the periphery of the retina being less sensitive to colour than the macula. A single series of observations must not include both direct-vision and indirect-vision items. This is a less effective method than (*c*); and see also section 17.3.

Two scales are in common use for the description of star colours, Hagen's (sometimes called Osthoff's) and Franks':

Hagen		Franks
-3^c	blue	B
-2^c	pale blue	WB
-1^c	bluish white	BW
0^c	white	W
1^c	yellowish white	YW
2^c	pale yellow	WY
3^c	pure yellow	Y
4^c	deep yellow	OY
5^c	pale orange	YO
6^c	orange	O
7^c	deep orange	RO
8^c	reddish orange	OR
9^c	red, with tinge of orange	R
10^c	spectrum red (never encountered	SR

17.6 Observational records

As usual, it is necessary to keep two separate records: an Observing Book for the direct recording of observations as they are made, and a Permanent Record made up from the Observing Book.

17.6.1 Observing Book: A convenient form of layout is as follows:

Date................................

(1)	(2)	(3)		(4)	(5)	(6)	(7)
	Time	*Sky State*					
Star	*GMAT*	*Gen-*	*Local*	*Instrument*	*Observations*	*Class*	*Notes*
	(0ʰ = noon)	*eral*					

Date: Day of week, GMAT date, and Julian date (see section 17.14), e.g. Sunday, 1950 Nov 12, JD 2433598.

Col. (1): The generally used name; in most cases this will be Argelander's designation (see section 17.12).

Col. (2): GMAT. Either to the nearest minute in all cases, or in the case of SPVs and Irregular Variables only; or to the nearest 5 mins, according to the variable's period. Even with LPVs it is useful to record the approximate time, since this gives an indication of the star's altitude in doubtful cases.

Col. (3): Clarity, rather than freedom from turbulence, is the important factor. The General State refers to the whole sky, and need be entered only at the beginning and end of each observing session. Local State refers to the immediate vicinity of the variable, and should be entered against each observation. A convenient scale is:

    1: good,
    2: fair, to fairly good,
    3: poor.

Other useful abbreviations are:

    C: clouds,
    H: haze, mist, or uniform high cloud,
    W: windy.

If the General State is apparently uniform, the mag of the faintest visible star near the zenith may be quoted. The following stars of UMi are also useful in this connexion:

    α . . . . mag 2.0
    γ 3.0
    δ 4.3
    θ 5.0
    19 . . . . 5.5

Col. (4): Abbreviations recognised by the B.A.A. include:

    NE: naked eye,
    B : binoculars,

T : telescope, followed by the magnification, e.g. T 30. If two telescopes are regularly employed, T is used for the larger and t for the smaller.

This scheme can be elaborated. For instance:

F : finder

RR: refractor ⎱ plus aperture and magnification, e.g. RL 15/40 indicates
RL: reflector ⎰ a 15 cm reflector, ×40.

Col. (5): See sections 16.2.1–17.2.4.

Col. (6): The degree of reliability of the observation, as judged by the observer at the time he makes it:

1: reliable, complete confidence,

2: fair,

3: unreliable.

It is important that this should be entered at the time of the observation, taking account of all relevant factors, and that it should under no circumstances be altered later.

Col. (7): Any notes on conditions in relation to which the observations should be considered, e.g. trouble from artificial lights or aurorae, observation hurried or otherwise suspect, etc. Suggested abbreviations:

M: Moon above horizon. If it is near the variable, the fact should be mentioned.

T: Twilight.

Z 70: Zenith Distance of variable $70°$; the ZD should be recorded if greater than $65°$.

L: Star low. (Used as an alternative to ZD, sometimes.)

17.6.2 Permanent Record: Made up daily from the Observing Book. Cards are more convenient than a book, unless loose-leaf. Separate card or page for each variable. It is assumed that the Permanent Record is kept in duplicate, one copy being sent annually to a central pool for the correlation and discussion of different observers' work. This copy will include a covering note giving the observer's locality, longitude and latitude, and any other general information that he may think relevant. It is most important, however, that *each individual sheet* sent to the B.A.A. or elsewhere should contain the observer's name, a statement of the apertures of T and t (see notes on column (4) of the Observing Book, above), and whether they are refractors or reflectors.

The form recommended by the V.S.S., and in use by all its members, is shown on the opposite page.

The currently high cost of publication means that observers working on their own (i.e., not as members of a group, such as the B.A.A. Variable Star Section or the A.A.V.S.O., with their recognised methods and procedures) will not normally be able, when publishing their results, to quote the observations in full as made and recorded in the Observing Book; but they should indicate clearly the method used—Argelander Step (e.g. $a(1)V$, $V(2)b$), Pogson's Step $(a-1, b+2)$, Fractional $(a,1,V,2,b)$, etc.— and should place a copy of the observations in a suitable archive for future reference.

VARIABLE STAR SECTION ESTIMATE REPORT FORM

VARIABLES

MONTH	DAY	TIME	METHOD	ESTIMATE	DED. MAG.	CLASS	INSTR.	COMMENTS	OFFICE USE ONLY
JAN	03	12·05	FM	(1)V(2)N	13·0	1	1		
	04	10·26	P	(2)V(1)R	13·6	1			
	05	10·40	PR	-2;5+1	13·9	1	N		
	06	10·37	R	-3;=5,7+3	14·0	1	N		
	10	11·12	R	(2)V(1)S	13·9	2	C		
	12	11·34	R	(1)v;v;v(2)s	13·8:3		MH		
	12	11·50	<S		<14·0		I		

Please do not write in boxes labelled 'Office Use Only'

STAR (1 2 3 4 5 6 7 8 9): S U

CONSTELLATION (10 11 12): U M A

SEQUENCE No (13 14 15 16 17 18): 0 1 8 · 0 1

YEAR (19 20 21 22): 1 9 8 5

OBSERVER'S NAME

OFFICE USE ONLY
OBSERVER ABBREVIATION (23 24 25):

See Instruction Sheet for Standard Abbreviations and Method of Completion

17.7 Observational errors

Two equally experienced observers, observing the same star under identical conditions, may differ systematically in their estimates by as much as a fifth of a magnitude, even though each series is homogeneous. The discrepancy may grow to a whole magnitude, or even more, when the star is very bright (e.g. nova at maximum).

Van der Bilt (B.17.2) found that accidental errors tend to cancel one another out in a large mass of observational material; in particular, no systematic correlations of error with type of instrument or aperture were possible. Readers interested in the problem of discussing a large quantity of heterogeneous material are recommended to refer to this paper.

The sources of accidental error are manifold, and include:

(*a*) Wrong identification of the variable, or of one or more of the comparison stars. Because of this danger, and also because the locating of his variables often wastes a great deal of time with the newcomer to variable-star work, a separate section (17.8) is devoted to this subject. The importance of correctly identifying the variable, and the comparison stars used, need hardly be stressed.

(*b*) Erroneous mags of the comparison stars. If not working from B.A.A. or A.A.V.S.O. charts (see section 17.4), B.23.10, 23.19, 25.6 are convenient sources, although if more recent photoelectric measures (e.g., B.23.2) are available, they should be used.

(*c*) The use of different comparison stars by different observers.

(*d*) Variations in atmospheric transparency, background brightness, etc.

(*e*) Different colour sensitivity of different instruments, leading to different estimates of the star's colour, and thence to divergent estimates of its brightness.

(*f*) The employment of different instruments or photometric methods in a single series of observations.

(*g*) Unsystematic use of direct and averted vision. Any variable whose observation requires, around minimum, the use of averted vision should be observed in this manner at all times, an effort being made to systematise the distance and p.a. of the fixation point relative to the observed point. If, however, the variable should be perceptible by direct vision, estimates should be made by both direct and averted vision, in order to secure a check 'overlap'. See section 17.3.

(*h*) Unsystematic dark adaptation. Always wait until adaptation is complete, or virtually so, before starting the night's work (not less than 15 mins). Any light at the observing site should be heavily red-screened;

screen out any external light shining on to the observing site (street lamps, etc).

(*i*) Decreased accuracy when the variable is very bright.

(*j*) Decreasing accuracy as the variable fades through about 1 mag above the instrument's threshold.

(*k*) Giving the observation an emotional bias by looking up, or thinking about, the star's expected magnitude beforehand; the telescope should be approached with a mind empty of preconceptions.*

(*l*) Position-angle error.** That the apparent brightness of a star depends upon its position in the field, and that the perceived relative brightness of two stars depends upon the inclination of the line joining them to that joining the observer's eyes, is well established. But in detail the evidence of different observers is conflicting. Most commonly it appears that when the line-of-stars and the line-of-eyes are parallel, equally bright stars appear equally bright; when the two lines are perpendicular,† maximum inequality in the apparent brightness of the two stars occurs. The rotation of the field through 90° may cause discrepancies of the order of 0.5 to more than 1.0 mag, the lower of the two stars usually, though not invariably, appearing the brighter. The quantity of the error appears to be independent of the brightness of the stars.

Even though the variable-star observer is, in principle, making successive observations of two stars, each at the centre of the field, the line-of-stars does not for that reason effectively cease to exist, nor the rule 'keep the line-of-eyes parallel to the line-of-stars' lose its validity. For star *b* does not instantaneously and miraculously replace star *a* at the centre of the field: in fact the telescope has to be moved so that a path can be traced out from *a* to *b* with the help of the known star patterns of the field, *b* being the point of aim but the observer also being unconsciously aware of *a* as the point of departure. One cannot in practice expunge from memory and awareness everything that happens between the moment that *a* ceases to be observed at the field centre and the moment when *b* occupies that position.

(*m*) Purkinje effect and colour-brightness equation.‡ The former begins to obtrude when the intensity has fallen to a level at which rod-vision is being used more than cone-vision. Probably because of different thresholds

---

* Comment of the Director of the V.S.S. [then W. M. Lindley – ed.], when he read this in manuscript: 'I would like you to print note (*k*) in block letters!'
** See also section 17.3, and *A.A.H.*, section 24.9.
† In one case, maximum divergence from equality occurred systematically when the line-of-stars was inclined at 30° to the vertical, the line-of-eyes being horizontal.
‡ See also section 17.5, and *A.A.H.*, section 24.

at which this change-over occurs, two observers of a red variable over an amplitude of 8 mags found a relative personal equation which varied with magnitude, being 0.0 at both ends of the range but 0.7 in the middle.

(*n*) Scale error.*

17.8 Locating variables

A great deal of time can be wasted locating a variable that has just been added to one's observing list unless a rational procedure is adopted. After it has been observed a number of times, of course, the observer's familiarity with the field and immediate surroundings, as well as with the particular course that has been adopted for locating the field, will reduce the time spent on finding to a matter of seconds in most cases, and at worst to two or three minutes.

Once this familiarity with the stars on the observing list has been gained, it is as quick to pick them up by the visual method described below as by using circles; and it has the additional advantage of gradually increasing one's familiarity with the whole face of the night sky.

Whether using circles or the visual method, what the observer is trying to locate is not the variable but the field. Hence the finding operation is not affected by the possible faintness or even the invisibility of the variable itself. For the same reason it is helpful to study the field chart beforehand and memorise the configuration of its brighter stars, so that it will be immediately recognised when it enters the field of the telescope or finder.

The visual method consists in effect of nothing more than the intelligent use of a star atlas, starting with the whole visible hemisphere of sky and narrowing down the position of the variable to within smaller and smaller areas. Suppose the variable sought is X Peg. Its position, 21^h $18^m.6$, + $14° 14'$, is plotted on *Norton's Star Atlas* (B.20.14). Then starting from the nearest well-known star group, which in this case may be the Great Square of Pegasus, work nearer to the variable: the southern side of the Square will lead to the rapid identification of ϵ Peg, which is only $8°$ Sf the variable. Using ϵ Peg as a springboard for further operations, note from the map that the variable lies midway between it and γ Del, and about $3°$ Nf the centre of the line joining them. At this stage it should be possible to direct the finder at the field with sufficient accuracy for it to be picked up with little or no sweeping; a preliminary examination of the vicinity with binoculars will prove a great help. The field having been located, identify all the visible comparison stars from the X Peg chart (the

* *A.A.H.*, section 24.3.14.

variable itself may or may not be visible), making yourself so familiar with the pattern of stars that the field's identification will be even quicker next time. The rule to remember when locating V.S. fields by this method is never to go on to the next stage before being quite sure of the last—it is a waste of time, for example, trying to narrow down the position of the X Peg field to within less than 8° if the star you are working from, in the belief that it is ε Peg, is in fact β Aqr.

17.9 Equipment

(a) Type of telescope:

Refractors and reflectors are equally suitable, the secondary spectrum of the former merely resulting in a systematic over-estimation of the brightness of red stars, when compared with white, amounting to about 0.1 mag or less. In the case of Newtonians it is essential that the flat should be uniformly coated, and also large enough to intercept the cone from the whole of the field.

(b) Aperture:

Binoculars and the smallest telescopes can do useful work in the field of variable stars; optical quality much inferior to that demanded by, for example, planetary observation will be tolerated. Moderately large apertures are required to reach many variables at minimum, and the owners of such instruments should concentrate on these, since they are in many cases beyond the reach of smaller apertures. The disadvantages of large apertures are (*i*) excessive brightness of the images; this can be overcome by stopping; (*ii*) small fields; (*iii*) weight and difficulty of handling make readjustments of position between the variable and the comparison star rather slow, with consequent deterioration of the accuracy of the estimates.

(c) Mounting:

An equatorial fitted with circles is a convenience, since it saves some time in the location of unfamiliar or difficult fields, but it is in no sense a necessity.

(d) Oculars:

Wide, flat fields with good definitions to the edge are the ideal, since they offer the greatest choice of comparison stars with minimum movement of the telescope. For the same reason, magnifications as low as possible are used. A bar in the field of the HP ocular used for variable-star work is useful for eliminating bright field stars which would otherwise render the comparisons difficult and the results unreliable.

(*e*) *Accessory instruments:*

Brightness estimations being most accurate when made in the range of roughly 2 to 4 mags above the instrument's threshold, it follows that a diversity of available aperture is required. The brighter variables can be wholly observed with the naked eye; those lying within the approximate magnitude range 4 to 7, wholly with binoculars; and those fainter than mag 7 at maximum, telescopically. In general, estimates made with binoculars tend to be more accurate than those made with the naked eye, and telescopic estimates than either. Variables of large amplitude cannot always be wholly observed with the same equipment, being too faint at minimum for accurate observation with binoculars or the naked eye, whilst at maximum the telescopic field is too restricted for any suitably bright comparison stars to be anywhere near it. The undesirable switch from monocular to binocular vision can be eliminated over a certain range by means of a predictor telescope or a small hand telescope of the coast-guard type.* Nevertheless, any attempt to combine observations made with different apertures inevitably introduces errors into a series of observations, since they can only be made consonant by reducing them to the values that would have been derived by the use of a standard aperture; and the nature of the necessary corrections is not known with any accuracy.

(*f*) *Photometers:*

See *A.A.H.*, section 21; especially the discussion of energy-conversion photometers, section 21.6.

17.10 Novae

17.10.1 Scope for work:

(*a*) Discovery (see section 17.10.2).

(*b*) Keeping past novae under survey, especially the recurrent novae which may brighten without warning at any time. This work requires moderate aperture, however.

(*c*) Photometric and spectroscopic observations, especially valuable being those made in the stages immediately following the discovery announcement. The early stages in the development of novae are still badly under-observed, since they are so often past before the nova is even discovered. (See also section 17.10.3.)

17.10.2 Searching: Naked-eye discoveries of bright novae have, in the past, often been made by regular meteor observers, giving a clue to one at

---

* See *A.A.H.*, section 22.2, for further details.

least of the desiderata for success in this direction: an intimate knowledge of the constellations, so that the appearance of even a 4-mag star in an unusual position catches the eye. Figure 36 shows the magnitude distribution of 162 novae discovered up to 1979, for which data are available, 26 of which were above mag 5.0 at maximum.

Given sufficient familiarity with the galactic region, it is worth while making a quick survey with binoculars at the commencement of each night's observing period. Even a naked-eye nightly survey of the zones bounding the Milky Way is likely to bring ultimate success, and is undertaken in a matter of minutes once the stars down to mag 4 or 5 are known.

There is a strong case to be made for cooperative searching: each observer is then responsible for a smaller section of the sky than when on his own, can learn it more thoroughly, and can get through his routine nightly survey more quickly. He should gradually learn the positions of all the stars in his zone, making mag 4 his first objective and the naked-eye limit his ultimate one. Allotted zones should be based on the galactic equator as median line. The subjoined Table of galactic distributions refers to 119 novae brighter than mag 10.0 at maximum, for which data are available:

Number of novae	Galactic latitude	Number of novae	Galactic longitude
2	$-39°$ – $-30°$	43	$0°$ – $39°$
2	-29 – -20	18	40 – 79
9	-19 – -10	9	80 – 119
71	-9 – 0	4	120 – 159
21	1 – 10	5	160 – 199
7	11 – 20	3	200 – 239
2	21 – 30	7	240 – 279
3	31 – 40	6	280 – 319
2	$\geqslant 41$	22	320 – 359

As regards galactic longitude, it should be noted that the preponderance of novae occurs in the quadrant containing the galactic centre $(0°)$.

Searches for novae, their early spectra being so distinctive, could be carried out with an objective prism. Whipple's Nova Monocerotis 1939, and Anderson's novae of 1891 and 1900, for example, were spotted by their spectra.

For photographic searching, see section 17.11.

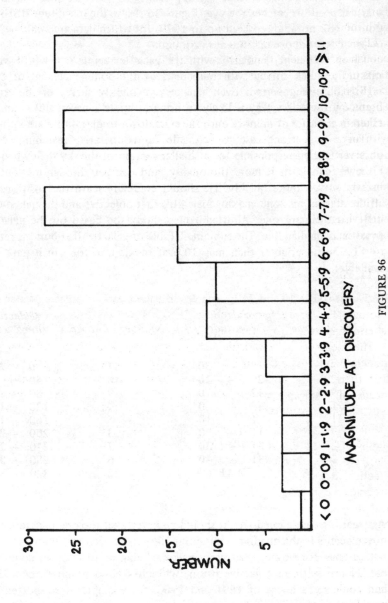

FIGURE 36

17.10.3 Spectroscopy: The complex spectroscopic changes concurrent with the brightness variations of novae are described in the textbooks. Early observations, in particular, are likely to be of value even when made with modest equipment. An ordinary slitless star spectroscope, used with a LP ocular and an aperture of about 10 cm, is capable of showing bright enough spectra down to about mag 4. No possible spectroscopic observations of novae, at any stage, should be neglected, however.

17.10.4 Supernovae: Only three supernovae have definitely been observed in the Galaxy in the past millennium: (a) what is now the Crab nebula, (b) N. Cassiopeiae 1572, which has not been certainly identified with any object now visible, (c) Kepler's nova of 1604, recently identified as a small fan-shaped nebulosity with an emission spectrum similar to that of the Crab nebula. However, an appreciable number of the supernovae observed professionally in the nearer external galaxies have been sufficiently bright for amateur detection (say, mag 11 and above), and both visual and photographic patrols of likely galaxies are currently in operation.

17.11 Photography

Sections 14, 16, 20 and 21 in *A.A.H.* deal with various aspects of stellar photography. In particular, section 21.5 discusses different ways of determining stellar magnitude from a photograph.

The fact that new variable stars of almost naked-eye brightness and moderate amplitude (up to 0.5 mag, say) are still being discovered, means that there is still some scope for the amateur use of photography as a means of discovery within the magnitude range that can be covered satisfactorily with simple equipment. However, the physical labour of comparing such photographs for new variables, and eliminating those candidates already in the catalogues, would make such a project a daunting one. Therefore, the use of photography as a weapon of discovery is, in practice, confined to the search for novae. Cooperative nova-hunting programmes have been under way for some time, in both the British Isles and the U.S.A., the object being to photograph given regions of the sky on a night-to-night basis, checking the negatives immediately for any sign of a suspicious object. The drawback to such work is the ever-increasing cost of the photographic material; the advantage, over visual sweeping, is that a large area of sky can be photographed in considerably greater detail than could be covered by an observer, and furthermore a record of the variable stars lying in the field of view is being accumulated for subsequent analysis.

In practice, photography of known variable stars is most usefully applied to the short-period types, where a run of exposures on a single night can lead to the derivation of the time of maximum or minimum brightness. It has the advantage over visual monitoring that observer bias, often a serious source of error, can be eliminated by measuring the series of negatives in random order. Another useful duty that the camera can perform is the establishment of magnitude sequences for new variables, using the techniques described in, e.g., B.17.6 and 17.27. B.17.25 describes the use of multiple-exposure photography on a single negative to monitor the behaviour of flare stars.

17.12 Nomenclature

(*a*) Standard system, originated by Argelander, adopted by the I.A.U., and now universally employed:

(*i*) If the variable had a previous name, this is retained.

(*ii*) In each constellation, 334 variables are designated by a capital Roman letter, or combination of two letters, followed by the constellation abbreviation,* and thereafter by the letter V and a number, this series of numbers opening at 335:

R, S, T	Z	9 stars	
RR, RS, RT	RZ	9 stars	
SS, ST, SU	SZ ⎱ ZZ ⎰	36 stars	334 stars
AA, AB, AC	AZ (omitting J)	25 stars	
BB, BC, BD	BZ ⎱ QZ ⎰ (omitting J)	255 stars	
V 335, V 336	(see paragraph (*d*) below).		

(*b*) Harvard College Observatory positional designations:
6 figures, the 1st two being the hrs of the star's RA (1900),

```
        „   2nd      „     mins     „    RA,
        „   last     „     degrees  „    Dec.
```

N and S Dec are distinguished by italicising or underlining the last two figures, or by preceding them with a minus sign, if the variable is in S Dec.

(*c*) Chandler's number: A 4-figure group, representing the star's RA (1900), expressed in secs and divided by 10. These are not used nowadays, but are to be found in some of the older catalogues and papers, as well as in Hagen's charts.

* If given in full, the genitive of the constellation name is used.

(*d*) Chambers-André-Nijland nomenclature: Variables in each constellation numbered according to order of discovery, the number being prefixed by V; it is this sytem which is now used from V 335 onward, in each constellation.

(*e*) Various provisional and 'private' nomenclatures, notably the series of numbers, each prefixed by the letters HV, allotted by the Harvard College Observatory to its own discoveries.*

17.13 Classification

The following list of variable-star classes and abbreviations has been abstracted from Information Sheet No. 3 of the B.A.A. Variable Star Section. It is based upon B.25.9 and later Supplements, and corresponds to the system adopted by the I.A.U.

1. PULSATING VARIABLES

C (or Cep): Cepheid type. High-luminosity stars, with period ranging from 1^d to 70^d, amplitude $0^m.1-2^m$. The period and form of the light-curve are generally constant. Spectral class at maximum light is F, with G–K at minimum, being later in class the longer the period.

Cδ: Classical long-period Cepheid variables belonging to the Galactic disc (Population I), obeying the well-known period-luminosity relationship. (δ Cep)

CW: Cepheids belonging to the spherical component of the Galaxy (Population II), otherwise known as W Virginis stars. Period-luminosity curve is similar to that of the Cδ stars, but at the same period the CW stars are $1^m.5-2^m$ fainter. Periods of 3^d to 10^d are practically absent. (W Vir)

L: Slow, irregular variables with periodicity absent or only intermittently traceable. Many insufficiently observed stars are erroneously assigned to this group.

Lb: Slowly varying stars (generally giants) of late spectral class. Red irregular stars are also assigned to class Lb when their spectral class and luminosity are unknown. (U Del)

Lc: Irregularly variable supergiants of late spectral class. (RW Cep)

---

* Similarly HN for nebulae discovered at Harvard.

M: Long-period giant type, of amplitude greater than $2^m.5$, having well-expressed periodicity. Period ranges from about 80^d to 1000^d. Usually of late spectral class. (o Cet)

SR: Semiregular type, consisting of giants and supergiants having an appreciable but often disturbed periodicity. Periods range from about 30^d to 1000^d or more; amplitude rarely exceeds 1^m-2^m.

SRa: Giants of late spectral class with fairly stable period, but the light-curve is of variable amplitude and form. (V Boo)

SRb: As above, but with poorly expressed periodicity due to temporary cessation of variation, or to change of cycle length. (AF Cyg)

SRc: Supergiants of late spectral class. (μ Cep)

SRd: Giants and supergiants of earlier spectral class (F, G and K). (RU Cep)

RR: RR Lyrae type (otherwise known as cluster variables or short-period Cepheids). Periods range from $0^d.05$ to $1^d.2$. As a general rule, the spectral class is A (more rarely F), and the amplitude does not exceed 1^m-2^m.

RRab: Stars having a sharply asymmetric light-curve, with a steep ascending branch. Period exceeds $0^d.21$. (RR Lyr)

RRc: Stars having an almost symmetrical light-curve, often sinusoidal. The mean period is $0^d.3$. (SX UMa)

RRs: Stars of luminosity 2^m-3^m fainter than the above type, belonging to the disc population and absent from clusters. Otherwise referred to as dwarf Cepheids. (CY Aqr)

RV: RV Tauri type variables. Supergiants of comparatively stable period, with total amplitude up to 3^m, characterised in the light-curve by a double wave consisting of alternate main and secondary minima of variable depth. The main minima are often replaced by the secondary, and vice versa. The 'formal period' (interval between two successive main minima) ranges from 30^d to 150^d.

RVa: Variables of constant mean brightness. (AC Her)

RVb: Variables whose mean brightness varies periodically. (RV Tau)

δC: δ Cephei type variables (often called δ Canis Majoris stars). A very homogeneous group of giant stars (spectral class from B0 to B3), period from $0^d.1$ to $0^d.6$. Brightness variation less than $0^m.1$. (β Cep) δ Sct: δ Scuti type stars. Variables of late A and F spectral class, amplitude generally less than $0^m.1$, period not exceeding $0^d.2$. Greatly resemble RRs type (δ Sct)

α CV: $α^2$ Canum Venaticorum stars (magnetic variables). Spectral class Ap, period 1^d–25^d, amplitude not exceeding $0^m.1$. They have powerful and variable magnetic fields. ($α^2$ CVn)

ZZ: ZZ Ceti variables. Short-period white dwarfs, some of which may be old novae; their period ranges from a few minutes to dozens of minutes. In general, variations with different periods are superimposed. (ZZ Cet)

2. ERUPTIVE VARIABLES

(a) *Irregular variables connected with diffuse nebulae, and rapid irregular stars*

Ia: Very heterogeneous class of irregular variables of early spectral class O–A). (BU Tau)

In: Orion variables. Stars connected with diffuse nebulae, mainly main-sequence subgiant objects. If rapid light variations have been observed, the suffix 's' is added. In (RU Ori); Ins (SU Aur)

Ina: Orion variables of early spectral class (O–A). The suffix 's' is added for rapid variation. Ina (AB Aur); Inas (T Ori)

Inb: Orion variables of intermediate or late spectral class (F–M). The suffix 's' is added for rapid variation. Inb (GW Ori); Inbs (NV Ori)

InT: T Tauri type. Membership of this class is determined exclusively by spectral features. Spectral class ranges from F to M. (T Tau)

Is: Rapid irregular variables, apparently not connected with diffuse nebulae, with brightness variations of $0^m.5$–$1^m.0$ during several hours or days. There is no sharp distinction between rapid irregular stars and Orion variables. (DI And)

Isa: Rapid irregular variables of early spectral class (O–A). (φ Per)

Isb: Rapid irregular variables of intermediate and late spectral class (F–M). (V377 Cas)

UV: UV Ceti type variables. These are dMe stars, sometimes subject to flares with amplitudes of 1^m–6^m lasting from a few seconds to an hour or so. (UV Cet)

UVn: Flash variables, connected with diffuse nebulae. These stars are, in fact, Orion variables of late spectral class. (V389 Ori)

BY: BY Draconis variables. These are emission stars of late spectral class, showing periodic light variations (up to $0^m.3-0^m.5$) with periods of from a fraction of a day to a few days. The form of the light-curve varies. The light variations are probably caused by the axial rotation of a star with anisotropic surface brightness which varies from time to time. (BY Dra)

SD: S Doradus type variables: high-luminosity stars of spectral class Bpeq–Fpeq, showing considerable irregular light variations. (P Cyg)

γC: γ Cassiopeiae type variables: irregular variables of spectral class Be III–V, usually rapidly rotating objects, whose light variations are connected with the ejection of shells from the equatorial zone. (γ Cas)

(*b*) *Novae and nova-like variables*

N: Novae. Hot dwarf stars showing spontaneous increase in brightness, with amplitude generally in the range 7^m-16^m, this increase taking place over a period of from one to dozens or even hundreds of days. The brightness afterwards decreases over years or decades, until the original brightness is reached. It is probable that all novae are members of close binary systems.

Na: Rapid development, characterised by a fast rise followed by a decrease of 3^m in 100^d or less after the maximum. (V1500 Cyg)

Nb: Slow development, with decrease of 3^m from maximum taking 150^d or more. (HR Del)

Nc: Particularly slow novae, remaining at maximum brightness for many years and fading extremely slowly. (RT Ser)

Nr: Recurrent novae. Distinguished from typical novae in that two or more outbursts have been observed. (RS Oph)

Nl: Nova-like variables. A very heterogeneous class resembling novae in their light variation or spectral properties, most of which apparently bear no actual relationship to the novae. (V Sge)

Z And: Z Andromedae type variables are a heterogeneous class of symbiotic objects, usually with composite spectra. (CH Cyg)

RCB: R Coronae Borealis variables. Stars of high luminosity, spectral class F–K and R, characterised by non-periodic fades in brightness of different amplitude (1^m–9^m) and duration (from dozens to several hundred days). (R CrB)

UG: U Geminorum or SS Cygni variables. Dwarf stars showing normally only small fluctuations, the brightness of which increases occasionally by 2^m–6^m during 1^d–2^d, returning to the original brightness in the course of several days or tens of days. The interval between such outbursts can vary greatly, but the mean cycle length ranges from ten to several thousand days. Many (possibly all) of these stars are close binary systems. Often referred to as dwarf novae. (U Gem)

Z Cam: Z Camelopardalis variables. In physical characteristics, spectral properties and light variations they are similar to UG stars; however, they are occasionally interrupted by periods of constant light, when the star maintains an intermediate brightness for several cycles. Values of the mean periods range from 10^d to 40^d; amplitudes from 2^m to 5^m. (Z Cam)

SN: Supernovae. Stars which suddenly increase in brightness by 20^m or more, and then slowly fade. Their light-curves are similar in general appearance to those of novae, but their emission bands are several times as wide. (CM Tau = SN 1054)

3. ECLIPSING BINARY STARS

E: Binary systems whose orbital plane is so close to the line of sight that the components undergo mutual occultation. The observer consequently experiences a variation in the apparent brightness of the system, the period of which coincides with the period of the orbital motion of the components.

EA: Algol type variables. The components are spherical or only slightly ellipsoidal, and their light-curves permit the moment of the beginning and end of occultation to be determined. Secondary minimum may be absent. Period ranges from 0^d2 to $10,000^d$ or more, and the amplitude may attain several magnitudes. (β Per)

EB: β Lyrae type variables. The components are ellipsoidal, and the light change is continuous, preventing the moments of the beginning and end of occultation from being determined. A secondary minimum is always observed. The period usually exceeds 1^d; amplitude is usually less than 2^m. The components are usually of an early spectral class. (GK Cep)

279

EW: Variables of the W Ursae Majoris type, having a period of less than 1^d, and consisting of ellipsoidal components almost in contact. The apparent total light varies with a period equal to that of the orbital motion, due not to mutual occultations but to the change in the luminous area that is presented to the observer. (b Per)

17.14 The Julian Date

In computations and records covering wide time-intervals it is inconvenient to have to deal in miscellaneous units, such as years, months, and days.

The Julian Calendar consists of a continuous series of numbered days, originating at a sufficiently distant past date to ensure that no date prior to JD 1 will be required. The date of the origin of the series is 4713 B.C., whence 1950 = Julian Year 6663.

In the following table are tabulated the last 4 figures of the Julian dates

	Jan.	Feb.	Mar.	Apr.	May	Jun.	Jul.	Aug.	Sep.	Oct.	Nov.	Dec.
1970	0587	0618	0646	0677	0707	0738	0768	0799	0830	0860	0891	0921
71	0952	0983	1011	1042	1072	1103	1133	1164	1195	1225	1256	1286
72	1317	1348	1377	1408	1438	1469	1499	1530	1561	1591	1622	1652
73	1683	1714	1742	1773	1803	1834	1864	1895	1926	1956	1987	2017
74	2048	2079	2107	2138	2168	2199	2229	2260	2291	2321	2352	2382
1975	2413	2444	2472	2503	2533	2564	2594	2625	2656	2686	2717	2747
76	2778	2809	2838	2869	2899	2930	2960	2991	3022	3052	3083	3113
77	3144	3175	3203	3234	3264	3295	3325	3356	3387	3417	3448	3478
78	3509	3540	3568	3599	3629	3660	3690	3721	3752	3782	3813	3843
79	3874	3905	3933	3964	3994	4025	4055	4086	4117	4147	4178	4208
1980	4239	4270	4299	4330	4360	4391	4421	4452	4483	4513	4544	4574
81	4605	4636	4664	4695	4725	4756	4786	4817	4848	4878	4909	4939
82	4970	5001	5029	5060	5090	5121	5151	5182	5213	5243	5274	5304
83	5335	5366	5394	5425	5455	5486	5516	5547	5578	5608	5639	5669
84	5700	5731	5760	5791	5821	5852	5882	5913	5944	5974	6005	6035
1985	6066	6097	6125	6156	6186	6217	6247	6278	6309	6339	6370	6400
86	6431	6462	6490	6521	6551	6582	6612	6643	6674	6704	6735	6765
87	6796	6827	6855	6886	6916	6947	6977	7008	7039	7069	7100	7130
88	7161	7192	7221	7252	7282	7313	7343	7374	7405	7435	7466	7496
89	7527	7558	7586	7617	7647	7678	7708	7739	7770	7800	7831	7861
1990	7892	7923	7951	7982	8012	8043	8073	8104	8135	8165	8196	8226
91	8257	8288	8316	8347	8377	8408	8438	8469	8500	8530	8561	8591
92	8622	8653	8682	8713	8743	8774	8804	8835	8866	8896	8927	8957
93	8988	9019	9047	9078	9108	9139	9169	9200	9231	9261	9292	9322
94	9353	9384	9412	9443	9473	9504	9534	9565	9596	9626	9657	9687
1995	9718	9749	9777	9808	9838	9869	9899	9930	9961	9991	10022	10052

of the zero day of each month from 1970 to 1995. The figures must be added to 2,440,000 to derive the full Julian date. Thus the JD of

$$
\begin{array}{ll}
\text{1970 Feb 10 is} & 2,440,000 \\
 & 618 \\
 & 10 \\ \hline
 & 2,440,628
\end{array}
$$

$$
\begin{array}{ll}
\text{1984 Apr 1 is} & 2,440,000 \\
 & 5,791 \\
 & 1 \\ \hline
 & 2,445,792
\end{array}
$$

17.15 Conversion of hours and minutes to decimals of a day

Observers are asked to report their observations to the B.A.A. timed in $^{h \ m}$ GMAT (see section 17.6.2), as in their Observational Record, and also to convert to decimals or a Julian Day. It is the usual practice, moreover, to quote times of variable-star observations by decimals of a day rather than in hours and minutes, and it is often convenient to have a conversion table handy. The following table gives results correct to the nearest $0^{d}.005$, or $\pm 7^{m}.2$.

To convert Hours and Minutes to Decimals of a Day:

h	m		h	m		h	m		h	m	
0	00										
		.00									
	07		2	02		3	57		5	52	
		.01			.09			.17			.25
	21			16		4	12		6	07	
		.02			.10			.18			.26
	36			31			26			21	
		.03			.11			.19			.27
0	50		2	45			40			36	
		.04			.12			.20			.28
1	04		3	00		4	55		6	50	
		.05			.13			.21			.29
	19			14		5	09		7	04	
		.06			.14			.22			.30
	33			28			24			19	
		.07			.15			.23			.31
1	48			43			38			33	
		.08			.16			.24			.32

h	m		h	m		h	m		h	m	
7	48	.33	12	07	.51	16	26	.69	20	31	.86
8	02	.34		21	.52		40	.70	20	45	.87
	16	.35		36	.53	16	55	.71	21	00	.88
	31	.36	12	50	.54	17	09	.72		14	.89
8	45	.37	13	04	.55		24	.73		28	.90
9	00	.38		19	.56		38	.74		43	.91
	14	.39		33	.57	17	52	.75	21	57	.92
	28	.40	13	48	.58	18	07	.76	22	12	.93
	43	.41	14	02	.59		21	.77		26	.94
9	57	.42		16	.60		36	.78		40	.95
10	12	.43		31	.61	18	50	.79	22	55	.96
	26	.44	14	45	.62	19	04	.80	23	09	.97
	40	.45	15	00	.63		19	.81		24	.98
10	55	.46		14	.64		33	.82		38	.99
11	09	.47		28	.65	19	48	.83		52	1.00
	24	.48		43	.66	20	02	.84	23	59	
	38	.49	15	57	.67		16	.85			
11	52	.50	16	12	.68						

In critical cases ascend.

Observations submitted to the A.A.V.S.O., and to most other national societies, are normally required to be timed in UT.

17.16 The discussion of observations

Statistical techniques for the analysis of variable-star observations are beyond the scope of this volume (see *A.A.H.*, section 29), and more can be learned from a study of papers in one of the journals than from textbooks; but certain simple graphical techniques for estimating the time of maximum and minimum of a variable star should be familiar to every observer.

17.16.1 Pogson's method of bisected chords: The observations are plotted on a graph of magnitude against time, and a smooth curve is drawn, by hand, consistent with the estimated accuracy of the observations and the known characteristics of the star. The instant of, say, maximum, is determined by the following procedure. Points on the rising and falling branches of the curve are read off at intervals of half a magnitude or less, and the dates or times of the two points for each magnitude are averaged. The points on the graph corresponding to each magnitude and averaged time are marked, and a smooth curve drawn through them to meet the light curve. This point of intersection defines the date and magnitude of maximum (see Figure 37).

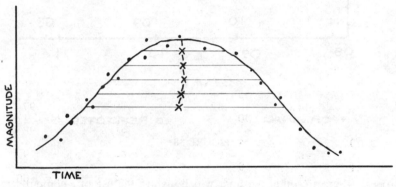

FIGURE 37

17.16.2 Tracing-paper method: When the maximum or minimum is known to be nearly symmetrical (as for example with EA stars), the preceding method is not the most efficient. In the so-called 'tracing-paper' method a graph of the observations is again plotted, but no curve is drawn. A tracing of the plot is made, including the time axis. The tracing-paper is now reversed left-to-right and the magnitude scale is lined up with the original; the traced plot is then moved along the time axis until the position of best fit is found. If time t_1 on the original plot falls under time t_2 on the traced plot, the time corresponding to the axis of symmetry of the curve is the mean of t_1 and t_2. Figure 38 illustrates the appearance of the result obtained using the estimates given in section 17.2.4 above, the deduced time of minimum being about $09^h 35^m$.

17.16.3 Other methods: The extraction of details of maxima and minima is part of the process of reducing a large mass of data to a smaller bulk that can be handled more easily; but information is lost in the

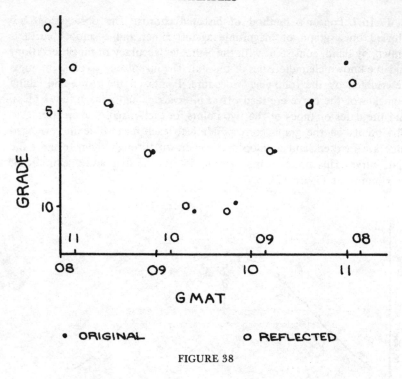

FIGURE 38

process. There are other methods which involve the use of a computer to extract useful information directly from the full set of data. An example is the use of periodogram analysis to identify secondary periods in the light curves of semiregular variables (B.17.16). For those who have access to a computer, there is vast scope for useful work in the discussion or rediscussion of existing observations in the literature and in the files or organisations such as the B.A.A. Variable Star Section and the A.A.V.S.O.

17.17 Correction for light-time

For variable stars with very rapid changes, such as eclipsing binaries with periods less than a few days, it is usual to correct the times of observation or of deduced minima for the difference in light-time between the Earth and the Sun, which can be as much as $\pm\ 8^{\mathrm{m}}\ 20^{\mathrm{s}}$. Times of the original observations are referred to (loosely!) as geocentric, while those of the corrected observations are heliocentric. Strictly speaking, the reduction should be to the centre of mass of the solar system.

Let α, δ be the star's RA and Dec in angular measure. The ecliptic longitude and latitude of the star are given by

$$\sin \beta = \sin \delta \cos \epsilon - \cos \delta \sin \epsilon \sin \alpha$$
$$\cos \beta \cos \lambda = \cos \delta \cos \alpha$$
$$\cos \beta \sin \lambda = \sin \delta \sin \epsilon + \cos \delta \cos \epsilon \sin \alpha$$

where ϵ is the obliquity of the ecliptic ($23° 26'$ for the year 2000).

If L☉ is the Sun's longitude (tabulated in the B.A.A. *Handbook*, for example), the correction to be added is, to sufficient accuracy,

$$- 498^{s}4 \ \cos (\lambda - \text{L}☉) \cos \beta$$

Note that this is positive when the Sun and the star are more than 90° apart, as is most frequently the case for stars observed in a dark sky. Care should be taken to ensure that the trigonometric terms are given the correct sign when the argument is greater than 90°. The calculation of such corrections is laborious, and graphical methods have been devised, such as the nomogram published by Zverev in B.17.34. Such corrections should not be made in observations reported to the B.A.A., unless specifically requested. Observers working on their own should make it clear whether times given in any publication of their results are geocentric or heliocentric.

SECTION 18

BINARIES

18.1 Work

Professional observatories hold all the trumps in this field, though double stars provide a field for gaining experience in the technique of micrometer measures and orbit computation, and there is a limited scope for amateur photography (see section 18.4). Primarily, however, binaries are of interest to the amateur as tests of instrumental performance.

As regards the discovery of new faint binaries, Jonckhere has recently denied the truth of Aitken's dictum that the need today is not for more discoveries, but for more numerous and more accurate measures of already known doubles.

Ephemerides of visual binary stars are given in B.18.20, 18.21.

18.2 Measures of position

With a 15 cm aperture or more, work of value can be done with micrometer or camera; but it should be remembered that unpublished measures are valueless. With apertures smaller than 15 cm it is a waste of time to concentrate on binaries. Dawes, the eagle-eyed, worked with instruments ranging from 96 mm to 21.5 cm, mostly 17.8–21.5 cm, all refractors.

With an ordinary filar micrometer a motor-driven 30 cm can make reliable measures of doubles with separations down to about $0''.5$, but by the method of matching ellipses this threshold has been considerably lowered, with still reasonable accuracy (θ to within $15°$ and r to $\pm0''.025$ on the average):

D	R (theoretical threshold)	Measurable threshold of r
75 mm	$1''.82$	$0''.75–1''.10$
12.5 cm	1.09	0.45–0.65
20 cm	0.68	0.3–0.4
28 cm	0.49	0.2–0.3

The method is to match as nearly as possible the observed elongated star image (unresolved) with one of a series of white ellipses photographed on a dark ground. A set of five such ellipses is employed, one axis being 1.0 cm in each case, the other progressing from 0.96 to 1.16 cm in 0.04 cm steps. The reliability of the measures falls off sharply outside the range of eccentricities 1.04–1.12; also if the magnitude difference exceeds 2. For this work a high magnification is recommended (30D), and good seeing is absolutely essential.

The choice of stars for observation is important:

(a) For reliable measures, the pair should be well above the instrument's theoretical resolution threshold; Aitken's opinion was that to secure the greatest accuracy possible with a given equipment, the separation should not be less than twice the instrument's theoretical threshold, given good seeing.

(b) Rapidly moving pairs are, generally, those most needing observation: those with separations of less than 0".5, say (implying an aperture of not less than 25 cm), for which one set of measures every 1 to 5 years is sufficient. For those with separations of 1"–5", measures every 20 years or so are adequate.

(c) Stars which are shown by the catalogues to have been poorly observed in the past or to be at a critical point in their orbit.

Separation is measured in " arc; position angle from 0° (N), through 90° (f), 180° (S), and 270° (p), where the hour circle of the primary passes through the N and S points of the field (see A.A.H., Figure 192).

For the measurement of θ and r by the filar micrometer, see A.A.H., section 18.6.

For the use of objective and exit-pupil diaphragms in the observation of binaries, see A.A.H., section 27.1.

The observer should be aware of the following precautions and sources of error in double-star measurements:

(a) Seeing is the all-important factor in the measurement of difficult pairs, and measures should not be attempted when the seeing is poor; inferior results are inevitable, and these are worse than merely useless. For this reason it is well to have prepared working lists, supplemented where necessary by charts, so that full use can be made of periods of good seeing when they occur.

(b) Each final measure should be the mean of about 3 nights' observations (more if the measures are discordant, or the pair is very difficult), each night's observations consisting of at least three measures of θ and three of the double distance. In this way errors due to seeing and the physiological condition of the observer may be to some extent reduced.

(*c*) Systematic errors are more difficult to eliminate. They are mainly dependent on the personal idiosyncrasy of the observer, the relative brightness of the two stars, their separation, and the inclination of the radius vector to the horizontal. Some mitigation of these can be achieved by taking the mean of measures made on either side of the meridian, and with and without a totally reflecting prism, placed between the eye and the ocular, which inverts the field. According to Voûte, more trouble is encountered when the radius vector is vertical than when horizontal.

(*d*) Tilt the head so that the line through the eyes is systematically either parallel or perpendicular (preferably the former) to the line through the components.

(*e*) See that the observing position (especially of the head and neck) is comfortable.

(*f*) With unequally bright pairs there is a tendency when measuring θ to set the web tangential to the discs instead of bisecting them or setting it parallel to their centres. Therefore bring the stars nearly into contact with the web, first from one side and then from the other, taking an equal number of readings on either side of the web.

(*g*) Between successive measures of θ, throw the web off in opposite directions. Also remove the eye from the telescope between measures and attempt to come to the next measure with no recollection of the last one.

(*h*) Separations, when small, tend to be over-estimated, owing to the dilation of the star image when the web is placed across it. Small instruments hence have a systematic tendency to over-estimate r, since they cannot use sufficient magnification to separate out the images (it being assumed, of course, that the instrument's resolving power is adequate in the first place). This factor becomes increasingly important as the value of r is reduced from about twice the theoretical resolution threshold. There are two alleviants: (*i*) a Barlow lens placed forward of the primary focus increases the separation of the images in the field without increasing the apparent thickness of the webs; (*ii*) before the measures are made, the appearance of each star when accurately bisected by the web should be carefully noted and memorised; when making the measures, a conscious attempt is made to reproduce these appearances accurately.

(*i*) Webs and stars must be simultaneously in accurate focus.*

(*j*) The accuracy with which θ is measurable depends, *inter alia*, upon the accuracy of the setting of the polar axis. The gravity of the errors from this cause increases with increasing Dec, and for satisfactory results to be obtained in the vicinity of the Pole itself extreme precision in the orienta-

---

* See, further, *A.A.H.*, section 18.2.

tion and elevation of the polar axis is demanded. For Innes's modification of the Struve method of dealing with errors of this nature, see B.18.5.

For a survey of micrometers more suited to amateur use than is the filar, see *A.A.H.*, section 18, and B.18.6.

18.3 Magnification

The general subject of magnification limits is discussed in *A.A.H.*, section 3; it is of relevance here since the measurement of double stars normally entails magnifications near the upper limit. The atmosphere is therefore frequently the limiting factor in this work; when turbulence forbids the use of adequate magnification it is advisable to restrict observation to pairs of sufficient separation to be observable with low magnifications. Aitken, for example, never attempted to observe the closer pairs with the Lick 36 in (91 cm) refractor unless a magnification of ×1500 (16D) was feasible. Burnham's invariable rule was to use the lowest magnification yielding sufficient separation for a good measure to be made, thus exposing himself as little as possible to the effects of inferior seeing.

Taking 28D as an average permissible magnification on good nights, at any rate over the small and moderate aperture range, and following Aitken's usual practice with the 36 in of using

    ×520 on pairs wider than 2″,
    ×1000 on pairs between 2″ and 1″,
    ×1000–×3000 on pairs closer than 0″.5,

we have, as a very rough guide, that

    a 20 cm will tackle pairs wider than 2″,
    a 35 cm will tackle pairs wider than 1″,
    a 50 cm will tackle pairs wider than 0″.5.

This agrees more or less with W. H. Pickering's dictum that for work of value, atmosphere and instrument must be capable of supporting a magnification of at least ×800. Most of Burnham's measures with the 36-in were made with ×1000–×1500 (11D–16D), ×2600 (29D) being reserved for the very closest pairs; Barnard employed the same magnifications, finding that ×2600 could be used very rarely indeed.

Lewis (B.18.11) carried out a valuable investigation into the magnifications actually employed by 36 double-star observers in all parts of the world, using instruments from 97 mm to 91 cm aperture. His results are summarised below. The actual magnifications most used with the different apertures in col. (1) are tabulated in col. (2), the equivalent magnification per centimetre of aperture in col. (3). Lewis found that these empirical figures could be nearly represented by the formula

$$M = 89\sqrt{D}$$

as can be seen from cols. (4) and (5) and Figure 39.

It will be seen that the maximum practicable magnification, when expressed in terms of aperture, is not constant over the range of apertures employed, but falls continuously with increasing aperture.*

From these empirical figures it might appear that $28\,D$ is in fact seldom realised, even with small instruments, and that the conclusions reached on the preceding page need fairly drastic amendment. It must be pointed out, however, that the figures derived by Lewis represent the most commonly employed magnification with each aperture, and were obtained by weighting each in proportion to the number of times it was employed over a period of time. Considerably higher magnifications can be used, all along

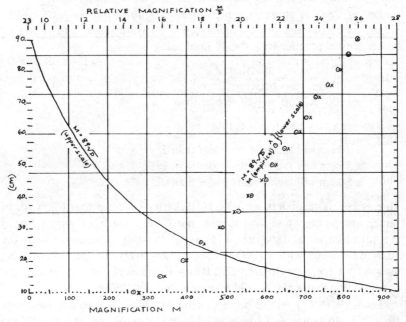

FIGURE 39

Magnification and aperture in double-star work

* See also, in this connexion, *A.A.H.*, section 3.

(1) D (cm)	(2) M	(3) Equivalent M per cm	(4) M = $89\sqrt{D}$	(5) Equivalent M per cm
10	264	26.4	281	28.1
15	333	22.2	345	23.0
20	392	19.6	398	19.9
25	440	17.6	445	17.8
30	492	16.4	487	16.2
35	532	15.2	527	15.1
40	563	14.1	563	14.1
45	604	13.4	597	13.3
50	620	12.4	629	12.6
55	651	11.8	660	12.0
60	682	11.4	689	11.5
65	707	10.9	718	11.0
70	731	10.4	745	10.6
75	765	10.2	771	10.3
80	790	9.9	796	10.0
85	816	9.6	821	9.7
90	840	9.3	844	9.4

the range of apertures, when the seeing is better than average; $28D$ is not infrequently practicable at least into the moderate-aperture range, though by the time D reaches 90 cm the experience of Burnham and Barnard indicates that it can be very rarely used.

18.4 Photography

The measurement of wide pairs can be undertaken more accurately from photographic plates than by direct observation with a micrometer. Given a moderate aperture the measurement of pairs wider than about 3″ may be undertaken, but the necessity for a plate-measuring machine and the fact that this field is covered by the professional observatories make the work somewhat ill-adapted to amateur requirements and qualifications.

18.5 Apparent orbits

The true orbit of the *comes* about the primary (it is assumed for simplicity that only the *comes* is in motion, the primary being used as the reference point for this motion) is observed as projected upon a plane perpendicular to the line of sight. Its projection in this plane is known as the apparent orbit. The determination of the apparent orbit can be made either from the observed position angles and separations at different times, or from the elements of the true orbit.

Two simple graphical methods of establishing the apparent orbit of a binary are described below. The first, modified from J. Herschel, is both the more elegant and the more accurate. It is instructive occasionally to use both methods on the same set of observational material and to compare the results obtained. For the construction of the apparent orbit from the elements of the true orbit, see B.18.1.

First method:

Given: θ (position angle) and r (angular separation) at a variety of dates, t.

The advantage of this method is that it employs θ only, which is in general a more accurately observable quantity than r, in establishing the form of the orbit.

1. Construct a pro-forma for the following Table:

(1)	(2)	(3)	(4)	(5)	(6)	(7)	(8)
θ	t	Δt	$\Delta\Delta t$	$\dfrac{1}{v}$	$\sqrt{\dfrac{\Delta t}{\Delta\theta}}$	r	r'

2. Plot the given values of θ against t. Through these points draw the best possible freehand curve; this is known as the interpolating curve.

3. In col. (1) set down a series of equally spaced position angles (say every $5°$ or $10°$ throughout the given series of values of θ), read off the corresponding values of t from the interpolating curve, and tabulate these in the left-hand side of col. (2). It will probably be necessary to smooth this curve.

4. In the left-hand side of col. (3) tabulate the difference between each pair of values of t; in the left-hand side of col. (4) tabulate the difference between each pair of values of Δt. The first differences in col. (3) should increase or decrease steadily; the existence of any sudden jumps is more easily seen in the column of second differences, which, however, can be omitted when some experience has been gained.

Any such sudden increase or decrease must be eliminated by adjusting the date in col. (2) which is responsible for it. A second interpolating curve is then constructed from the values in col. (1) and the adjusted values which have been entered in the right-hand side of col. (2). This in turn is tested for smoothness and when satisfactory is known as the smoothed interpolating curve.

5. Take the mean of the values of Δt on either side of each value in col. (2), divide it by the constant difference between the values in col. (1), and enter in col. (5).

6. The square roots of the values in col. (5) are taken and entered in col. (6); these values are now, by Kepler's Second Law, proportional to the radii vectores at each position angle in col. (1).

7. From a central point on a sheet of graph paper lay off the radii vectores in col. (1). Using the largest scale that the sheet allows, cut off on each radius vector a length proportional to the corresponding value in col. (6).

8. Through these points draw the best possible ellipse (ellipsograph or pins and cotton loop).

9. If any points lie far from any possible curve, work back to the smoothed interpolating curve, make the necessary correction, and draw a new ellipse.

To find the scale of this apparent orbit:

10. Plot the observed values of r against the times, t, at which the observations were made.

11. Draw the interpolating curve through these points.

12. Tabulate in col. (7) the values of r, read off this curve, corresponding to the values of t in col. (2).

13. In col. (8) tabulate the scale values of the radii vectores of the apparent orbit corresponding to the values of θ in col. (1).

14. The value in $''$ arc of one scale division is then given by dividing the sum of the values in col. (7) by the sum of those in col. (8).

Second method:

Given: a set of values of θ and r at different times, as before.

1. Construct a pro-forma for the Table on page 294:

2. Enter the observational data in cols. (1)–(3).

3. In col. (4) tabulate the intervals separating the dates tabulated in col. (1).

4. In col. (5) tabulate the differences between the values of θ in col. (2).

5. The angular velocity at each time in col. (1) is obtained by dividing

293

(1)	(2)	(3)	(4)	(5)	(6)	(7)	(8)
t	θ	r	Δt	$\Delta \theta$	$\dfrac{\Delta t}{\Delta \theta}$	$\sqrt{\dfrac{\Delta t}{\Delta \theta}}$	$\sqrt{\dfrac{\Delta t}{\Delta}}\, \times$

each corresponding value of Δt by that of $\Delta \theta$. These are entered in col. (6).

6. Take the square roots of these values and enter them in col. (7).

7. From a central point, representing the primary, lay off the position angles tabulated in col. (2); mark each with the corresponding date from col. (1); lay off on each, using the largest scale that the paper will allow, a distance proportional to the corresponding value in col. (3). The points so plotted will probably show a considerable scatter, but draw through them the best ellipse that is possible.

8. From this first rough orbit estimate as nearly as possible the period (p), and measure the major and minor axes (a, b), converting them (the scale being known) to " arc.

9. The radius of the circle whose area is equal to that of the orbit is given by $\frac{1}{2}\sqrt{ab}$. Calculate the value of $\frac{1}{2}\sqrt{ab}. \sqrt{\dfrac{360}{p}}$; call this k.

10. Insert this value at the head of col. (8), thus: $\sqrt{\dfrac{\Delta t}{\Delta \theta}} \times [k]$, and tabulate the values.

11. Each of these values, in " arc, represents the radius of a circle whose segment bounded by the corresponding radii vectores is equal in area to the segment of the orbital ellipse bounded by the same radii vectores. On the diagram of the approximate orbit draw the arcs of these circles, using in turn the radii tabulated in col. (8). By their means the necessary corrections can be applied to the rough orbit. A surprising degree of accuracy can be obtained, though this stage of the process requires practice.

18.6 True orbits

If one of the foci of the apparent orbit coincides with the primary, it is also the true orbit. If, however, the plane of the true orbit is not perpendicular to the line of sight, this coincidence will not occur.

The true orbit is defined by the following elements:

P: Period (mean solar years),

T: Time of periastron passage,

e: Eccentricity,

a: Semimajor axis ($''$ arc),

Ω (or ω):   Position angle of the node which lies between $0°$ and $180°$,

i (or γ):   Inclination of the orbital plane to the plane of projection,

λ (or ω):   Angle between the line of nodes and the major axis, measured in the plane of the true orbit.

There are numerous methods of computing the elements of the true orbit, among them:

(a) The Thiele-Innes method (B.18.1).
(b) Kowalsky's analytical method: derivation of the elements from the constants of the general equation of the apparent orbit (B.18.1, 18.18).
(c) Zwiers' graphical method: involves the prior construction of an accurate apparent orbit (B.18.1).
(d) Russell's method (B.18.14).

18.7 Spectroscopic binaries

The orbit of a spectroscopic binary may be determined by a variety of methods:

(a) The Lehmann-Filhés method (B.18.1).
(b) Schwarzschild's method (B.18.1).
(c) Zurhellen's methods (B.18.1).
(d) King's method (B.18.1, 18.9).
(e) Russell's short method (B.18.1, 18.15).

18.8 Eclipsing binaries

See section 17.

For methods of computing the orbits of eclipsing binaries, see B.18.1, 18.17; also B.18.16.

SECTION 19

NEBULAE AND CLUSTERS

The scope for original and valuable work in this field lies with the large instruments of the professional observatories. Nebulae and clusters are nevertheless among the more exciting discoveries of the telescopic beginner who is making his first survey of the heavens. It should not, therefore, be supposed that to deal with them summarily is to imply that their observation is worthless.

Just as lunar cartography has an educational value out of all proportion to its scientific usefulness, so is observation of 'deep-sky' objects equally valuable training in learning how to see, interpret, and draw. In addition, there is no better way of learning the night sky. Finally, many amateurs seek nebulae, clusters, and the more pictorial double stars as ends in themselves, without any thought of an end-product other than the satisfaction of locating, studying and recording them, either visually or photographically.

The amateur embarking on deep-sky observation requires, essentially, nothing more than an aperture of say 30 mm and upwards (numerous catalogued objects can be discerned using binoculars), an atlas such as B.20.14, or B.20.2 when he has become more proficient, and a manual, e.g. B.19.1, 19.5, or 19.8.

As an indication of what can be achieved photographically in this field, the pages of B.1.3 regularly include both black-and-white and colour photographs taken by amateurs.

SECTION 20

BIBLIOGRAPHY

Abbreviations used in this Bibliography are listed below. Bibliographical data are enclosed in round brackets ( ); my own comments, if any, in square brackets [ ]. Each division of the Bibliography is arranged alphabetically by authors unless some other arrangement (e.g. chronological) is specified.

A.A. Astronomical Almanac.
A.J. Astronomical Journal.
A.N. Astronomische Nachrichten.
Ann. Cape Obs. Annals of the Cape Observatory.
Ann. d'Astrophys. Annales d'Astrophysique.
Ann. Lowell Obs. Lowell Observatory Annals.
Ann. Obs. Strasbourg. Annales de l'Observatoire de Strasbourg.
Ap. J. Astrophysical Journal [previously 'Sidereal Messenger', 1883–92; 'Astronomy and Astro-Physics', 1892–95].
Astr. Mitt. Göttingen. Astronomische Mittheilungen der Königlichen Sternwarte zu Göttingen.
Astr. Pap., Wash. Astronomical Papers of the American Ephemeris & Nautical Almanac.
B.A.A.H. Handbook of the British Astronomical Association.
Brit. J. Physiol. Optics. British Journal of Physiological Optics.
B.S.A.F. Bulletin de la Société Astronomique de France ['L'Astronomie'].
Connais. Temps. Connaissance des Temps.
Contr. Dunsink Obs. Dunsink Observatory Contributions.
Contr. Lick Obs. Contributions from the Lick Observatory.
Contr. Perkins Obs. Perkins Observatory Contributions.
Contr. Princeton Obs. Princeton University Observatory Contributions.
C.R. Comptes Rendus des Séances de l'Académie des Sciences.
Engl. Mech. English Mechanics [formerly 'The English Mechanic'].
Harv. Ann. Annals of Harvard College Observatory.
Harv. Bull. Harvard Bulletins.
Harv. Circ. Harvard College Observatory Circulars.
Harv. Monogr. Harvard Observatory Monographs.
Harv. Repr. Harvard Reprints.
H.d.A. Handbuch der Astrophysik (Julius Springer, Berlin).
I.A.U. International Astronomical Union.

Internat. Ass. Acad. International Association of Academies.
Irish Astr. J. Irish Astronomical Journal.
J. Appl. Phys. Journal of Applied Physics.
J.B.A.A. Journal of the British Astronomical Association.
J. Can. R.A.S. Journal of the Royal Astronomical Society of Canada.
J. Opt. Soc. Amer. Journal of the Optical Society of America.
J. Sci. Instr. Journal of Scientific Instruments.
L.O.B. Lick Observatory Bulletins.
Lund. Medd. Meddelande fran Lunds Astronomiska Observatorium.
Mem. Amer. Acad. Arts Sci. Memoirs of the American Academy of Arts and Sciences.
Mem. B.A.A. Memoirs of the British Astronomical Association.
Mem. R.A.S. Memoirs of the Royal Astronomical Society and Memoirs of the Astronomical Society of London [consecutive volume numbering].
M.N. Monthly Notices of the Royal Astronomical Society.
M.W.C. Mount Wilson Contributions.
N.A. The Nautical Almanac and Astronomical Ephemeris (H.M.S.O., annually).[Superseded by the Astronomical Almanac.]
Nat. Nature.
Obs. The Observatory.
Perkins Obs. Repr. Perkins Observatory Reprints.
Phil. Trans. Philosophical Transactions of the Royal Society.
Photogr. J. Photographic Journal.
Phys. Rev. Physical Review [U.S.A.].
Pop. Astr. Popular Astronomy.
Proc. Amer. Acad. Arts Sci. Proceedings of the American Academy of Arts and Sciences.
Proc. Amer. Phil. Soc. Proceedings of the American Philosophical Society.
Proc. Phys. Soc. Lond. Proceedings of the Physical Society.
Proc. R. Irish Acad. Proceedings of the Royal Irish Academy.
Proc. Roy. Soc. Proceedings of the Royal Society.
Publ. A.A.S. Publications of the American Astronomical Society.
Publ. A.S.P. Publications of the Astronomical Society of the Pacific.
Publ. Tartu. Publications de l'Observatoire Astronomique de l'Université de Tartu (Dorpat).
Publ. Astrophys. Obs. Potsdam. Publikationen des Astrophysikalischen Observatoriums zu Potsdam.
Publ. Carneg. Instn. Carnegie Institution Publications, Washington.
Publ. Cincinn. Obs. Cincinnati Observatory Publications.
Publ. Dom. Astrophys. Obs. Dominion Astrophysical Observatory Publications.
Publ. Lick Obs. Lick Observatory Publications.
Publ. L. McC. Obs. Publications of the Leander McCormick Observatory of the University of Virginia.
Publ. Michigan Obs. Michigan Observatory Publications.
Publ. Univ. Pa. Pennsylvania University Publications (Astronomical Series).
Publ. Yale Obs. Yale University Observatory Publications.
Rep. Obs. Circ. Republic Observatory Circular.
Result. Obs. Nac. Argent. Resultados del Observatorio Nacional Argentino.

Rev. d'Opt. Revue d'Optique Théorique et Instrumentale.
Rev. Mod. Phys. Review of Modern Physics.
Rev. Sci. Instr. Review of Scientific Instruments.
Sci. Amer. Scientific American.
Sci. Progr. Science Progress.
Sky & Tel. Sky and Telescope [amalgamation of 'Sky' and 'Telescope'].
Smithson. Contr. Smithsonian Contributions to Knowledge.
Trans. I.A.U. Transactions of the International Astronomical Union.
Veröff. Univ. Berlin-Babelsberg. Veröffentlichungen der Universitätssternwarte zu Berlin-Babelsberg.
Veröff. König. Astron. Rechen-Insit. Veröffentlichungen des Königlichen Astronomischen Rechen-Instituts zu Berlin.
Wash. Nat. Ac. Proc. Proceedings of the National Academy of Science of the United States of America.

1. OBSERVATIONAL AIDS

1.1 *Astronomical Almanac* (H.M.S.O., annually). [General ephemerides for Sun, Moon and major planets and their brighter satellites, as well as much other information; identical in content to the earlier *Nautical Almanac*.]

1.2 *Handbook* of the British Astronomical Association (B.A.A., annually). [Ephemerides for Sun, planets and satellites; lunar phases and occultations, as well as risings and settings; expected comet returns; meteors showers, etc.]

1.3 *Sky & Telescope* (Sky Publishing Corp., 49 Bay State Road, Cambridge, MA 02138, USA, monthly). [Articles, news notes, observations, telescope-making forum, etc.]

1.4 *The Astronomer* (The Old Wheel, Sutton, Norwich, England, monthly). [Rapid publications of observational reports from UK and overseas amateurs.]

1.5 *Circulars* of the British Astronomical Association (B.A.A., at irregular intervals). [Immediate announcement of comet and nova discoveries, etc.]

1.6 *I.A.U. announcement cards* (Central Bureau for Astronomical Telegrams, Smithsonian Astrophysical Observatory, Cambridge, MA 02138, USA). [Telegram, telex and airmail circulars detailing discoveries, etc.]

1.7 *The Observer's Handbook* (R.A.S. of Canada, 124 Merton Street, Toronto, annually). [Similar in scope to B.A.A. *Handbook*.]

2. AMATEUR OBSERVATIONAL ASTRONOMY

2.1 J. GRIBBIN, *Astronomy for the Amateur* (Macmillan, 1976).

2.2 R. H. LAMPKIN, *Naked Eye Stars* (Gall & Inglis, 1972).

2.3 D. H. MENZEL, *Field Guide to the Stars and Planets* (Collins, 1977).

2.4 H. G. MILES, *Artificial Satellite Observing* (Faber & Faber, 1974).

2.5 P. A. MOORE (ed), *Practical Amateur Astronomy* (Lutterworth, 1978).

2.6 P. A. MOORE (ed), *Yearbook of Astronomy* (Eyre & Spottiswoode, annually).

2.7   J. MUIRDEN, *Astronomy with Binoculars* (Faber & Faber, 1976).
2.8   G. D. ROTH, *Astronomy: A Handbook* (Springer-Verlag, Berlin, 1975).
2.9   T. W. WEBB, *Celestial Objects for Common Telescopes* (Dover Books, New York, 1962).

3. Sun

3.1   G. ABETTI (trans. J. B. SIDGWICK), *The Sun* (Faber, 1963).
3.2   G. ABETTI, Solar Physics (*H.d.A.*, **4**, 57).
3.3   F. ADDEY, The Wilson Effect (*J.B.A.A.*, **74**, No. 2, 43).
3.4   A. F. ALEXANDER, Longitudinal Distribution of Sunspot Areas, 1925–36 (Rotations 964–1113): Relative Positions of the Most and Least Spotted Regions (*J.B.A.A.*, **54**, No. 2, 31).
3.5   A. F. O'D. ALEXANDER, Area, Distribution and the Sunspot Cycle (*J.B.A.A.*, **55**, No. 2, 43).
3.6   R. G. ATHAY, A Model of the Chromosphere from Radio and Optical Data, p. 98 (*Radio Astronomy*, ed.: R. N. Bracewell, Stanford Univ. Press).
3.7   H. D. BABCOCK & C. E. MOORE, *The Solar Spectrum, λ6600 to λ13495* (*Public. Carneg. Instn.*, 579, 1947).
3.8   V. BAROCAS, Sudden Enhancement of Atmospherics (*J.B.A.A.*, **83**, No. 2, 98).
3.9   V. BAROCAS, The Observation of Solar Flares, and the Work of the Solar Section (*J.B.A.A.*, **86**, No. 3, 232).
3.10   B. F. BAWTREE, Solar Eclipses (*Mem. B.A.A.*, **30**, No. 3, 94).
3.11   B. F. BAWTREE, Tables for the Calculation of Astronomical Quantities (*Mem. B.A.A.*, **34**, No. 4, 1).
3.12   W. M. BAXTER, *The Sun and the Amateur Astronomer* (David & Charles, 1973).
3.13   K. BISPHAM & H. HILL, Observations of Polar Faculae (*J.B.A.A.*, **79**, No. 3, 200).
3.14   E. J. BLUM & A. BOISCHOT, Occultation of the Crab Nebula by the Solar Corona (*Obs.*, **77**, 206, 1957).
3.15   R. J. BRAY & R. F. LOUGHHEAD, *The Solar Granulation* (Chapman & Hall, 1967).
3.16   R. J. BRAY & R. F. LOUGHHEAD, *Sunspots* (Dover Books, 1979).
3.17   E. BUDDING, The Assessment of 'White Light' Observations of the Sun (*J.B.A.A.*, **80**, No. 2, 125).
3.18   C. M. CHERNAN, *The Handbook of Solar Flare Monitoring and Propagation Forecasting* (Tab Books, U.S.A., 1978).
3.19   W. N. CHRISTIANSEN & J. A. WARBURTON, The Distribution of Radio Brightness over the Solar Disc at a Frequency of 21 cms. (*Aust. J. Sci. Res.* **6**, 262, 1953).
3.20   W. N. CHRISTIANSEN & D. S. MATHEWSON, The Origin of the Slowly Varying Component (*Radio Astronomy*, ed.: R. N. Bracewell, Stanford Univ. Press, 1959).
3.21   A. L. CORTIE, The Stonyhurst discs for Measuring the Positions of Sunspots (*J.B.A.A.*, **18**, No. 1, 26).

BIBLIOGRAPHY

3.22 A. L. CORTIE, On the Types of Sun-Spot Disturbances (*Ap.J.*, **13**, No. 4, 260).

3.23 R. E. COX, Hints on Eclipse Instrumentation for Photography (*Sky & Tel.*, **39**, No. 2, 124).

3.24 H. E. DALL, Filter Type Solar Prominence Telescopes for Amateurs (*J.B.A.A.*, **77**, No. 2, 94).

3.25 H. E. DALL & M. R. WHIPPEY, Filter Type Solar Prominence Telescopes for Amateurs (*J.B.A.A.*, **89**, No. 2, 122).

3.26 M. DAVIDSON, The Computation of Total Solar Eclipses (*J.B.A.A.*, **49**, No. 8, 299).

3.27 L. M. DOUGHERTY, Solar Graticules (*J.B.A.A.*, **87**, No. 6, 582).

3.28 L. M. DOUGHERTY, Metallized Polyester Film as a Solar Filter (*J.B.A.A.*, **89**, No. 5, 450).

3.29 J. DRAGESCO, A Folded Refractor Exclusively for the Sun (*Sky & Tel.*, **49**, No. 5, 323).

3.30 J. A. EDDY, *A New Sun* (NASA, 1979).

3.31 B. EDLÉN, The Identification of the Coronal Lines (*M.N.*, **105**, No. 6, 323). [George Darwin Lecture, 1945.]

3.32 M. A. ELLISON, The Construction of an Auto-collimating Spectrohelioscope (*J.B.A.A.*, **50**, No. 3, 107).

3.33 M. A. ELLISON, Problems of the Motions of Solar Prominences (*Nat.*, **147**, 662).

3.34 M. A. ELLISON, *The Sun and Its Influence*, Routledge & Kegan Paul.

3.35 M. A. ELLISON, The Recording of Sudden Enhancements of Atmospherics, (S.E.A's) for Purposes of Flare Patrol (*J.B.A.A.*, **69**, No. 3, 1959).

3.36 A. FITTON, Markings at the Solar Poles near Sunspot Minimum (*J.B.A.A.*, **75**, No. 4, 236).

3.37 A. GABRIËL, A Narrow-band H-α Telescope for Visual and Photographic Solar Observations (*J.B.A.A.*, **86**, No. 2, 140).

3.38 F. F. GARDNER, The Effect of S.I.D's on 2.28 mcs Pulse Reflections from the Lower Ionosphere (*Aust. J. Phys.*, **12**, 42, 1959).

3.39 R. G. GIOVANELLI & J. A. ROBERTS, Optical Observations of the Solar Disturbances causing Type II Radio Bursts (*Aust. J. Phys.*, **11**, 353, 1958).

3.40 G. E. HALE, The Spectrohelioscope and its Work, I (*Ap. J.*, **70**, No. 5, 265 = *M.W.C.*, 388).

3.41 G. E. HALE, The Brightness of Prominences as Shown by the Spectrohelioscope (*M.N.*, **95**, No. 5, 467).

3.42 G. E. HALE, A Simple Solar Telescope and Spectrohelioscope (*A.T.M.*, **1**, 192).

3.43 R. A. HAM, A Versatile Radio Instrument (*J.B.A.A.*, **81**, No. 3, 204).

3.44 F. J. HARGREAVES, Camera Work in the Eclipse (*J.B.A.A.*, **37**, No. 6, 212).

3.45 L. D. J. HARRIS, Solar Microwave Bursts and their Association with Solar Proton Flares (*J.B.A.A.*, **84**, No. 1, 9).

3.46 A. HEWISH, Radio Observations of the Solar Corona at Sunspot

Minimum, I.A.U. Symposium No. 4 (*Radio Astronomy* p. 298. Cambridge, 1957).

3.47 A. HEWISH, The Occultation of the Crab Nebula by the Solar Corona (*Obs.*, **77**, 152, 1957).

3.48 A. HEWISH, The Scattering of Radio Waves in the Solar Corona (*Mon. Notices of R.A.S.*, **118**, No. 6 (1958), p. 534).

3.49 J. S. HEY & V. A. HUGHES, Centimetre Wave Observations of the Solar Eclipse of June 30 1954 (*Obs.*, **76**, 226, 1956).

3.50 H. HILL, Observations of a Solar Flare on 1966 August 28 (*J.B.A.A.*, **77**, No. 1, 28).

3.51 H. Hill, An Amateur's Spectrohelioscope (*J.B.A.A.*, **78**, No. 5, 342).

3.52 A. HUNTER, Recent Advances in Astronomy (*Sci. Progr.*, **34**, No. 136, 751).

3.53 C. DE JAGER, The Structure of the Chromosphere and the Low Corona (*Radio Astronomy*, ed.: R. N. Bracewell, Stanford Univ. Press, 1959).

3.54 M. KOMESAROFF, Polarization Measurements of the Three Special Types of Solar Radio Burst (*Aust. J. Phys.*, **11**, 201, 1958).

3.55 M. LAMMERER, German Amateur's Solar Telescope (*Sky & Tel.*, **58**, 1, 78).

3.56 R. C. MAAG, Building a Recorder for Sudden Enhancement of Atmospherics (*Sky & Tel.*, **45**, No. 6, 392).

3.57 R. C. MAAG *et al.* (Eds.), *Observe and Understand the Sun* (Astronomical League, Washington, U.S.A., 1977).

3.58 J. C. MABY, A Programme for the Observation and Photography of Sunspots (*J.B.A.A.*, **41**, No. 4, 190).

3.59 J. C. MABY, Sunspot Photography with a small Visual Refractor (*J.B.A.A.*, **47**, 321).

3.60 E. W. MACKMAN, An Easily Adjusted Radio Telescope (*J.B.A.A.*, **89**, No. 4, 356).

3.61 S. W. MATHERS & G. A. J. FERRIS, Aluminised Filters for Solar Photography (*J.B.A.A.*, **79**, No. 5, 376).

3.62 E. W. MAUNDER, Eclipse Suggestions (*J.B.A.A.*, **15**, No. 8, 317).

3.63 E. W. MAUNDER, Carrington's Method of Determining the Positions of Sun-spots (*J.B.A.A.*, **21**, No. 2, 94).

3.64 E. W. MAUNDER, Carrington's Method of Observing Sun-spots (*Mem. B.A.A.*, **23**, No. 2, 65).

3.65 D. J. McCLEAN, Solar Radio Emission of Spectral Type 4 and its Association with Geomagnetic Storms (*Aust. J. Phys.*, **12**, 404, 1959).

3.66 A. MAXWELL, Radio Emissions from Solar Flares (*Sky & Tel.*, **46**, No. 7, 4).

3.67 A. MAYNE, Stonyhurst Discs—How to Make Them (*J.B.A.A.*, **78**, No. 5, 356).

3.68 K. J. MEDWAY, An Observation of a Solar Flare in White Light on 1975 July 3 (*J.B.A.A.*, **85**, No. 5, 423).

3.69 J. MEEUS, The Next Decade of Solar Eclipses (*Sky & Tel.*, **52**, No. 2, 97).

3.70 D. H. MENZEL, *Our Sun* (Blakiston, Philadelphia, 1949).

3.71 M. MINNAERT, G. F. W. MULDERS & J. HOUTGAST, *Photometric Atlas of the Solar Spectrum from* λ*6312 to* λ*8771, with an Appendix from* λ*3332 to* λ*3637* (Amsterdam, 1940). [The Utrecht Atlas.]

3.72 S. K. MITRA, *The Upper Atmosphere* (Royal Asiatic Society of Calcutta, Second edn: 1952).

3.73 H. W. NEWTON, Early Visual Observations of Bright Chromospheric Eruptions on the Sun's Disk (*J.B.A.A.*, **50**, No. 8, 273).

3.74 H. W. NEWTON, A Famous Sunspot and its Epilogue (*J.B.A.A.*, **54**, No. 9, 244).

3.75 H. W. NEWTON, Sunspots, Bright Eruptions and Magnetic Storms (*Obs.* **62**, No. 787, 318).

3.76 J. B. NEWTON, A Spectroscope Attachment for Viewing Solar Prominences (*Sky & Tel.*, **39**, No. 2, 120).

3.77 A. A. NEYLAN, An Association between Solar Radio Bursts at Metre and Centimetre Wavelengths (*Aust. J. Phys.*, **12**, 399, 1959).

3.78 T. VON OPPOLZER, *Canon der Finsternisse* (Dover, 1963). [Data of 8000 solar and 5000 lunar eclipses from B.C. 1205 to A.D. 2152.]

3.79 E. PETTIT, The Properties of Solar Prominences as Related to Type (*Ap. J.*, **98**, 6).

3.80 E. W. PIINI, A Three-Way Camera for Solar Eclipse Photography (*Sky & Tel.*, **46**, No. 3, 187).

3.81 R. PIKE, A White-Light Solar Telescope (*Sky & Tel.*, **60**, No. 3, 245).

3.82 H. PINNOCK, Photographing the Sun (*J.B.A.A.*, **55**, 124).

3.83 J. G. PORTER, The Heliographic Co-ordinates of Sunspots (*J.B.A.A.*, **53**, No. 2, 63).

3.84 K. RAWER, *The Ionosphere* (Crosby-Lockwood, 1958).

3.85 J. A. ROBERTS, Evidence of Echoes in the Solar Corona from a New Type of Radio Burst (*Aust J. Phys.*, **11**, 215, 1958).

3.86 F. ROUVIÈRE, A Telescope for Solar Photography (*Sky & Tel.*, **56**, No. 4, 356).

3.87 W. SCHMIEDECK, A Simple Technique for Recording the Sun's Spectrum (*Sky & Tel.*, **57**, 4, 395).

3.88 F. J. SELLERS, A Note on the Spectrohelioscope and a Description of a Vibrating Slit Mechanism (*J.B.A.A.*, **48**, No. 6, 243).

3.89 F. J. SELLERS, Estimation of Sunspot Areas (*J.B.A.A.*, **49**. No. 7, 256).

3.90 F. J. SELLARS, Co-operative Sunspot Observation (*J.B.A.A.*, **49**, No. 2, 75).

3.91 F. J. SELLERS, Motions of Solar Prominences (*J.B.A.A.*, **51**, No. 6, 221).

3.92 C. A. SHAIN & C. S. HIGGINS, Location of the Sources of 19 mc/s Solar Bursts (*Aust. J. Phys.*, **12**, 357, 1959).

3.93 O. B. SLEE, The Occultation of a Radio Star by the Solar Corona (*Obs.*, **76**, 228, 1956).

3.94 J. R. SMITH, Radio Observations of the Sun at 136 Mc/s for the Period 1963 May to 1964 December (*J.B.A.A.*, **75**, No. 2, 87).

3.95 J. R. SMITH, A Radio Telescope for 136 Mc/s (*J.B.A.A.*, **75**, No. 4, 250).

3.96 H. T. STETSON, *Sunspots in Action* (N.Y., 1947).

3.97 H. T. STETSON, *Sunspots and Their Effects* (McGraw-Hill, N.Y., 1937).

3.98 G. SWARUP & R. PARTHASARATHY, Solar Brightness Distribution at a Wavelength of 60 cms. (*Aust. J. Phys.*, **11**, 338, 1958).

3.99 F. N. VEIO, A Miniaturized Spectrohelioscope (*J.B.A.A.*, **85**, No. 3, 242).

3.100 D. J. WADSWORTH, Some Observations on the White Light Solar Faculae (*J.B.A.A.*, **88**, No. 5, 444).

3.101 J. P. WILD, K. V. SHERIDAN, & A. A. NEYLAN, An Investigation of the Speed of the Solar Disturbances responsible for Type III Radio Bursts (*Aust. J. Phys.*, **12**, 369, 1959).

3.102 W. REES WRIGHT, Solar Observations with Unsilvered Mirrors (*J.B.A.A.*, **54**, No. 4/5, 103).

3.103 A. T. YOUNG, The Problem of Shadow Band Observations (*Sky & Tel.*, **43**, No. 5, 291).

3.104 *Publ. Astrophys. Obs., Potsdam*, **2**, Tables 33, 34; 1881.

3.105 *Publ. Astrophys. Obs., Potsdam*, **1**, 1879.

3.106 *Quarterly Bulletin on Solar Activity* (I.A.U., Zurich, 1939–). (Formerly *Bulletin for Character Figures of Solar Phenomena* (Zurich, 1917–1938).]

3.107 Solar Flares in White Light (*Obs.*, **67**, No. 839, 156).

3.108 [VARIOUS] Some Hints for Photographers of Total Solar Eclipses (*Sky & Tel.*, **45**, No. 5, 322).

4. MOON

[*a*. Maps and specialised works; arranged chronologically.]

4.1 W. BEER & J. H. MÄDLER, *Der Mond nach seinen kosmischen und individuellen Verhältnissen oder Allgemeine vergleichende Selenographie* (Berlin, 1837). [Marks the birth of modern selenography.]

4.2 W. BEER & J. H. MÄDLER, *Mappa Selenographica* (Berlin, 1837). [Accompanying B.4.1.]

4.3 J. F. J. SCHMIDT, *Der Mond* (Leipzig, 1856).

4.4 E. NEISON, *The Moon, and the Condition and Configurations of its Surface* (Longmans, Green, 1876. [Comprehensive but rather out of date. Map (diameter 24 ins) included in sections in text.]

4.5 J. F. J. SCHMIDT, *Charte der Gebirge des Mondes* (Reimer, Berlin, 1878). [Schmidt's great contribution to selenography.]

4.6 W. G. LOHRMANN: *Mondcharte* (Leipzig, 1878). [25 charts, with descriptions.]

4.7 R. A. PROCTOR, *The Moon – Her Motions, Aspect, Scenery, and Physical Condition* (Longmans, Green, 1878). [Includes a 12-in map based on B.4.2. One of the standard works in English, though now rather badly dated.]

4.8 T. G. ELGER, *The Moon – A Full Description and Map of its Principal Physical Features* (Philip, Lond., 1895). [Map, diameter 18 ins, in 4 quadrants, with index of formations; new edition of Map, edited by H. P. Wilkins (Philip, 1950).]

4.9 M. M. LOEWY & M. P. PUISEUX, *Atlas Photographique de la Lune* Paris Observatory, 1896–1900). [Issued in 12 parts.]

4.10 N. S. SHALER, A Comparison of the Features of Earth and the Moon (*Smithson. Contr.*, 1903).

4.11 J. NASMYTH & J. CARPENTER, *The Moon considered as a Planet, a World, and a Satellite* (Murray, 4th edn., 1903; 1st edn., 1874). [Rather badly out of date.]

4.12 W. H. PICKERING, *The Moon – A Summary of Existing Knowledge of Our Satellite, with a Complete Photographic Atlas* (Doubleday, Page, N.Y., 1903). [Moon divided into 16 areas, each photographed 5 times (sunrise, morning, noon, evening, sunset), plus key photographs and charts of the 4 quadrants.]

4.13 P. FAUTH, *The Moon in Modern Astronomy* (Owen, Lond., 1907).

4.14 P. PUISEUX, *La Terre et la Lune – Forme Extérieure et Structure Interne* (Gauthier-Villars, Paris, 1908).

4.15 W. GOODACRE, *Lunar Map* (1910). [One of the most accurate; based on 1433 surveyed points. Original 77 ins diameter; published in 25 sections reduced to 60 ins diameter. Reproduced in B.4.20.]

4.16 J. N. KRIEGER, *Mond-Atlas* (Vienna, 1912) [Vol. 1, text; vol. 2, atlas.]

4.17 M. A. BLAGG, Collated List of Lunar Formations Named or Lettered in the Maps of Neison, Schmidt and Mädler. (*Internat. Ass. Acad.*, Edinburgh, 1913).

4.18 H. P. WILKINS, *A New Map of the Moon* (Lond., 1924) [24 sections plus index map; diameter about 60 ins.]

4.19 K. ANDĚL, *Mappa Selenographica* (Prague, 1926) [Single sheet, diameter about 24 ins, with key map of same size.]

4.20 W. GOODACRE, *The Moon, with a Description of its Surface Formations* (Bournemouth, 1931) [One of the finest modern works on the Moon. The lunar map, B.4.15 is reproduced here on reduced scale.]

4.21 M. A. BLAGG & K. MÜLLER, *Named Lunar Formations* (Lund, Humphreys, 1935).

4.22 W. H. WESLEY & M. A. BLAGG, *I.A.U. Map of the Moon* (Lund, Humphreys, 1935). [Being vol. 2 of B.4.21. 14 charts; lunar diameter about 90 cm.]

4.23 P. FAUTH, *Unser Mond* (Breslau, 1936).

4.24 P. FAUTH, *Übersichtskarte des Mondes* (Breslau, 1936). [6 charts measuring about 30 x 40 cm.]

4.25 F. C. LAMÈCH, *Carte Générale de la Lune* (Gizard, Barrère et Thomas, Paris, 1946). [Single sheet, diameter about 24 ins.]

4.26 H. P. WILKINS, *Map of the Moon* (Lond., 3rd edn. 1951). [The great 300-in diameter map; in 22 sections and 3 special Libratory Maps.]

4.27 R. B. BALDWIN, *The Face of the Moon* (University of Chicago Press, 1949).

4.28 D. W. G. ARTHUR, *The Diameters of Lunar Craters* (pub. by the author, 1951). [Part 1 (including 1000 craters) of a catalogue of crater diameters.]

4.29 H. P. WILKINS & P. A. MOORE, *The Moon* (Faber & Faber, 1955).

4.30 G. P. KUIPER (Ed.), *Photographic Lunar Atlas* (University of Chicago Press, 1960).

4.31 R. B. BALDWIN, *The Measure of the Moon* (University of Chicago Press, 1963).

4.32 E. A. WHITAKER *et al.*, *Rectified Lunar Atlas* (University of Arizona Press, 1963).

4.33 D. W. G. ARTHUR & A. P. AGNIERAY, *Lunar Designation and Positions*, University of Arizona Press, 1964).

4.34 S. MIYAMOTO & A. HATTORI, *Photographic Atlas of the Moon* (University of Kyoto, Japan, 1964).

4.35 R. B. BALDWIN, *A Fundamental Survey of the Moon* (McGraw-Hill, U.S.A., 1965).

4.36 D. ALTER (Ed.), *Lunar Atlas* (Dover, U.S.A., 1968).

4.37 Z. KOPAL, *The Moon* (Reidel, 1969).

4.38 P. MOORE, *Moon Flight Atlas* (G. Philip & Son, 1970).

4.39 Z. KOPAL, *A New Photographic Atlas of the Moon* (R. Hall & Co., 1971).

4.40 P. MOORE, *A Survey of the Moon* (Norton, U.S.A., 1974).

4.41 Z. KOPAL & R. W. CARDER, *Mapping of the Moon* (Reidel, 1974).

4.42 H. R. POVENMIRE, *Graze Observer's Handbook* (Vantage Press, U.S.A., 1975).

4.43 D. ALTER, *Pictorial Guide to the Moon* (T. Y. Crowell, U.S.A., 1979).

4.44 J. MEEUS & H. MUCKE, *Canon of Lunar Eclipses −2002 to +2526* (Astronomisches Büro Wien, Vienna, 1979).

[*b*. Papers and miscellaneous works referred to in the text; arranged alphabetically.]

4.45 J. ASHBROOK, Some Very Thin Lunar Crescents (*Sky & Tel.*, **42**, No. 2, 78).

4.46 J. ASHBROOK, More About the Visibility of the Lunar Crescent (*Sky & Tel.*, **43**, No. 2, 95).

4.47 R. BARKER, Physical Change on the Moon (*J.B.A.A.*, **48**, No. 9, 347).

4.48 L. F. BALL, The Lunar Mare Marginalis (*J.B.A.A.*, **47**, No. 7, 260).

4.49 L. J. COMRIE, *A Short Semi-Graphical Method of Predicting Occultations* (H.M.S.O., 1926). [Copies obtainable from Scientific and Computing Service, 23 Bedford Square, London, W.C.1.]

4.50 M. DAVIDSON, The Reduction of Occultations for Stars Fainter than Magnitude 7.5 (*J.B.A.A.*, **48**, No. 3, 120). [See also B.105, 93.]

4.51 M. DAVIDSON, Note on Mr F. Robbins' Paper (*J.B.A.A.*, **48**, No. 4, 175). [Refers to B.4.76.]

4.52 L. E. FITTON, Transient Lunar Phenomena—a New Approach (*J.B.A.A.*, **85**, No. 6, 511).

4.53 L. E. FITTON, A Photoelectric Lunar Scanner (*J.B.A.A.*, **89**, No. 5, 465).

4.54 J. T. FOXELL, Lunar Occultation Maps (*Mem. B.A.A.*, **30**, No. 3, 107).

4.55 W. H. HAAS, Colour Changes on the Moon (*Pop. Astr.*, **45**, No. 6, 337).

4.56 W. H. HAAS, Does Anything Ever Happen on the Moon? (*J. Can. R.A.S.*, **36**, No. 6, 237; No. 7, 317; No. 8, 361; No. 9, 397).

4.57 C. H. HAMILTON, Lunar Changes (Eratosthenes region) (*Pop. Astr.*, **32**, 327).

4.58 H. R. HATFIELD, Lunar Photography for Beginners (*J.B.A.A.*, **76**, No. 2, 90).

4.59 J. J. HILL, Some Notes on Lunar Photography with a 4¼-inch Refractor (*J.B.A.A.*, **55**, No. 3, 77).

4.60 R. T. A. INNES, Reduction of Occultations of Stars by the Moon (*A.J.*, **35**, No. 835, 155).

4.61 B. LYOT, Recherches sur la polarisation de la lumière des planètes . . . . (*Ann. d'Obs. de Paris, sect. de Meudon*, **8**, part 1, 38).

4.62 N. A. KOSYREV, Luminescence of the Lunar Surface and the Intensity of Corpuscular Radiation from the Sun (*Publ. Crimean Astrophys. Obs.*, **16**, 148).

4.63 N. A. KOSYREV, Observation of a Volcanic Process on the Moon (*Sky and Tel.*, **18**, 184).

4.64 P. MOORE, Areas on the Moon Suspected of Variability (*J.B.A.A.*, **75**, No. 2, 119).

4.65 P. MOORE, Report on Transient Phenomena (*J.B.A.A.*, **77**, No. 1, 47).

4.66 P. MOORE, Transient Lunar Phenomena: A Review, 1967 (*J.B.A.A.*, **78**, No. 2, 138).

4.67 P. MOORE, Extension of the Chronological Catalogue of Reported Lunar Events: October 1967—June 1971 (*J.B.A.A.*, **81**, No. 5, 365).

4.68 T. L. MACDONALD, The Altitudes of Lunar Craters (*J.B.A.A.*, **39**, No. 8, 314).

4.69 T. L. MACDONALD, Studies in Lunar Statistics (*J.B.A.A.*, **41**, 172, 228, 288, 367; **42**, 291).

4.70 E. PETTIT & S. B. NICHOLSON, Lunar Radiation and Temperatures (*Ap.J.*, **71**, No. 2, 102 = *M.W.C.*, 392).

4.71 E. PETTIT, Radiation Measurements on the Eclipsed Moon (*Ap.J.*, **91**, No. 4, 408 = *M.W.C.*, 627).

4.72 W. H. PICKERING, Eratosthenes, I: A Study for the Amateur (*Pop. Astr.*, **27**, 579); No. 2 (*ibid.*, **29**, 404); No. 3 (*ibid.*, **30**, 257); No. 4 (*ibid.*, **32**, 69); No. 5 (*ibid.*, **32**, 302); No. 6 (*ibid.*, **32**, 393).

4.73 T. RACKHAM, *Astronomical Photography at the Telescope* (Faber, London, 1959). 3rd edn. 1972.

4.74 C. L. RICKER, Lunar Transient Phenomena: the A.L.P.O. Programme (*J.B.A.A.*, **78**, No. 3, 217).

4.75 W. F. RIGGE, *The Graphic Construction of Eclipses and Occultations* (Loyola University Press, Chicago, 1924).

4.76 F. ROBBINS, Remarks on Dr Davidson's Paper, (*J.B.A.A.*, **48**, No. 4, 171). [See also B. 4.50, 4.51.]

4.77 D. H. SADLER & M. W. P. RICHARDS, The Prediction and Reduction of Occultations (H.M.S.O., Supplement to *N.A.*, 1938).

4.78 P. K. SARTORY, A Method of Rendering Obvious Small Differences of Colour or Contrast Observations (*J.B.A.A.*, **75**, No. 2, 98).

4.79 P. K. SARTORY, Report on Transient Phenomena (*J.B.A.A.*, **77**, No. 1, 47).

4.80 S. A. SAUNDER, The Determination of Selenographic Positions and the Measurement of Lunar Photographs (*M.N.*, **60**, No. 3, 174).

4.81 S. A. SAUNDER, The Determination of Selenographic Positions . . . Determination of a first group of Standard Points by Measures made at the Telescope and on Photographs (*M.N.*, **62**, No. 1, 41).

4.82 S. A. SAUNDER, The Determination of Selenographic Positions . . . Results of the Measurement of Four Paris Negatives (*Mem. R.A.S.*, **57**, 1).

4.83 G. E. TAYLOR, Occultations (*J.B.A.A.*, **81**, No. 1, 16).

4.84 V. VAND, A Theory of the Evolution of the Surface Features of the Moon (*J.B.A.A.*, **55**, 47).

4.85 R. L. WATERFIELD, The Reappearance of Stars from Occultation (*J.B.A.A.*, **82**, No. 1, 46).

4.86 C. T. WHITMELL, Stellar Occultation (*J.B.A.A.*, **23**, No. 4, 182).

4.87 H. P. WILKINS, The Lunar Mare 'X' (*J.B.A.A.*, **48**, No. 2, 80).

4.88 H. P. WILKINS, The Lunar Mare Australe (*J.B.A.A.*, **50**, No. 9, 304).

4.89 H. P. WILKINS, Total Lunar Eclipse, 1942 March 2 (*J.B.A.A.*, **52**, No. 3, 108).

4.90 H. P. WILKINS, The Total Eclipse of the Moon, 1942 Aug 26 (*J.B.A.A.*, **52**, No. 9, 297).

4.91 H. P. WILKINS, Lunar Thermal Researches (*J.B.A.A.*, **53**, No. 2, 86).

4.92 H. P. WILKINS, A Thermal Eyepiece (*J.B.A.A.*, **54**, No. 2, 38).

4.93 J. YOUNG, A Statistical Investigation of the Diameters and Distribution of Lunar Craters (*J.B.A.A.*, **50**, No. 9, 309).

4.94 – Lunar Occultations (*N.A.*, 1950, pp. 592–597).

4.95 – Lunar Memoirs of the B.A.A. (*Mem. B.A.A.*, 1, 2, 3, 7, 10, 13, 20, 23, 32, 34, 36).

4.96 – Report of the General Ordinary Meeting (*J.B.A.A.*, **54**, No. 8/9, 150).

5. PLANETS:MISCELLANEOUS

5.1 R. M. BAUM, *The Planets: Some Myths and Realities* (David & Charles, 1973).

5.2   J. H. BOTHAM, Planetary Photography with a 9-inch Refractor (*J.B.A.A.*, **71**, No. 4, 152).

5.3   C. R. CHAPMAN, *The Inner Planets* (Charles Scribner's Sons, U.S.A., 1977).

5.4   F. J. HARGREAVES, Presidential Address to the B.A.A. (*J.B.A.A.*, **55**, No. 1, 1).

5.5   W. K. HARTMAN, *Moons and Planets* (Bodger & Quigley, U.S.A., 1972).

5.6   F. W. HYDE & R. FULFORD-JONES, An Investigation into the Use of Colour Filters in Visual Observation (*J.B.A.A.*, **72**, No. 4, 163). Discussion, *J.B.A.A.*, **72**, No. 6, 289.

5.7   Z. KOPAL, *The Realm of the Terrestrial Planets* (Wiley, U.S.A., 1979).

5.8   G. KUIPER & B. M. MIDDLEHURST (Eds.), *Planets and Satellites,* (University of Chicago Press, 1961).

5.9   S. W. MILBOURN, Explanation of the Symbols used in Orbital Elements (*J.B.A.A.*, **85**, No. 4, 329).

5.10 R. B. MINTON, Hints on Planetary Photography for Amateurs (*Sky & Tel.*, **40**, No. 1, 56; and **40**, No. 2, 116).

5.11 P. A. MOORE, *Guide to the Planets* (Eyre & Spottiswoode, 1955).

5.12 R. S. PRICE, Planetary Photography at High Resolution (*Sky & Tel.*, **52**, No. 3, 220).

5.13 J. HEDLEY ROBINSON, The Use of Colour Filters in Visual Planetary Observation (*J.B.A.A.*, **90**, No. 5, 434).

5.14 G. D. ROTH, *Handbook for Planet Observers* (Faber & Faber, 1970).

5.15 W. SANDNER, *Satellites of the Solar System* (Faber & Faber, 1965).

5.16 E. C. SLIPHER, *Photographic Study of the Brighter Planets* (National Geographical Society, U.S.A., 1964).

5.17 S. WHEATCRAFT, Finding Faint Planetary Satellites (*Sky & Tel.*, **54**, No. 3, 243) [photography].

5.18 F. L. WHIPPLE, *Earth, Moon and Planets* (Harvard University Press, 1963).

5.19 J. A. WOOD, *The Solar System* (Prentice-Hall, U.S.A., 1979).

5.20 – *The Solar System* (W. H. Freeman & Co., U.S.A., 1976) [reprints of features from *Scientific American*].

6. MERCURY

6.1   E. M. ANTONIADI, *La Planète Mercure et la Rotation des Satellites* (Gauthier-Villars, 1934).

6.2   P. LOWELL, New Observations of Mercury (*Mem. Amer. Acad. Arts Sci.*, **12**, 1897).

6.3   H. McEWEN, Mercury, Part III (*J.B.A.A.*, **39**, No. 8, Plates VII, VIII).

6.4   H. McEWEN, The Markings of Mercury (*J.B.A.A.*, **46**, No. 10, 382).

6.5   H. McEWEN, *Mercury and Venus Section Observing Notes* (B.A.A. Sectional Notes, No. 3).

6.6 H. C. NIGHTINGALE, A Suspected Phase Anomaly of Mercury (*J.B.A.A.*, **78**, No. 1, 45).

6.7 G. V. SCHIAPARELLI, Sulla rotazione di Mercurio (*A.N.*, **123**, No. 2944, 242).

6.8 R. G. STROM, The Planet Mercury as Viewed by Mariner 10 (*Sky & Tel.*, **47**, No. 6, 360).

6.9 — Planispheres of Mercury (*J.B.A.A.*, **46**, No. 10, 357, Plate I).

7. VENUS

7.1 E. M. ANTONIADI, The Markings and Rotation of the Planet Venus (*J.B.A.A.*, **44**, No. 9, 341).

7.2 J. ASHBROOK, Some Naked-Eye Observations of Venus (*Sky & Tel.*, **53**, No. 1, 12).

7.3 C. BOYER, The Four-day Rotation of Venus' Atmosphere (*J.B.A.A.*, **83**, No. 5, 363).

7.4 V. A. BRONSHTEN, Investigations of Schroter Effect in the U.S.S.R. (*J.B.A.A.*, **81**, No. 3, 181).

7.5 R. H. CHAMBERS & J. TAYLOR, An Investigation into the Phase Anomaly of Venus (*J.B.A.A.*, **76**, No. 5, 310).

7.6 B. F. DAVIES, Photometric Considerations in Observing Venus (*J.B.A.A.*, **73**, No. 5, 188).

7.7 F. V. DAVIES, The Phase Anomaly of Venus. Does Refraction Play a Part? (*J.B.A.A.*, **82**, No. 5, 341).

7.8 P. DEVADAS, Bands and Belts on Venus (*J.B.A.A.*, **73**, No. 4, 165).

7.9 E. L. ELLIS, The Dichotomy of Venus (*J.B.A.A.*, **84**, No. 5, 351).

7.10 C. FLAMMARION, *La Planète Vénus – discussion générale des observations* (Paris, 1897).

7.11 A. W. HEATH *et al.*, Filter Observations of Venus in 1959 (*J.B.A.A.*, **71**, No. 6, 242).

7.12 D. C. HEGGIE, On the Phase Anomaly of the Inner Planets (*J.B.A.A.*, **80**, No. 4, 288).

7.13 J. HISCOTT, Ultraviolet Observations of Venus in 1969 (*J.B.A.A.*, **82**, No. 3, 198).

7.14 W. J. LEATHERBARROW, The Rotation of Venus – Observations Made during the Eastern Elongation of Spring 1964 (*J.B.A.A.*, **81**, No. 3, 177).

7.15 J. MEEUS, The Inferior Conjunctions of Venus, 1960 to 2023 (*J.B.A.A.*, **81**, No. 2, 114).

7.16 P. MOORE, *The Planet Venus* (Faber, London, 1959).

7.17 P. MOORE, The Future of the Mercury and Venus Section (*J.B.A.A.*, No. 4, pp. 167–173, 1960).

7.18 T. RACKHAM, Photography of Venus, 1956 (*J.B.A.A.*, **67**, No. 5, 160) [photography in ultra-violet light].

7.19 J. HEDLEY ROBINSON, Why Observe Venus? (*J.B.A.A.*, **90**, No. 1, 36).

7.20 B. WARNER, The Observation of Detail on the Planet Venus (*J.B.A.A.*, **71**, No. 5, 202).

7.21 B. WARNER, The Phase Anomaly of Venus (*J.B.A.A.*, **73**, No. 2, 65).

7.22 A. T. YOUNG & L. G. YOUNG, Observing Venus Near the Sun (*Sky & Tel.*, **43**, No. 3, 140).

8. MARS

8.1   E. M. ANTONIADI *et al.*, Reports of the Mars Section of the B.A.A. (*Mem. B.A.A.*, **2**, 1892; **4** & **7**, 1896; **27**, 1919–20; **37**, 1941).

8.2   E. M. ANTONIADI, *La Planète Mars, 1659–1929* (Hermann, Paris, 1930).

8.3   E. M. ANTONIADI, La Planète Mars en 1935 (*B.S.A.F.*, **49**, 401).

8.4   M. ATCHISON, A Major Change on Mars – Developments in the Region of Aethiopis, 1958–1963 (*J.B.A.A.*, **73**, No. 7, 265).

8.5   J. BLUNCK, *Mars and Its Satellites* (Exposition Press, U.S.A., 1977).

8.6   C. F. CAPEN, Martian Yellow Clouds – Past and Future (*Sky & Tel.*, **41**, No. 2, 117).

8.7   C. F. CAPEN & L. J. MARTIN, Mars' Great Storm of 1971 (*Sky & Tel.*, **43**, No. 5, 276).

8.8   W. W. COBLENZ, Climatic Conditions on Mars (*Pop. Astr.*, **33**, No. 5, 310; No. 6, 363).

8.9   W. W. COBLENZ & Co. O. LAMPLAND, Radiometric Measurements on Mars in 1924 (*Pop. Astr.*, **32**, No. 9, 570).

8.10 C. FLAMMARION, *La Planète Mars* (Gauthier-Villars, Paris, vol. 1 1892, vol. 2 1909).

8.11 E. GOFFIN & J. MEEUS, Mars' Closest Approaches to Earth (*Sky & Tel.*, **56**, No. 2, 106).

8.12 N. E. GREEN, Observations of Mars at Madeira, August and September 1877 (*M.N.*, **38**, No. 1, 38).

8.13 N. E. GREEN, Observations of Mars at Madeira, in August and September 1877 (*Mem. R.A.S.*, **44**, No. 3, 123).

8.14 A. W. HEATH *et al.*, Mars Through Colour Filters, 1958 (*J.B.A.A.*, **70**, No. 6, 270).

8.15 J. L. INGE *et al.*, A New Map of Mars from Planetary Patrol Photographs (*Sky & Tel.*, **41**, No. 6, 336).

8.16 A. P. LENHAM, Observations of Mars in 1958 (*J.B.A.A.*, **74**, No. 4, 128).

8.17 P. LOWELL, *Mars* (Longmans, Green, 1896).

8.18 P. LOWELL, *Mars and its Canals* (NY ., 1906).

8.19 P. LOWELL, *Mars as the Abode of Life* (Macmillan, N.Y., 1909).

8.20 J. T. MELKA & R. MELVIN, Suggestions for a Photographic Patrol of Mars (*Sky & Tel.*, **50**, No. 6, 424).

8.21 P. MOORE, *Guide to Mars* (W. W. Norton & Co., U.S.A., 1977).

8.22 E. PETTIT & S. B. NICHOLSON, Measurements of the Radiation from the Planet Mars (*Pop. Astr.*, **32**, No. 10, 601).

8.23 W. H. PICKERING, Mars Reports: Nos. 1–7 (*Pop. Astr.*, **22**, 1914); Nos. 8–12 (*ibid.*, **23**, 1915); Nos. 13–17 (*ibid.*, **24**, 1916); No. 18 (*ibid.*, **25**, 1917); Nos. 19–20 (*ibid.*, **26**, 1918); No. 21 (*ibid.*, **27**, 1919); No. 22 (*ibid.*, **28**, 1920); No. 23 (*ibid.*, **29**, 1921); No. 24 (*ibid.*, **30**, 1922); Nos. 25–26 (*ibid.*, **31**, 1923);

Nos. 27–28 (*ibid.*, **32**, 1924); Nos. 29–31, (*ibid.*, **33**, 1925); Nos. 32–37 (*ibid.*, **34**, 1926); Nos. 38–40 (*ibid.*, **35**, 1927); Nos. 41–42 (*ibid.*, **36**, 1928); No. 43 (*ibid.*, **37**, 1929); No. 44 (*ibid.*, **38**, 1930).

8.24 G. DE VAUCOULEURS (trans., P. A. MOORE), *The Planet Mars* (Faber, 1950).

8.25 G. DE VAUCOULEURS, A New Variation on Mars (*Sky & Tel.*, **49**, No. 4, 222).

8.26 — *Ann. Lowell Obs.*, **3**, Supplement, 24, 1905.

9. JUPITER

9.1 L. E. ALSOP *et al.*, in *Radio Astronomy*, p. 69 (ed: R. N. Bracewell, Stanford Univ. Press, 1959).

9.2 C. H. BARROW, Thesis, London Univ. (1958).

9.3 C. H. BARROW, The Latitudes of Radio Sources on Jupiter (*J.B.A.A.*, **69**, 211, 1959).

9.4 C H. BARROW & T. D. CARR, A Radio Investigation of Planetary Radiation (*J.B.A.A.*, **67**, 200, 1957).

9.5 D. P. BAYLEY, Colour of Jupiter's Polar Regions (*J.B.A.A.*, **55**, No. 5, 116).

9.6 C. R. CHAPMAN, Dependence of the Prominence of Jupiter's Red Spot on Differential Atmospheric Flow (*J.B.A.A.*, **78**, No. 5, 371).

9.7 R. DOEL, An Update on South Equatorial Belt Disturbance Analysis (*Strolling Astronomer*, No. 26, 254).

9.8 D. DUTTON, Naked-eye Observations of Jupiter's Moons (*Sky & Tel.*, **52**, No. 6, 482).

9.9 W. E. FOX, The Great Red Spot on Jupiter (*J.B.A.A.*, **78**, No. 1, 16).

9.10 F. J. HARGREAVES, How to Observe Jupiter, and Why (*J.B.A.A.*, **60**, No. 7, 187).

9.11 A. W. HEATH *et al.*, Observations of Jupiter Using Colour Filters (*J.B.A.A.*, **75**, No. 1, 45).

9.12 A. W. HEATH & J. HEDLEY ROBINSON, Jupiter Through Colour Filters, 1975/1976 (*J.B.A.A.*, **87**, No. 5, 485).

9.13 A. E. LEVIN, [Presidential Address to the B.A.A., 1931] (*J.B.A.A.*, **42**, No. 1).

9.14 A. E. LEVIN, Mutual Eclipses and Occultations of Jupiter's Satellites (*Mem. B.A.A.*, **30**, No. 3, 149).

9.15 B. LYOT, Observations Planétaires au Pic du Midi en 1941 (*B.S.A.F.*, April 1943).

9.16 E. F. McCLAIN & R. M. SLOANAKER, *Radio Astronomy*, p. 61 (ed.: R. N. Bracewell, Stanford Univ. Press, 1959).

9.17 R. A. McINTOSH, Colour Variation in Jupiter's Equatorial Zone (*J.B.A.A.*, **46**, No. 8, 285).

9.18 R. A. McINTOSH, Disturbance on Jupiter's South Equatorial Region (*J.B.A.A.*, **60**, No. 8, 247).

9.19 D. W. MILLAR, Variations of Surface Features on Jupiter (*J.B.A.A.*, **54**, No. 8/9, 162).

BIBLIOGRAPHY

9.20 R. B. MINTON, Initial Development of the 1971 June South Equatorial Belt Disturbance on Jupiter (*J.B.A.A.*, **83**, No. 4, 263).

9.21 B. M. PEEK, A Hint to Observers of Jupiter (*J.B.A.A.*, **47**, No. 4, 154).

9.22 B. M. PEEK, [Presidential Address to the B.A.A., 1939] (*J.B.A.A.*, **50**, No. 1).

9.23 B. M. PEEK, [Presidential Address to the B.A.A., 1940] (*J.B.A.A.*, **51**, No. 1).

9.24 B. M. PEEK, *The Planet Jupiter* (Faber, London, 1958).

9.25 B. M. PEEK, The Determination of the Longitudes of Sources of Emission of Radio Noise on Jupiter (*J.B.A.A.*, **69**, 70, 1959).

9.26 B. M. PEEK, On the Life and Continuity of Jupiter's South Tropical Disturbance (*J.B.A.A.*, **73**, No. 3, 109).

9.27 J. H. ROGERS, Disturbances and Dislocations on Jupiter (*J.B.A.A.*, **90**, No. 2, 132).

9.28 T. SATO, A Possible Interpretation of the Changeable Aspect of the Great Red Spot on Jupiter (*J.B.A.A.*, **81**, No. 1, 30).

9.29 W. DE SITTER, Jupiter's Galilean Satellites (*M.N.*, **91**, No. 7, 706). [George Darwin Lecture, 1931.]

9.30 C. F. O. SMITH, The Colour of Jupiter's Polar Regions (*J.B.A.A.*, **55**, No. 1, 23).

9.31 C. F. O. SMITH, Jupiter in Retrospect (*J.B.A.A.*, **57**, No. 1, 37).

9.32 F. G. SMITH, A Search for Radiation from Jupiter at 38 mc/s and at 85 mc/s. (*Obs.*, **75**, 252, 1955).

9.33 A. STANLEY WILLIAMS, *Zenographical Fragments – The Motions and Changes of the Markings of Jupiter in the Apparition 1886–87* (Mitchell & Hughes, Lond., 1889).

9.34 A. STANLEY WILLIAMS, *Zenographical Fragments II – The Motions and Changes of the Markings on Jupiter in 1888* (Taylor & Francis, Lond., 1909).

9.35 A. STANLEY WILLIAMS, On the Observed Changes in the Colour of Jupiter's Equatorial Zone (*M.N.*, **80**, No. 5, 467).

9.36 A. STANLEY WILLIAMS, Periodic Variation in the Colours of the Two Equatorial Belts of Jupiter (*M.N.*, **90**, No. 7, 696).

9.37 A. STANLEY WILLIAMS, The Colour Variations of Jupiter's Equatorial Zones (*J.B.A.A.*, **47**, No. 2, 68).

9.38 A. STANLEY WILLIAMS, The Colour Variations of Jupiter's Equatorial Belts (*Obs.*, **60**, No. 754, 74).

9.39 W. K. WACKER, Large-scale Disturbances on Jupiter (*Strolling Astronomer*, No. 25, 145).

9.40 W. H. WRIGHT, On Photographs of the Brighter Planets by Light of Different Colours (*M.N.*, **88**, No. 9, 709). [George Darwin Lecture, 1928.]

9.41 −33 Reports of the Jupiter Section of the B.A.A., 1891 −(*Mem. B.A.A.*, 1–35), especially Reports No. 27, 29, 32, 33.

9.42 −*J.B.A.A.*,**44**, 219; **43**, 86, 404; **42**, 205, 362; **37**, 62; *Mem. B.A.A.*, **34**, No. 2.

10. SATURN

10.1   M. A. AINSLIE, Photographs of Saturn by the Lick Observatory (*J.B.A.A.*, **55**, No. 5, 125).

10.2   A. F. O'D. ALEXANDER, The Future of the Saturn Section [of the B.A.A.] (*J.B.A.A.*, **57**, No. 1, 45).

10.3   A. F. O'D. ALEXANDER, Intensities of Belts, Zones and Rings, 1946–48 (*J.B.A.A.*, **59**, No. 7, 207).

10.4   A. F. O'D. ALEXANDER, *The Planet Saturn* (Faber & Faber, 1962).

10.5   E. M. ANTONIADI, Observations of Saturn in 1936–1937, when his Ring was seen more or less edgewise (*J.B.A.A.*, **47**, No. 7, 252).

10.6   A. APPLEYARD, The Equatorial Zone of Saturn (*J.B.A.A.*, **79**, No. 1, 40).

10.7   A. APPLEYARD, Saturnicentric Latitude Tables (*J.B.A.A.*, **81**, No. 3, 186).

10.8   R. BARKER, Saturn in 1932 (*J.B.A.A.*, **43**, No. 2, 56).

10.9   R. BARKER, Notes on Saturn in 1933 (*J.B.A.A.*, **44**, No. 2, 74).

10.10   R. BARKER, Notes on Saturn in 1934 (*J.B.A.A.*, **45**, No. 1, 41).

10.11   R. BARKER, Saturn's Satellites in 1936 (*J.B.A.A.*, **47**, No. 4, 152).

10.12   J. L. BENTON, Latitudes of Saturn's Features by Visual Methods (*J.B.A.A.*, **86**, No. 5, 383).

10.13   L. J. COMRIE, Phenomena of Saturn's Satellites (*Mem. B.A.A.*, **30**, No. 3, 97).

10.14   K. DELANO, Magnitude Variations of Saturn's Satellites (*J.B.A.A.*, **79**, No. 2, 124).

10.15   F. O'B. ELLISON, Some Observations on Saturn near the time of the Ring Plane passing through the Sun and the Earth, 1936–1937 (*J.B.A.A.*, **50**, No. 6, 213). [See also *J.B.A.A.*, **50**, No. 7, 230.]

10.16   [W. HAY], Interim Report of the Saturn Section [dealing with observations of Will Hay's prominent white spot on Saturn] (*J.B.A.A.*, **44**, No. 6, 220).

10.17   M. B. B. HEATH, Latitude of Saturn's South Equatorial Belt (*J.B.A.A.*, **62**, No. 6, 202).

10.18   P. H. HEPBURN, The Diameters and Densities of the Six Inner Satellites of Saturn (*J.B.A.A.*, **33**, No. 6, 244).

10.19   P. H. HEPBURN, Note and Correction to [the above paper] (*J.B.A.A.*, **33**, No. 7, 284).

10.20   R. G. HODGSON, Resolving the Iapetus Problem (*J.B.A.A.*, **81**, No. 3, 191).

10.21   H. M. JOHNSON, The White Spot on Saturn's Rings (*J.B.A.A.*, **51**, No. 9, 309).

10.22   A. P. LENHAM, An Analysis of Saturn Intensity Observations (*J.B.A.A.*, **76**, No. 4, 258).

10.23   P. A. MOORE, The Magnitude of Saturn's Satellites (*J.B.A.A.*, **79**, No. 2, 121).

10.24   J. B. MURRAY, Visual Observations of Saturn's Ring D (*J.B.A.A.*, **89**, No. 3, 250).

10.25 R. W. PAYNE, Visual Photometry of Titan (*J.B.A.A.*, **81**, No. 2, 123).

10.26 R. W. PAYNE, Photometry of Iapetus (*J.B.A.A.*, **81**, No. 3, 193).

10.27 K. J. H. PHILLIPS, Recent Latitude Measurements of Features on Saturn (*J.B.A.A.*, **79**, No. 2, 113) [correction, **79**, No. 3, 243].

10.28 T. E. R. PHILLIPS, Micrometer Measures of the Rings and Ball of Saturn (*J.B.A.A.*, **34**, No. 5, 185).

10.29 T. E. R. PHILLIPS, The Rotation of Saturn (*J.B.A.A.*, **44**, No. 1, 29).

10.30 T. E. R. PHILLIPS, Report of Saturn Section [of the B.A.A.] (*J.B.A.A.*, **46**, No. 10, 361).

10.31 J. G. PORTER, Phenomena of Saturn's Satellites (*J.B.A.A.*, **84**, No. 3, 209).

10.32 R. A. PROCTOR, *Saturn and its System* (Longmans, Green, Lond., 1865).

10.33 J. HEDLEY ROBINSON, Colour on Saturn? (*J.B.A.A.*, **85**, No. 1, 34). [See also **85**, No. 4, 360.]

10.34 R. E. SCHMIDT, Disappearances of Saturn's Rings, 1600–2100 (*Sky & Tel.*, **58**, 6, 500).

10.35 E. D. SHERLOCK, Colour of Saturn's Polar Regions (*J.B.A.A.*, **56**, No. 1, 16).

11. URANUS

11.1 A. F. O'D. ALEXANDER, *The Planet Uranus* (Faber & Faber, 1965).

11.2 W. H. STEAVENSON, The Satellites of Uranus (*J.B.A.A.*, **74**, No. 2, 54).

12. ASTEROIDS

12.1 J. ASHBROOK, Amateurs Observe the Rotation of Eros (*Sky & Tel.*, **49**, No. 5, 331).

12.2 N. T. BOBROVNIKOFF, The Spectra of Minor Planets (*L.O.B.*, **14**, No. 407).

12.3 C. R. CHAPMAN & D. MORRISON, The Minor Planets – Sizes and Mineralogy (*Sky & Tel.*, **47**, No. 2, 92).

12.4 T. GEHRELS, *Asteroids* (University of Arizona Press, 1979).

12.5 A. O. LEUSCHNER, Research Surveys of the Orbits of Minor Planets 1–1091, from 1801.0 to 1929.5 (*Publ. Lick Obs.*, **19**, 1935).

12.6 P. D. MALEY, Exploring for Satellites of Minor Planets (*J.B.A.A.*, **90**, No. 1, 90).

12.7 J. MEEUS, Oppositions of Ceres, Pallas, Juno and Vesta, 1970 to 1999 (*J.B.A.A.*, **84**, No. 1, 36).

12.8 D. MORRISON, Diameters of Minor Planets (*Sky & Tel.*, **53**, No. 3, 181).

12.9 F. PILCHER & J. MEEUS, *Table of Minor Planets* [privately published, 1973; available from Geoffrey Falworth, 11 Wimbledon Avenue, Blackpool].

12.10 D. N. WALLENTINE, The A.L.P.O. Minor Planets Section (*Strolling Astronomer*, No. 26, 18).

13. ZODIACAL LIGHT, ETC

13.1 D. E. BLACKWELL, The Zodiacal Light and its Interpretation (*Endeavour*, **19**, 14).
13.2 R. B. BOUSFIELD, The Zodiacal Band (*M.N.*, **94**, No. 9, 824).
13.3 T.A. CLARK & F. BABOTT, The Zodiacal Light Observed from Latitude 52° North (*Sky & Tel.*, **46**, No. 2, 130).
13.4 A. COLEMAN, The Photography of the Zodiacal Light (*J.B.A.A.*, **44**, No. 7, 262).
13.5 A. E. DOUGLASS, Zodiacal Light and Counterglow and the Photography of Large Areas and Faint Contrasts (*Photogr. J.*, New Series, **40**, No. 2, 44).
13.6 C. T. ELVEY, Photometry of the Gegenschein (*Ap. J.*, **77**, No. 1, 56).
13.7 M. V. GAVIN, Atmospheric Phenomena and Zodiacal Light (*J.B.A.A.*, **82**, No. 5, 353).
13.8 C. HOFFMEISTER, Beitrag zur Photometrie der Südlichen Milchstrasse und des Zodiakallichts (*Veröff. Univ. Berlin-Babelsberg*, **8**, No. 2, 1930).
13.9 C. HOFFMEISTER & J. PATON, *Visual Observation of the Airglow and Other non-Auroral Luminosities of the Sky* (London, 1956).
13.10 W. B. HOUSMAN, *Aurora and Zodiacal Light Section Observing Notes* (B.A.A. Sectional Notes, No. 4).
13.11 M. F. INGHAM [Meeting Report] (*Obs.*, **87**, No. 958, 104).
13.12 R. J. LIVESEY, Some Tropical Astronomy (*J.B.A.A.*, **80**, No. 5, 378).
13.13 H. C. NIGHTINGALE, Some Notes on the Zodiacal Light, Zodiacal Band and Gegenschein (*J.B.A.A.*, **85**, No. 5, 417).
13.14 R. O. REDMAN, Dust and Gas Between the Earth and the Sun (*Obs.*, **79**, No. 912, 172).
13.15 F. E. ROACH & P. M. JAMNICK, The Sky and Eye (*Sky & Tel.*, **17**, 4, 164).
13.16 R. G. ROOSEN, The Light of the Night Sky (*Sky & Tel.*, **47**, 4, 231).
13.17 L. RUDAUX & G. DE VAUCOULEURS, *Larousse Encyclopedia of Astronomy*, (Hamlyn, 1959, p. 292).
13.18 W. SANDNER, Leuchtende Nachtwolken, Polarlichter, Zodiakallicht (*Handbuch für Sternfreunde*, Berlin, 1967, p. 352).
13.19 H. SPENCER JONES, *General Astronomy* (London, 1956, p. 282).
13.20 A. THOM, The Zodiacal Light (*J.B.A.A.*, **49**, No. 3, 103).
13.21 —Reports of the Aurora and Zodiacal Light Section of the B.A.A. (*Mem. B.A.A.*, **19**, 1914–35, 1944).

14. AURORAE

14.1 S.-I. AKASOFU, Magnetospheric Substorms (*Q.J.*, **18**, No. 2, 170).

14.2   S.-I. AKASOFU, A Search for the Interplanetary Quantity Control-
       ling the Development of Geomagnetic Storms (*Q.J.*, **20**, No. 2,
       119).
14.3   S.-I. AKASOFU, *Polar and Magnetospheric Substorms* (Dordrecht,
       1972).
14.4   J. VAN ALLEN, Interplanetary Particles and Fields (*Sci. Amer.*,
       No. 233, 161).
14.5   B. BURRELL, Photographs of the Great Aurora, 17th April 1947
       (*J.B.A.A.*, **57**, No. 5, 205).
14.6   J. R. CAPRON, *Aurorae – Their Characters and Spectra* (London,
       1879).
14.7   T. G. COWLING, Interstellar and Interplanetary Plasma (*Q.J.*, **8**,
       No. 2, 130).
14.8   T. G. COWLING, Solar Wind (*Q.J.*, **12**, No. 4, 447).
14.9   E. DOYLERUSH, Correlation Between Aurorae and Solar Radio
       Bursts at 185 MHz (*J.B.A.A.*, **81**, No. 6, 449).
14.10 J. W. DUNGEY, Some Remaining Mysteries in the Aurora (*Q.J.*,
       16, No. 2, 117).
14.11 R. H. EATHER & S. B. MENDE, *The Radiating Universe* (Dord-
       recht, 1971, p. 255).
14.12 D. S. EVANS & G. R. JESSOP, *VHF-UHF Manual* (Radio Society
       of Great Britain, 1978).
14.13 R. A. HAM, Radio Observations of the Aurora Borealis (*J.B.A.A.*,
       87, No. 3, 252).
14.14 L. HARANG, *The Aurorae* (Chapman & Hall, 1951).
14.15 J. K. HARGREAVES, *The Upper Atmosphere and Solar-Terrestrial
       Relations* (London, 1979).
14.16 *I. T. T. Reference Data for Radio Engineers* (Indianapolis, 1977).
14.17 R. J. LIVESEY, The Distribution of Cloud and Its Effect on
       Observing (*J.B.A.A.*, **81**, No. 4, 292).
14.18 R. J. LIVESEY, Comparison of Solar and Auroral Activity at
       Glasgow, 1959–1973 (*J.B.A.A.*, **85**, No. 5, 424).
14.19 R. J. LIVESEY, The Aurora (*J.B.A.A.*, **89**, No. 2, 144).
14.20 R. J. LIVESEY, The Location of the Polar Aurora (*J.B.A.A.*, **90**,
       No. 3, 253).
14.21 *Manual No. 3, I.Q.S.Y. Aurora Committee* (Comité International
       de Géophysique, London, 1963).
14.22 H. M. MASSEY, *Space Physics* (Cambridge University Press, 1964).
14.23 H. W. NEWTON, *The Face of the Sun* (London, 1958).
14.24 J. PATON, Proposed Survey of the Frequency of Aurorae over the
       British Isles (*J.B.A.A.*, **62**, No. 7, 226).
14.25 J. PATON, Aurora Borealis: Photographic measurements of height
       (*Weather*, **1**, No. 6, 8, 1946).
14.26 J. PATON, Aurora Borealis (*Science News*, **11**, No. 9, 15, 1949).
14.27 J. A. RATCLIFFE, *Sun, Earth, and Radio* (London, 1970).
14.28 G. M. C. STONE, *Bull. R.S.G.B.*, **34**, pp. 13 and 15 (I.G.Y. V.H.F.
       Programme and Progress Report, 1958).
14.29 G. M. C. STONE, *Bull. R.S.G.B.*, **35**, p. 395 (Amateur Radio
       Participation in I.G.Y., 1950).

14.30 — *Photographic Atlas of Auroral Forms and Scheme for Visual Observations of Aurorae* (International Geodetic & Geophysical Union, Oslo, 1930).

15. METEORS

15.1 I. S. ASTAPOWITSCH, On The Cosmic Nature of Telescopic Meteors (*Obs.*, **60**, No. 762, 285).

15.2 A. BEER, Meteors (*Obs.*, **63**, No. 786, 229).

15.3 S. L. BOOTHROYD, Results of the Arizona Expedition: IV. Telescopic Observations of Meteor Velocities. (*Harv. Circ.*, 390). [See also B.15.58, 15.43–46.]

15.4 B. BOYD, The Accuracy of Estimated Meteor Durations (*Pop. Astr.*, **44**, No. 1, 39).

15.5 H. CHRÉTIEN, Les Perséids en 1904 (*B.S.A.F.*, **18**, 482).

15.6 E. H. COLLINSON & J. P. M. PRENTICE, The Photography of Meteors (*J.B.A.A.*, **37**, No. 7, 266).

15.7 E. H. COLLINSON, An Automatic Meteor Camera (*J.B.A.A.*, **39**, No. 5, 150).

15.8 E. H. COLLINSON, An Improved Automatic Meteor Camera (*J.B.A.A.*, **44**, No. 4, 157).

15.9 E. H. COLLINSON, Meteor Photography, 1928–31 (*J.B.A.A.*, **46**, No. 3, 116).

15.10 M. DAVIDSON, The Computation of the Real Paths of Meteors (*J.B.A.A.*, **46**, 292).

15.11 M. DAVIDSON, Variation in the Number of Meteors Observed for Different Hours and Different Times of the Year (*J.B.A.A.*, **24**, 352, 411).

15.12 M. DAVIDSON, Variation in the Number of Meteors observed for different Periods of the Year (*J.B.A.A.*, **24**, 477).

15.13 M. DAVIDSON, Computation of the Orbit of a Meteor Stream (*J.B.A.A.*, **44**, 116, 146).

15.14 M. DAVIDSON, A Simple Method of determining the Orbit of a Meteor by means of a Celestial Globe (*J.B.A.A.*, **24**, 307).

15.15 M. DAVIDSON, Note on Mr Porter's Paper (*J.B.A.A.*, **47**, 120). [See B.15.52.]

15.16 W. F. DENNING, General Catalogue of the Radiant Points of Meteoric Showers and of Fireballs and Shooting Stars observed at more than one Station (*Mem. R.A.S.*, **53**, 203). [A classic work, but now obsolete and not always reliable.]

15.17 W. J. FISHER, The Newton-Denning Method for Computing Meteor Paths with a Celestial Globe (*Proc. Nat. Acad. Sci.*, **19**, No. 2, 209 = *Harv. Repr.*, 85).

15.18 G. A. HARVEY, Four Years of Meteor Spectra Patrol (*Sky & Tel.*, **47**, No. 6, 378).

15.19 K. B. HINDLEY, The 11 Canis Minorid Meteor Stream (*J.B.A.A.*, **79**, No. 2, 138).

15.20 K. B. HINDLEY, Modern Meteor Observing (*J.B.A.A.*, **79**, No. 5, 393).

BIBLIOGRAPHY

15.21　K. B. HINDLEY, Taurid Meteor Stream Fireballs (*J.B.A.A.*, **82**, No. 4, 287).

15.22　K. B. HINDLEY, An International Centre for Meteor Observations (*J.B.A.A.*, **82**, No. 6, 459).

15.23　K. B. HINDLEY, The Quadrantid Meteor Stream (*Sky & Tel.*, **43**, No. 3, 162).

15.24　K. B. HINDLEY, The Meteor Section's Fireball Analysis and Meteorite Recovery Programme (*J.B.A.A.*, **85**, No. 2, 150).

15.25　C. HOFFMEISTER, *Die Meteore* (Leipzig, 1937).

15.26　J. S. HOPKINS, A Tentative Identification of Lines in a Meteor Spectrum (*Pop. Astr.*, **45**, No. 4, 214).

15.27　A. S. KING, A Spectroscopic Examination of Meteorites (*Ap. J.*, **84**, No. 5, 507).

15.28　L. KRESAK (Ed.), *Physics and Dynamics of Meteors* (Reidel, 1968).

15.29　F. A. LINDEMANN & G. M. B. DOBSON, Note on the Photography of Meteors (*M.N.*, **83**, No. 3, 163).

15.30　R. LLOYD EVANS & T. LLOYD EVANS, Observations of Meteor Spectra (*J.B.A.A.*, **76**, No. 4, 231).

15.31　B. LOVELL & J. A. CLEGG, *Radio Astronomy* (Chapman & Hall, 1952).

15.32　R. E. McCROSKY, The Lost City Metorite Fall (*Sky & Tel.*, **39**, No. 3, 154).

15.33　P. M. MILLMAN, The Theoretical Frequency Distribution of Photographic Meteors (*Selected Papers from Wash. Nat. Ac. Proc.*, **19**, 34, 1933).

15.34　P. M. MILLMAN, Amateur Meteor Photography (*Pop. Astr.*, **41**, No. 6, 298).

15.35　P. M. MILLMAN, An Analysis of Meteor Spectra (*Harv. Ann.*, **82**, No. 6, 113).

15.36　P. M. MILLMAN & D. HOFFLEIT, A Study of Meteor Photographs Taken through a Rotating Shutter (*Harv. Ann.*, **105**, No. 31, 601).

15.37　P. M. MILLMAN & D. W. R. McKINLEY, Three-Station Radar and Visual Triangulation of Metcors (*Sky & Tel.*, 8, No. 5, 114).

15.38　D. MILON, Observing the 1966 Leonids (*J.B.A.A.*, **77**, No. 2, 89).

15.39　C. P. OLIVIER, Calculation of the Heights of Meteors observed by H. L. Alden and C. P. Olivier (*Pop. Astr.*, **32**, No. 10, 591).

15.40　C. P. OLIVIER, Methods for Computing the Heights and Paths of Fireballs and Meteors (*Supplt. to the Pilot Chart of the North Atlantic Ocean for 1931*, Washington, 1931).

15.41　C. P. OLIVIER, *Meteors* (Williams & Wilkins, Baltimore, 1925).

15.42　E. ÖPIK, On the Visual and Photographic Study of Meteors (*Harv. Bull.*, 879, 5).

15.43　E. ÖPIK, Results of the Arizona Expedition: II. Statistical Analysis of Group Radiants (*Harv. Circ.*, 388). [See also B. 15.3, 15.44–46, 15.58.]

15.44　E. ÖPIK, Results of the Arizona Expedition: III. Velocities of

BIBLIOGRAPHY

Meteors observed visually (*Harv. Circ.*, 389). [See also B.15.58, 15.43, 15.3, 15.45–46.]

15.45 E. ÖPIK, Results of the Arizona Expedition: V. On the Distribution of Heliocentric Velocities of Meteors (*Harv. Circ.*, 391). [See also B.15.58, 15.43–44, 15.3, 15.46.]

15.46 E. ÖPIK, Results of the Arizona Expedition: VI. Analysis of Meteor Heights (*Harv. Ann.* **105**, No. 30, 1937). [See also B.15.58, 15.45–44, 15.3, 15.45.]

15.47 E. ÖPIK, A Statistical Method of Counting Shooting Stars and its Application to the Perseid Shower of 1920 (*Publ. Tartu,* **25**, No. 1, 1922).

15.48 E. ÖPIK, Results of Double-Count Observations of the Perseids in 1921 (*Publ. Tartu,* **25**, No. 4, 1923).

15.49 E. ÖPIK, Telescopic Observations of Meteors at the Tartu Observatory (*Publ. Tartu,* **27**, No. 2, 1930).

15.50 F. A. PANETH, *The Origin of Meteorites* (Oxford, 1940). [Halley Lecture, 1940.]

15.51 J. G. PORTER, The Reduction of Meteor Observations (*Mem. B.A.A.,* **34**, No. 4, 37).

15.52 J. G. PORTER, The Reduction of Meteor Observations (*J.B.A.A.,* **47**, 118). [See B.15.15.]

15.53 J. G. PORTER, An Analysis of British Meteor Data (*M.N.,* **103**, No. 3, 134).

15.54 J. P. M. PRENTICE, The Giacobinids, 1946 (*J.B.A.A.,* **57**, No. 2, 86).

15.55 J. P. M. PRENTICE, Fatigue and the Hourly Rate of Meteors (*J.B.A.A.,* **52**, No. 3, 98).

15.56 H. B. RIDLEY, The Phoenicid Meteor Shower of 1956 December 5 (*J.B.A.A.,* **72**, No. 6, 266).

15.57 D. W. SEARS, *The Nature and Origin of Meteorites* (Oxford University Press, 1978).

15.58 H. SHAPLEY, E. J. ÖPIK & S. L. BOOTHROYD, The Arizona Expedition for the Study of Meteors (*Wash. Nat. Ac. Proc.,* **18**, No. 1, 16 = *Harv. Repr.,* 74). [See also B.15.43–46, 15.3.]

15.29 J. SKYORA, La Photographie des Étoiles Filantes (*B.S.A.F.,* **38**, 64).

15.60 K. SUZUKI *et al.*, Recording Meteor Echoes by Radio (*Sky & Tel.,* **51**, No. 5, 359).

15.61 H. H. TURNER, On the Measurement of a Meteor Trail on a Photographic Plate (*M.N.,* **67**, No. 9, 562).

15.62 H. H. WATERS, The Photography of Meteors (*J.B.A.A.,* **46**, No. 4, 152).

15.63 F. WATSON, The Luminosity Function of the Giacobinids (*Harv. Bull.,* 895, 9).

15.64 F. WATSON, A Study of Telescopic Meteors (*Proc. Amer. Phil. Soc.,* **81**, No. 4, 493).

15.65 F. G. WATSON, *Between the Planets* (Cambridge, Mass., 1941).

15.66 F. WATSON & E. M. COOK, The Accuracy of Observations by Inexperienced Meteor Observers (*Pop. Astr.,* **44**, No. 5, 258).

15.67 F. L. WHIPPLE, The Harvard Photographic Meteor Programme (*Sky & Telesc.*, 8, No. 4, 90 = *Harv. Repr.*, 319).

15.68 B. S. WHITNEY, New Methods for Computing Meteor Heights (*M.N.*, 96, 544).

15.69 W. T. WHITNEY, The Determination of Meteor Velocities (*Pop. Astr.*, 45, No. 3, 162).

15.70 J. D. WILLIAMS, Binocular Observations of 718 Meteors (*Proc. Amer. Phil. Soc.*, 81, No. 4, 505).

15.71 E. WILLIS, *Tables for the Computation of Meteor Orbits* (privately printed, 1939).

15.72 C. C. WYLIE, The Calculation of Meteor Orbits (Formulas) (*Pop. Astr.*, 47, No. 8, 425).

15.73 C. C. WYLIE, The Calculation of Meteor Orbits (Tables) (*Pop. Astr.*, 47, No. 9, 478).

15.74 C. C. WYLIE, The Calculation of Meteor Orbits (Examples) (*Pop. Astr.*, 47, No. 10, 549).

15.75 C. C. WYLIE, Psychological Errors in Meteor Work (*Pop. Astr.*, 47, No. 4, 206).

15.76 C. C. WYLIE, The Relation of Group to Solo Counts in Meteor Work (*Pop. Astr.*, 42, No. 3, 157).

15.77 —*American Meteor Society Bulletins* (Flower Observatory, University of Pennsylvania).

15.78 — Reports of the Meteor Section of the B.A.A., 1893–1937 (*Mem. B.A.A.*, 1–32).

15.79 — Report of Meeting (*J.B.A.A.*, 68, 7, 229).

See also B. 16.48, 16.52.

16. COMETS

16.1 C. E. ADAMS, Calculation of a Comet's Coordinates (*J.B.A.A.*, 32, No. 6, 231). [See B.16.14, 16.29.]

16.2 G. E. D. ALCOCK, Comet and Nova Sweeps – a Summary (*J.B.A.A.*, 76, No. 1, 52).

16.3 J. BAUSCHINGER, *Die Bahnbestimmung der Himmelskörper* (Engelmann, Leipzig, 1928).

16.4 N. T. BOBROVNIKOFF, Physical Theory of Comets in the Light of Spectroscopic Data (*Rev. Mod. Phys.*, 14, No. 2–3 = *Perkins Obs. Repr.*, 31).

16.5 N. T. BOBROVNIKOFF, On the Systematic Errors in the Photometry of Comets (*Contr. Perkins Obs.*, 19).

16.6 N. T. BOBROVNIKOFF, On the Organisation of Physical Observations of Comets (*Pop. Astr.*, 42, No. 1, 2).

16.7 N. T. BOBROVNIKOFF, Investigations of the Brightness of Comets (Part I, *Contr. Perkins Obs.*, 15, 1941; Part II, *ibid.*, 16, 1942).

16.8 N. T. BOBROVNIKOFF, Observation of the Brightness of Comets (*Pop. Astr.*, 49, No. 9, 467 = *Perkins Obs. Repr.*, 28).

16.9 N. T. BOBROVNIKOFF, On the Brightness of Comets (*Pop. Astr.*, 50, No. 9, 473 = *Perkins Obs. Repr.*, 29).

16.10 N. T. BOBROVNIKOFF, The Brightness of Comet 1942*g* (Whipple) (*Pop. Astr.,* **51**, No. 9, 481 = *Perkins Obs. Repr.,* 32).

16.11 S. C. CHANDLER, Note on a Practical Problem for Beginners in Cometary Computation (*Pop. Astr.,* **6**, 459).

16.12 T. CLOSE, To Find the Parabolic Orbit of a Comet by a Graphical Method (*J.B.A.A.,* **49**, No. 6, 216).

16.13 L. J. COMRIE, Telegraphic and Published Positions of Comets and Asteroids (*Pop. Astr.,* **33**, No. 6, 382).

16.14 L. J. COMRIE, Note on Dr Adams's Paper and the Computation of Ephemerides (*J.B.A.A.,* **32**, No. 6, 234). [See B.16.1, 16.29.]

16.15 L. J. COMRIE, On Transferring Solar Co-ordinates etc from One Annual Equinox to Another (*J.B.A.A.,* **43**, No. 4, 158). [See B.16.28.]

16.16 R. T. CRAWFORD, *Determination of Orbits of Comets and Asteroids* (McGraw-Hill, N.Y., 1930).

16.17 A. C. D. CROMMELIN, Comet Catalogue – Sequel to Galle's 'Cometenbahnen' (*Mem. B.A.A.,* **26**, No. 2).

16.18 A. C. D. CROMMELIN, Continuation of Comet Catalogue (*Mem. B.A.A.,* **30**, No. 1).

16.19 A. C. D. CROMMELIN, On the Computation of a Comet's Ephemeris (*J.B.A.A.,* **6**, No. 3, 105).

16.20 A. C. D. CROMMELIN, On the Determination of the Orbit of a Comet or Planet by three Observations made at Intervals of a few days (*J.B.A.A.,* **7**, No. 3, 121; No. 5, 260; No. 6, 327).

16.21 A. C. D. CROMMELIN, The Computation of an Ephemeris of a Body moving in an Ellipse (*J.B.A.A.,* **26**, No. 4, 150).

16.22 A. C. D. CROMMELIN, Simplification of the Computation of an Ephemeris of a Comet moving in a Parabola (*J.B.A.A.,* **32**, No. 8, 305).

16.23 A. C. D. CROMMELIN & M. PROCTOR, *Comets* (Technical Press, Lond., 1937).

16.24 M. DAVIDSON, Variation in the Brightness of Comets (*J.B.A.A.,* **53**, No. 6, 175).

16.25 M. DAVIDSON, The Determination of the Parabolic Orbit of a Comet (*Mem. B.A.A.,* **30**, No. 1). [See also B.16.26.]

16.26 M. DAVIDSON, The Determination of the Parabolic Orbit of a Comet (*J.B.A.A.,* **43**, No. 2, 84). [See also B.16.27, 16.43.]

16.27 M. DAVIDSON, Remarks on the Relative Advantages of the Equator and the Ecliptic as Planes of Reference in Computing Cometary Orbits (*J.B.A.A.,* **43**, No. 3, 114).

16.28 M. DAVIDSON, Comet Whipple, 1933*f* (*J.B.A.A.,* **44**, No. 7, 267). [See also B.16.15.]

16.29 M. DAVIDSON, The Computation of Ephemerides (*J.B.A.A.,* **44**, No. 5, 185). [See also B.16.1; 16.14.]

16.30 M. DAVIDSON, A Method for Computing Approximate Elements in the Case of a General Orbit (*Mem. B.A.A.,* **30**, No. 1).

16.31 M. DAVIDSON, The Range of Solution in the Orbits of Comets (*J.B.A.A.,* **44**, No. 2, 68).

16.32　M. DAVIDSON, Note on Comet 1935a (Johnson) (*J.B.A.A.*, **45**, No. 5, 183). [See also B.16.31.]

16.33　M. DAVIDSON, Remarks on Mr Kellaway's Paper (*J.B.A.A.*, **53**, No. 4/5, 160). [Refers to B.16.47.]

16.34　A. H. DELSEMME (Ed.), *Comets, Asteroids, Meteorites* (University of Toledo, Ohio, 1977).

16.35　B. DONN *et al.* (Eds.), *The Study of Comets* (NASA, 1976).

16.36　E. DOOLITTLE, A Simple Method devised by F. C. Penrose for Finding the Orbit of a Heavenly Body by a Graphical Process (*Pop. Astr.*, **17**, 65, 138, 200, 292, 365). [See B.16.61.]

16.37　A. D. DUBYAGO, *The Determination of Orbits* (Macmillan, U.S.A., 1961).

16.38　P. DUFFETT-SMITH, *Practical Astronomy With Your Calculator* (Cambridge University Press, 1979).

16.39　J. GRIGG, A Graphic Method of Computing a Search Ephemeris for a Periodic Comet (*J.B.A.A.*, **9**, No. 8, 382).

16.40　W. P. HENDERSON, The Computation of the Perturbations of a Periodic Comet by Jupiter and Saturn (*Mem. B.A.A.*, **34**, No. 4, 21).

16.41　M. J. HENDRIE, An Analysis of Comet Discoveries (*J.B.A.A.*, **72**, No. 8, 384).

16.42　K. B. HINDLEY, On the Discovery of New Comets by Amateurs (*J.B.A.A.*, **75**, No. 3, 206).

16.43　R. T. A. INNES, Comments on Dr Davidson's Paper (*J.B.A.A.*, **43**, No. 3, 115). [Refers to B.16.26.]

16.44　R. T. A. INNES, Comet 1927f (Gale) (*Mem. B.A.A.*, **30**, No. 1). [See B.16.54, 16.36.]

16.45　L. C. JACCHIA, The Brightness of Comets (*Sky & Tel.*, **47**, No. 4, 216).

16.46　A. JONES, *Mathematical Astronomy with a Pocket Calculator* (Wiley, U.S.A., 1978).

16.47　G. F. KELLAWAY, A Note on the Orbit of Comet Whipple-Fedtke (1942g) (*J.B.A.A.*, **53**, No. 4/5, 159) [See B.16.33.]

16.48　A. KOPFF, Kometen und Meteore (*H.d.A.*, **4**, 426).

16.49　A. O. LEUSCHNER *et al.*, A Short Method of Determining Orbits from Three Observations [and other papers] (*Publ. Lick Obs.*, **7**, Nos. 1–3, 7–10).

16.50　A. E. LEVIN, Cometary Ephemerides: Correction Coefficients for change in date of perihelion (*J.B.A.A.*, **43**, No. 10, 429). [See also B.16.51.]

16.51　A. E. LEVIN, Cometary Ephemerides (*J.B.A.A.*, **44**, No. 1, 41). [See also B.16.50.]

16.52　A. F. LINDEMANN, A Revolving Eyepiece, electrically warmed (*M.N.*, **59**, 362).

16.53　G. MERTON, Comets and their Origin (*J.B.A.A.*, **62**, No. 1, 1951). [Also contains useful bibliography.]

16.54　G. MERTON, A Modification of Gauss's Method for the Determination of Orbits (*M.N.*, **85**, No. 8, 693). [See B.16.55.]

16.55 G. MERTON, A Modification of Gauss's Method for the Determination of Orbits (*M.N.*, **89**, No. 5, 451). [See B.16.54, 16.44.]

16.56 G. MERTON, Note on the above Paper of Mr Wood (*J.B.A.A.*, **36**,
No. 5, 151). [Refers to B.16.73.]

16.57 P. A. MOORE, *Comets* (Charles Scribner's Sons, 1978).

16.58 C. P. OLIVIER, *Comets* (Baillière, Tindall & Cox, 1930).

16.59 J. H. OORT, The Origin and Development of Comets (*Obs.*, **71**,
129). [1951 Halley Lecture.]

16.60 T. VON OPPOLZER, *Lehrbuch zur Bahnbestimmung der Kometen
und Planeten* (Engelmann, Leipzig, 1882; trans. PACQUIER,
Paris, 1886).

16.61 F. C. PENROSE, On a Method for Finding the Elements of the
Orbit of a Comet by a Graphical Process (*M.N.*, **42**, No. 2, 68).
[See also B.16.36.]

16.62 G. B. PETTER, The Determination of a Cometary Orbit (*Mem.
B.A.A.*, **21**, No. 2).

16.63 J. G. PORTER, *Comets and Meteor Streams* (Chapman & Hall,
1952).

16.64 J. RAHE & B. DONN, The Visual Observation of Comets (*Sky &
Tel.*, **41**, No. 4, 214).

16.65 W. C. RAND, A Nomogram for Comet Data (*J.B.A.A.*, **52**, No. 3,
104).

16.66 R. G. ROOSEN & B. G. MARSDEN, Observing Prospects for
Halley's Comet (*Sky & Tel.*, **49**, No. 6, 363).

16.67 Z. SEKANINA, The Prediction of Anomalous Tails of Comets
(*Sky & Tel.*, **47**, No. 6, 374).

16.68 R. W. SINNOTT, W. A. Bradfield and his Comet Seeker (*Sky &
Tel.*, **53**, No. 4, 306).

16.69 M. C. TRAYLOR, The Computation of an Ephemeris of a Planet
or a Comet (*Pop. Astr.*, **9**, 311).

16.70 Y. VÄISÄLÄ & L. OTERMA, Formulae and Directions for Computing the Orbits of Minor Planets and Comets (*Publ. Astron.
Obs., Turku*, 1951).

16.71 J. VINTER-HANSEN, The Orbits of Comets (*Pop. Astr.*, **52**,
370).

16.72 K. P. WILLIAMS, *The Calculation of the Orbits of Asteroids and
Comets* (Principia Press, Indiana, 1934; Williams & Norgate,
Lond., 1934).

16.73 H. E. WOOD, A Rapid Method of extending an Ephemeris (*J.B.A.A.*,
36, No. 5, 149). [See also B.16.56.]

See also B.15.65.

17. VARIABLES

17.1 R. G. ANDREWS, The Classification of Variable Stars (*J.B.A.A.*,
70, No. 5, 214).

17.2 F. M. BATESON *et al.*, Changing Trends in Variable Star Research
(*I.A.U. Colloquium 46*, 1979).

17.3 J. VAN DER BILT, The Light Curve of U Cygni (*Mem. B.A.A.*,
33, 1937).

17.4   A. N. BROWN & F. DE ROY, The Observation of Variable Stars
        (*J.B.A.A.*, **33**, No. 4, 143).
17.5   D. S. BROWN, The Photographic Observation of Variable Stars
        (*J.B.A.A.*, **73**, No. 8, 340).
17.6   D. S. BROWN, Some Experiments in Photographic Photometry
        using Comparison Prisms (*J.B.A.A.*, **75**, No. 2, 78).
17.7   J. T. BRYAN, Hunting for Supernovae (*J.B.A.A.*, **87**, No. 5, 457).
17.8   L. CAMPBELL, One Hundred Important Variable Stars (*Trans.
        I.A.U.* **6**, 237; Cambridge University Press, 1939).
17.9   L. CAMPBELL & L. JACCHIA, *The Story of Variable Stars*
        (Blakiston, Philadelphia, 1941).
17.10   R. S. DUGAN, A Finding List for Observers of Eclipsing Variables
        (*Contr. Princeton Obs.*, **15**). [Data on 269 eclipsing vari-
        ables.]
17.11   M. E. J. GHEURY, Notes pratiques sur l'Observation visuelle des
        Étoiles variables (*Ciel et Terre*, 1913, pp. 287, 319, 351; 1914–
        1919, pp. 1, 35, 78, 153, 180, 246; 1920, p. 36; Brussels).
17.12   J. B. GLASBY, The Future of the Variable Star Section (*J.B.A.A.*,
        79, No. 1, 70).
17.13   J. S. GLASBY, *Variable Stars* (Constable, 1968).
17.14   J. S. GLASBY, *The Nebular Variables* (Pergamon, 1974).
17.15   F. M. HOLBORN, The Methods of the Variable Star Section
        (*J.B.A.A.*, **68**, No. 8, 1958).
17.16   I. D. HOWARTH, Periodogram Analysis of Semi-regular Variable
        Stars (*J.B.A.A.*, **86**, 210, 379).
17.17   J. E. ISLES, Some Interesting Bright Variable Stars, 1967–70
        (*J.B.A.A.*, **84**, No. 1, 39).
17.18   Z. KOPAL, *The Computation of Elements of Eclipsing Binary
        Systems* (*Harv. Monogr.*, 8, 1950).
17.19   Z. KOPAL, *An Introduction to the Study of Eclipsing Variables*
        (*Harv. Monogr.*, 6, 1946).
17.20   R. J. LIVESEY, Control Charts and Small Amplitude Variable
        Stars (*J.B.A.A.*, **81**, No. 3, 196).
17.21   J. A. MATTEI *et al.*, Variable Stars and the A.A.V.S.O. (*Sky &
        Tel.*, **60**, No. 3, 180).
17.22   S. A. MITCHELL, Observations of 204 Long Period Variables
        (*Publ. L. McC. Obs.*, **60**, No. 1, 1935).
17.23   S. A. MITCHELL, Magnitudes and Coordinates of Comparison
        Stars (*Publ. L. McC. Obs.*, **60**, No. 2, 1935).
17.24   C. P. OLIVIER *et al.* (*Publ. Univ. Pa.*, **5**, No. 3). [Comparison stars
        for 52 variables.]
17.25   A. PAGE & B. PAGE, Multiple Exposure Techniques in Flare-star
        Photography (*J.B.A.A.*, **79**, No. 1, 26).
17.26   C. PAYNE-GAPOSCHKIN & S. GAPOSCHKIN, *Variable Stars*
        (*Harv. Monogr.*, 5, 1938).
17.27   W. E. PENNELL, Use of Modern 35 mm Panchromatic Films for
        Magnitude Determination (*J.B.A.A.*, **80**, No. 5, 371).
17.28   G. O. RAWSTRON, Automatic Photography of Eclipsing Variables
        (*Sky & Tel.*, **39**, No. 6, 397).

17.29 F. DE ROY, Tenth Report of the Variable Star Section of the
       B.A.A., 1920–1924 (*Mem. B.A.A.*, **28**, 1929).
17.30 J. SHEPHERD, A Photometer for Variable Star Observation
       (*J.B.A.A.*, **75**, No. 2, 105).
17.31 F. J. M. STRATTON, Novae (*H.d.A.*, **6**, 251).
17.32 W. STROHMEIER, *Variable Stars* (Pergamon, 1972).
17.33 A. D. THACKERAY, The Long-period Variables (*Obs.*, **59**, No.
       737, 285).
17.34 V. P. TSESEVICH (Ed.), *Eclipsing Variable Stars* (John Wiley &
       Sons, 1973).
17.35 M. ZVEREV (*Publ. Sternberg State Astr. Inst.*, No. 8, 1, 1936).
[See also entries under PHOTOMETRY in *A.A.H.*]

18. BINARIES

18.1   R. G. AITKEN, *The Binary Stars* (McGraw-Hill, 1935).
18.2   R. G. AITKEN, What we know about Double Stars (*M.N.*, **92**,
       No. 7, 596).
18.3   A. H. BATTEN, *Binary and Multiple Systems of Stars* (Pergamon,
       1973).
18.4   E. CROSSLEY, J. GLEDHILL & J. W. WILSON, *A Handbook of
       Double Stars, with a Catalogue of 1200 Double Stars and Exten-
       sive Lists of Measures* (Macmillan, Lond., revised edn., 1880).
18.5   M. A. ELLISON, A Note upon the Measurement of Double-Star
       Position-Angles near the Pole (*J.B.A.A.*, **54**, No. 8/9, 169).
18.6   K. GLYN JONES (Ed.), *Webb Society Deep-Sky Observer's Hand-
       book*, Vol. 1: *Double Stars* (Lutterworth, 1979).
18.7   W. D. HEINTZ, *Double Stars* (Reidel, 1978).
18.8   F. C. HENROTEAU, Double and Multiple Stars (*H.d.A.*, **6**, 299).
18.9   W. F. KING, Determination of the Orbits of Spectroscopic Binaries
       (*Ap. J.*, **27**, 125).
18.10 T. LEWIS, *Double Star Astronomy* (Taylor & Francis, Lond.,
       1908). [Reprint of *Obs.*, **31**, 88, 125, 162, 205, 242, 279, 307,
       339, 379.]
18.11 T. LEWIS, Double Star Astronomy (*Obs.*, **36**, 426).
18.12 J. MEEUS, Some Bright Visual Binary Stars (*Sky & Tel.*, **41**,
       No. 1, 21; **41**, No. 2, 88).
18.13 C. M. PITHER, Measuring Double Stars with a Grating Micrometer
       (*Sky & Tel.*, **59**, No. 6, 519).
18.14 H. N. RUSSELL, A Rapid Method for Determining Visual Binary
       Orbits (*M.N.*, **93**, 599).
18.15 H. N. RUSSELL, A Short Method of Determining the Orbit of a
       Spectroscopic Binary (*Ap. J.*, **40**, 282).
18.16 H. N. RUSSELL, On the Determination of the Orbital Elements of
       Eclipsing Variable Stars (*Ap. J.*, **35**, 315; **36**, 54).
18.17 B. W. SITTERLEY, A Graphical Method for Obtaining the Ele-
       ments of Eclipsing Variables (*Pop. Astr.*, **32**, 231).
18.18 W. M. SMART, On the Derivation of the Elements of a Visual
       Binary Orbit by Kowalsky's Method (*M.N.*, **90**, 534).

18.19 O. STRUVE, Spectroscopic Binaries (*M.N.*, **109**, No. 5, 487).
       [George Darwin Lecture, 1949.]
18.20 W. WEPNER, *291 Doppelstern-Ephermeriden für die Jahre 1975–2000* (Treugesell Verlag, Düsseldorf, 1976).
18.21 C. E. WORLEY & W. S. FINSEN, Third Catalogue of Orbits of Visual Binary Stars (*Rep. Obs. Circ.* 129, 1970).
[See also entries under MICROMETERS in *A.A.H.*]

19. NEBULAE AND CLUSTERS

19.1 R. BURNHAM, *Burnham's Celestial Handbook* (Dover, 1978, 3 vols).
19.2 R. J. BUTA, Observing Galaxies Visually (*Sky & Tel.*, **55**, No. 5, 482; **55**, No. 6, 595).
19.3 K. GLYNN JONES, *Messier's Nebulae and Star Clusters* (Faber & Faber, 1968).
19.4 K. GLYN JONES, *The Search for the Nebulae* (Neale Watson Academic, U.S.A., 1975).
19.5 K. GLYN JONES (Ed.), *Webb Society Deep-sky Observer's Handbook*, Vol. 2, *Planetary and Gaseous Nebulae*, Vol. 3, *Open and Globular Clusters* (Lutterworth, 1979).
19.6 J. H. MALLAS & E. KREIMER, *The Messier Album* (Sky Publishing Corp., U.S.A., 1978).
19.7 D. OVERBYE, Filters to Pierce the Night-time Veil (*Sky & Tel.*, **57**, No. 3, 231).
19.8 T. W. WEBB, *Celestial Objects for Common Telescopes* (Dover, 1962).

20. STAR ATLASES AND CHARTS
[See also various entries under 'Star Catalogues'.]

20.1 F. W. A. ARGELANDER, *Atlas des Nordlichen Gestirnten Himmels 1855.0* (Bonn, 1863; Ed. KUSTNER, Bonn, 1899). [140 charts covering the sky N of Dec $-2°$ to mag 9 on scale of 2 cm/degree.]
20.2 *Atlas of the Heavens* (Sky Publishing Corp., Cambridge, Mass.). [Previously known as *Atlas Coeli*; covers the whole sky to about mag 7.7 on a scale of 7.5 mm/degree. A standard atlas used by many amateurs.]
20.3 *Atlas Catalogue* (Sky Publishing Corp., Cambridge, Mass.). [Catalogue of interesting objects shown in B.20.2.]
20.4 *Atlas Borealis, Eclipticalis & Australis* (Sky Publishing Corp., Cambridge, Mass.). [On a uniform scale of 2 cm/degree, these three atlases cover the sky between Decs $+90°$ and $+30°$, $+30°$ and $-30°$, and $-30°$ and $-90°$ respectively. Stars to about mag 9.5 shown, spectral classes colour-coded. No nebulae, clusters, etc.]
20.5 E. E. BARNARD (Ed. E. B. FROST & M. R. CALVERT), *A Photographic Atlas of Selected Regions of the Milky Way* Publ. Carneg. Instn. 247, 1927). [Part I, photographs and descriptions; Part II, charts and tables.]

20.6   M. BEYER & K. GRAFF, *Stern-Atlas* (Hamburg, 1925, repr. 1952). [27 charts covering the whoke sky N of Dec $-23°$ to mag 9 (plus brighter nebulae and clusters) on scale of 1 cm/degree, epoch 1855. *B.A.A.H.*, 1926 contains a Table of Precession Corrections, 1855 to 1926, for application to the positions of the Beyer-Graff Atlas.]

20.7   G. BISHOP, *Ecliptic Chart* (London, 1848, etc). [24 charts, to mag 10, extending $3°$ on each side of the ecliptic, epoch 1825.]

20.8   K. F. BOTTLINGER, *Galaktischer Atlas* (Julius Springer, Berlin, 1937). [8 charts showing stars and nebulae to about mag 5.5 in galactic coordinates based on Pole at $12^h 40^m$, $+28°$.]

20.9   E. DELPORTE, *Atlas Céleste* (I.A.U., Cambridge, 1930). [26 maps covering whole sky to mag 6; I.A.U. constellation boundaries; lists of mags, spectroscopic types, and positions at 1875 and 1925 of all stars to mag 4.5, and principal doubles, variables, nebulae, etc.]

20.10 E. DELPORTE, *Délimitation Scientifique des Constellations* (*Tables et Cartes*) (I.A.U., Cambridge, 1930). [Virtually a quotation from B.20.9.]

20.11 J. FRANKLIN-ADAMS, *Photographic Chart of the Sky* R.A.S., 1914). [206 sheets, each $16°$ square, covering the whole sky to mag 15.5 on scale of about 1 in/$1°$.36. Charts 1-67, Dec $-90°$ to $-30°$; charts 68-139, $-15°$ to $+15°$; charts 140-206, $+30°$ to $+90°$. See also *M.N.*, **64**, 608, 1904; *ibid.*, **97**, 89, 1936.]

20.12 E. HEIS, *Atlas Coelestis Eclipticus: Octo Continens Tabulas ad Delineandum Lumen Zodiacale* (1878). [8 charts of the zodiacal region, especially made for the Zodiacal Light observer.]

20.13 W. J. LARSEN, Some Selected Sky Charts for the Amateur (*Sky & Tel.*, **56**, No. 6, 507).

20.14 A. P. NORTON, *A Star Atlas and Reference Handbook, Epoch 1950* (Gall & Inglis, 17th edn., 1978). [The stand-by of every amateur; some 7000 stars to mag 6.5, plus nebulae, clusters, etc, and a mass of general information and reference material of use to the amateur.]

20.15 J. PALISA & M. WOLF, *Palisa-Wolf Charts* (Vienna, 1908-31). [210 photographic charts in 11 Series; epoch 1875; scale 36 mm/ degree.]

20.16 *Palomar Sky Atlas.* [Two-colour photographic survey of the whole sky visible from Mt Palomar, made with the 122-cm Schmidt.]

20.17 C. PAPADOPOULOS, *True Visual Magnitude Photographic Star Atlas* (Pergamon, 1979).

20.18 F. E. ROSS & M. R. CALVERT, *Atlas of the Northern Milky Way* (University of Chicago Press, 1934). [39 plates, each about 13¼ ins square, exposed in a D = 5, F = 35-in camera at Mt Wilson and Flagstaff.]

20.19 E. SCHONFELD, *Atlas der Himmelszone zwischen $1°$ und $23°$ sudlicher Declination, 1855* (*Bonner Sternkarten, Zweite Serie*) (Bonn, 1887). [24 charts continuing B.20.1 to Dec $-23°$.]

BIBLIOGRAPHY

20.20   P. STUKER, *Sternelas für Freunde der Astronomie* (Stuttgart, 1925). [Photographic; to mag 7.5; epoch 1900.]

20.21   D. THOMAS, *Atlas der Sternbilder* (Salzburg, 1945). [32 main charts; also useful section on objects of interest.]

20.22   H. VEHRENBERG, *Photographic Star Atlas* (Hamburg; privately published). [In two volumes of loose photographs taken in blue light, to about mag 14; *Northern* covers Decs +90° to −26°; *Southern* from −14° to −90°. Scale 15 mm/degree, 464 maps altogether.]

20.23   H. VEHRENBERG, *Atlas of Deep Sky Splendours* (Sky Publishing Corp., U.S.A., 1978).

20.24   H. B. WEBB, *Atlas of the Stars* (privately printed, N.Y., 2nd edn., 1945). [110 charts covering the sky N of Dec −23° to about mag 9.5 on scale 1 cm/degree; epoch 1920.0, with coordinate intersections for 2000.0. Very useful supplement to B.20.14.]

21.   STAR CATALOGUES − POSITIONAL

21.1   A. KOPF, Star Catalogues, especially those of Fundamental Character (*M.N.*, **96**, No. 8, 714). [George Darwin Lecture, 1936; an extremely useful summary of 19th- and 20th-century work.]
[The following selection of the catalogues of the last 120 years, devoted primarily to star positions, is arranged chronologically.]

21.2   F. BAILY, *A Catalogue of Those Stars in the Histoire Celeste Francaise of J. Lalande* [etc] (British Association, London, 1837). [47,390 stars reduced to epoch 1800.0 from Lalande's observations.]

21.3   S. GROOMBRIDGE (ed. G. B. AIRY), *A Catalogue of Circumpolar Stars* (Murray, London, 1838). [4243 stars, epoch 1810.0; see also B.21.20.]

21.4   V. BAILY, *British Association Catalogue* (London, 1845). [8377 stars, epoch 1850.]

21.5   M. WEISSE, *Positiones mediae stellarum fixarum* [etc] (Petropoli, 1846). [Weisse's reductions of 31,085 stars within 15° of the equator to epoch 1825.0, using Bressel's observations; see also B.21.6. Abbrev: *WB*.]

21.6   M. WEISSE, *Positiones mediae stellarum fixarum* [etc] (Petropoli, 1863) [Continuation of B.21.5; 31,445 stars between Dec +15° and +45°. Abbrev: *WB2*.]

21.7   W. OELTZEN, *Argelanders Zonen-Beobachtungen vom 45° bis 80° nördlicher Declination in mittleren Positionen für 1842.0* (Wien, 1851-52). [Oeltzen's reductions of 26,425 stars from Argelander's observations. Abbrev: *OA(N)*.]

21.8   W. OELTZEN, *Argelanders Zonen-Beobachtungen vom 15° bis 31° südlicher Declination in mittleren Positionen für 1853.0* (Wien, 1857−58). [Continuation of B.21.7. Abbrev: *OA(S)*.]

21.9   F. W. A. ARGELANDER, *Bonner Durchmusterung des nördlichen Himmels* (Bonn, 1859−62; reprint 1903; 3 vols). [Approximate positions and visual mags to nearest 0.1 mag, to about mag 9.5. Arranged in successive 1°-wide Dec zones from +90° to −2°.

Charts, scale 2 cms/degree, have been issued in photostat. abbrev. *BD*. See B.21.12, 15.]

21.10 HEIS, *Atlas [and] Catalogus Coelestis Novus* (Cologne, 1872). [Abbrev: H'.]

21.11 J. BIRMINGHAM, The Red Stars: Observations and Catalogue (*Trans. Roy. Irish Acad.*, 26, 1877; Ed. T. E. ESPIN, Dublin, 1890). [Abbrevs: *B* and *E-B* respectively.]

21.12 E. SCHONFELD, *Durchmusterung* (1886). [Continuations of B.21.9 to Dec −23°. Abbrev: *SD*.]

21.13 B. A. GOULD, The Argentine General Catalogue (*Result. Obs. Nac. Argent.*, 14, Cordoba, 1886). [32,448 southern stars plus additional stars in clusters. Abbrevs: *CGA, AGC*.]

21.14 *Astronomische Gesellschaft Katalog* (Leipzig, 1890 etc). [The most complete catalogue of precision; epoch 1875. Abbrevs: *AG, AGC, CAG*. See B.21.38.]

21.15 J. M. THOME, Cordoba Durchmusterung (*Result. Obs. Nac. Argent.*, 16, 1892 etc.) [Continuation of B.21.12 to Dec −52° in 10° vols. To mag 10 approximately; with visual mags and charts. Abbrev: *CD*.]

21.16 *The Astrographic Catalogue*. [Initiated at the Paris Congress, 1887; work shared by observatories over the world, began 1892. Positions taken from photographic charts, scale 6 cm/degree. Abbrev: *AC*.]

21.17 D. GILL & J. C. KAPTEYN, The Cape Photographic Durchmusterung, 1875.0 (*Ann. Cape Obs.*, 3, 1896- 5, 1900). [Mags and approximate positions from Dec −19° to −90°, to mag 9. Abbrev: *CPD*.]

21.18 J. SCHEINER, Photographische Himmelskarte (*Publ. Astrophys. Obs. Potsdam*, 1899−1903). [Mags and positions, epoch 1900.0.]

21.19 S. NEWCOMB, Catalogue of Fundamental Stars for the Epochs 1875 and 1900 (*Astr. Pap., Wash.*, 8, No. 2, 77, 1905).

21.20 F. W. DYSON & W. G. THACKERAY, *New Reduction of Groombridge's Circumpolar Catalogue, Epoch 1810.0* (H.M.S.O., 1905). (See B.21.3.]

21.21 H. B. HEDRICK, Catalogue of Zodiacal Stars, for the Epochs 1900 and 1920 (*Astr. Pap., Wash.*, 8, No. 3, 405, 1905). [1607 stars to mag 7.5. Abbrev: *WZC*. See also B.21.37.]

21.22 J. BOSSERT, *Catalogue d'Etoiles Brillantes. 1900.0* (Gauthier-Villars, Paris, 1906). [3799 stars in 1° NPD zones.]

21.23 J. & R. AMBRONN, *Sternverzeichnis enthaltend alle Sterne bis zur 6.5 Grösse* (Julius Springer, Berlin, 1970). [Positions of 7796 stars to mag 6.5 for epoch 1900, and proper motions of 2226 stars.]

21.24 A. AUWERS, Neue Fundamentalkatalog des Berliner Astronomischen Jahrbuchs (*Veröff. König. Astron. Rechen-Instit.*, 33, 1910). [925 stars; one of the best fundamental catalogues; abbrev. *NFK*. See also B.21.33, 40.]

21.25 L. BOSS, *Preliminary General Catalogue of 6188 Stars for the Epoch 1900* (Publ. Carneg. Instn., 115, 1910). [Accurate posi-

BIBLIOGRAPHY

tions and proper motions of all naked-eye stars; abbrev: *PGC*.]

21.26 T. W. BACKHOUSE, *Catalogue of 9842 Stars, or all Stars Very Conspicuous to the Naked Eye* (Sunderland, 1911). [Epoch 1900; useful for amateurs.]

21.27 *Geschichte des Fixstern-Himmels* (Karlsruhe, 1922–23). [Collection of pre-1900 observations reduced to 1875.0.]

21.28 E. C. PICKERING & J. C. KAPTEYN, Durchmusterung of Selected Areas between δ = 0 and δ = +90° (*Harv. Ann.*, **101**, 1918).

21.29 E. C. PICKERING, J. C. KAPTEYN & P. J. VAN RHIJN, Durchmusterung of Selected Areas between δ = −15° and δ = −30° (*Harv. Ann.*, **102**, 1923).

21.30 E. C. PICKERING, J. C. KAPTEYN & P. J. VAN RHIJN, Durchmasterung of Selected Areas between δ = −45° and δ = −90° (*Harv. Ann.*, **103**, 1924).

21.31 L. BOSS & B. BOSS, *San Luis Catalogue of 15,333 Stars for the Epoch 1920.0* (Publ. Carneg. Instn., 386, 1928). [To mag 7; mostly southern.]

21.32 R. SCHORR & W. KRUSE, *Index der Sternörter 1900-25* (Bergedorf, 1928). [Monumental analysis of over 400 catalogues. Vol. 1, northern stars; vol. 2, southern.]

21.33 *Dritter Fundamentalkatalog des Berliner Astronomischen Jahrbuchs* (Berlin, 1934). [Abbrev: *FK3*. See B.21.24.]

21.34 F. SCHLESINGER, Catalogue of Bright Stars containing all important data known in June 1930 (*Publ. Yale Obs.*, 1930) [9110 stars to visual mag 6.5, and some fainter. Abbrev: *BS*.]

21.35 B. BOSS, *Albany Catalogue of 20,811 Stars for the Epoch 1910* (Carnegie Institution of Washington, 1931).

21.36 B. BOSS, *General Catalogue of 33,342 Stars for the Epoch 1950* (Carnegie Institution of Washington, 1937, 5 vols.).

21.37 J. ROBERTSON, Catalogue of 3539 Zodiacal Stars for the Equinox 1950.0 (*Astr. Pap., Wash.*, **10**, No. 2, 175, 1940). [Revision and enlargement of B.21.21.]

21.38 R. SCHORR & A. KOHLSCHÜTTER, *Zweiter Katalog der Astronomische Gesellschaft, Aquinoktium 1950* (Hamburg-Bergedorf, 1951). [Vols 1–4, Dec +90° to +50°. See B.21.14. Abbrev: *AGK2*.]

21.39 *Apparent Places of Fundamental Stars* (H.M.S.O., annually since 1941). [Mean and Apparent places of the 1535 stars of the *FK3* (B.21.33) and its Supplement.]

21.40 W. FRICKE & A. KOPFF, *FK4* (Karlsruhe-Brown, 1963). [The most recent revision of B.21.24; contains positions and proper motions for 1535 stars, mostly brighter than mag 7.5, over the whole sky.]

21.41 *Star Atlas of References Stars and Non-Stellar Objects* (MIT Press, Cambridge, U.S.A., 1969). [Based on the *Smithsonian Astrophysical Observatory Catalogue* (*SAOC*); consists of 152 charts covering the whole sky to mag 9, on a scale of 6.95′ = 1 mm. Abbrev: *SAOA*.]

331

22. STAR CATALOGUES – MOTIONS AND PARALLAXES

22.1 W. S. ADAMS & A. H. JOY, The Radial Velocities of 1013 Stars (*AP. J.*, **57**, No. 3, 149 = *M.W.C.*, 258).

22.2 J. BOSSERT, *Catalogue des mouvements propres des 5671 étoiles* (Paris, 1919).

22.3 W. W. CAMPBELL & J. H. MOORE, Radial Velocities of Stars brighter than Visual Magnitude 5.51 (*Publ. Lick Obs.*, **16**, 1928).

22.4 W. S. EICHELBERGER, Positions and Proper Motions of 1504 Standard Stars, 1925.0 (*Astr. Pap.*, *Wash.*, **10**, Part 1, 1925).

22.5 H. KNOX-SHAW & H. G. SCOTT BARRETT, *The Radcliffe Catalogue of Proper Motions in Selected Areas 1 to 115* (Oxford University Press, 1934). [To mag 14 in the Selected Areas on and N of the equator.]

22.6 J. H. MOORE, General Catalogue of Radial Velocities of Stars, Nebulae and Clusters (*Publ. Lick Obs.*, **18**, 1932).

22.7 J. S. PLASKETT & J. A. PEARCE, A Catalogue of the Radial Velocities of O and B Type Stars (*Publ. Dom. Astrophys. Obs.*, **5**, No. 2, 99).

22.8 J. G. PORTER, E. I. YOWELL & E. S. SMITH, A Catalog of 1474 Stars with proper motion exceeding four-tenths of a second per year (*Publ. Cincinn. Obs.*, **20**).

22.9 F. SCHLESINGER, *General Catalogue of Stellar Parallaxes* (Yale, 1924). [Includes all determinations available in 1924 Jan.]

22.10 F. SCHLESINGER & L. F. JENKINS, General Catalogue of Stellar Parallaxes (*Publ. Yale Obs.*, 1935). [B.22.9 revised to 1935.]

22.11 R. SCHORR, *Eigenbewegungs-Lexikon* (Hamburg Observatory, Bergedorf, 1 vol, 1923; 2 vols, 1936). [2nd edn. includes all proper motions available at the end of 1935; 94,741 stars, with mags and spectral types. Vol. 1, N stars; vol. 2, S stars. Abbrev: *EBL*.]

23. STAR CATALOGUES – PHOTOMETRIC

23.1 S. I. BAILEY, A Catalogue of 7922 Stars observed with the Meridian Photometer, 1889–91 (*Harv. Ann.*, **34**). [The 'Southern Meridian Photometry', abbrev: *SMP*. Continuation of B.23.16 to the S Pole.]

23.2 V. M. BLANCO *et al.*, *Photoelectric Catalogue* (Publ. U.S. Naval Observatory, Second Series, vol. XXI, Washington, 1968). [U, B, V photometry of more than 2000 stars, some as faint as mag 15.]

23.3 A. BRUNN, *Atlas photométrique des Constellations de +90° à −30°* (privately printed, France, 1949). [55 sheets showing all *BD* stars to mag 7.5, scale 1 cm/degree, epoch 1900. Against each star is printed its visual mag, to 0.01 mag for stars of mag 6.50 and brighter, to 0.1 mag for those fainter than 6.50. Other data include photometric mags of all extragalactic nebulae brighter than mag 12.0 photographic.]

BIBLIOGRAPHY

23.4   S. CHAPMAN & P. J. MELOTTE, Photogrpahic Magnitudes of 262 Stars within 25′ of the North Pole (*M.N.*, **74**, No. 1, 40).

23.5   B. G. FESSENKOFF, *Photometrical Catalogue of 1155 Stars* (Kharkow, 1926).

23.6   W. FLEMING, Spectra and Photographic Magnitudes of Stars in Standard Regions (*Harv. Ann.*, **71**, No. 2, 27).

23.7   B. A. GOULD, *Uranometria Argentina* (Buenos Aires, 1879). [Visually determined mags, and positions, of stars to mag 7 South of Dec +10°. Abbrev: *UA, G.*]

23.8   [Harvard], Stars near the North Pole (*Harv. Ann.*, **48**, No. 1, 1).

23.9   Harvard Standard Regions (*Harv. Ann.*, **71**, No. 4, 233).

23.10   D. HOFFLEIT, *Catalogue of Bright Stars* (Yale University Observatory, Conn., 1964). [Photometric data to stars of about mag 7.]

23.11   B. IRIARTE *et al.*, *Arizona-Tonantzintla Catalogue* (*Sky & Tel.*, **30**, No. 1, 21). [Five-colour photometry (U, B, V, R, I) of 1325 bright stars north of about Dec −50°, carried out between 1963 and 1965.]

23.12   R. H. LAMPKIN, *Naked Eye Stars* (Gall & Inglis, 1972). [Position, magnitude and spectrum for all stars to about mag 5.5.]

23.13   H. S. LEAVITT, The North Polar Sequence (*Harv. Ann.*, **71**, No. 3, 47).

23.14   Magnitudes of Stars of the North Polar Sequence (*B.A.A.H.*, 1926, 32).

23.15   G. MULLER & P. KEMPF, Photometrische Durchmusterung des nördlichen Himmels (*Publ. Astrophys. Obs. Potsdam*, Nos. 31, 43, 44, 51, 52, 1894–1907).

23.16   E. C. PICKERING, Adopted Photographic Magnitudes of 96 Polar Stars (*Harv. Circ.*, 170).

23.17   E. C. PICKERING, Observations with the Meridian Photometer, 1879–82 (*Harv. Ann.*, **14**, 1884). [The Harvard Photometry (*HP*): magnitudes of 4260 stars, including all brighter than mag 6 N of Dec −30°.]

23.18   E. C. PICKERING, Revised Harvard Photometry (*Harv. Ann.*, **50**). [Positions, visual mags, and spectral types of 9110 stars, mostly mag 6.5 and brighter. Abbrevs: *HR, RHP.*]

23.19   E. C. PICKERING, A Catalogue of 36,682 Stars Fainter than Magnitude 6.50 . . . forming a Supplement to the Revised Harvard Photometry (*Harv. Ann.*, **54**).

23.20   C. PRITCHARD, *Uranometria Nova Oxoniensis* (Oxford, 1885). [Wedge-photometer redeterminations of Argelander's *Uranometria Nova* magnitudes; 2784 entries. Abbrev: *UO.*]

23.21   F. H. SEARES, Magnitudes of the North Polar Sequence. (Report of the Commission de photométrie stellaire) (*Trans. I.A.U.*, **1**, 69).

23.22   F. H. SEARES, J. C. KAPTEYN & P. J. VAN RHIJN, *Mt Wilson Catalogue of Photographic Magnitudes in Selected Areas 1–139* (Carnegie Institution, Washington, 1930).

23.23　F. H. SEARES, F. E. ROSS & M. C. JOYNER, *Magnitudes and Colours of Stars North of +80°* (Publ. Carneg. Instn., 532, 1941).
23.24　R. W. SINNOTT [Chart of part of North Polar Sequence] (*Sky & Tel.*, 51, No. 5, 356).

24. STAR CATALOGUES — SPECTROSCOPIC

[Many of the catalogues mentioned elsewhere in this Bibliography quote spectroscopic types. Specially demanding mention, however, are the following.]
24.1　E. C. PICKERING, The Draper Catalogue of Stellar Spectra (*Harv. Ann.*, 27, 1890).
24.2　A. C. MAURY, Spectra of Bright Stars (*Harv. Ann.*, 28, 1897). [Together with B.24.1 constitutes the 'old' Draper Catalogue of over 10,000 stars.]
24.3　A. J. CANNON & E. C. PICKERING, The Henry Draper Catalogue of Stellar Spectra (*Harv. Ann.*, 91–99, 1918–24).
24.4　A. J. CANNON, The Henry Draper Extension (*Harv. Ann.*, 100). [Together with B.24.3 constitutes the 'new' Draper Catalogue (*HD*): mags, spectral types and positions of 225,000 stars to mag 10 approximately.]
25.5　A. SCHWASSMANN & P. J. VAN RHIJN, *Bergedorfer Spektral-Durchmusterung* (Bergedorf, 1935, 1938). [To mag 13 (photographic) in the northern Kapteyn areas.]

25. STAR CATALOGUES — VARIABLE STARS

25.1　A. J. CANNON,* Second Catalogue of Variable Stars (*Harv. Ann.*, 55, 1907). ['Second' with reference to B.25.12; 1957 variables.]
25.2　S. C. CHANDLER,* Catalogue of Variable Stars (*A.J.*, 8, No. 11/12, 82, 1888). [225 variables.]
25.3　S. C. CHANDLER,* Second Catalogue of Variable Stars (*A.J.*, 13, No. 12, 89, 1893). [260 variables.]
25.4　S. C. CHANDLER,* Third Catalogue of Variable Stars (*A.J.*, 16, No. 9, 145, 1896). [393 variables.]
25.5　S. C. CHANDLER,* Revision of Elements of the Third Catalogue of Variable Stars (*A.J.*, 24, No. 1, 1, 1904).
25.6　J. G. HAGEN, *Atlas [et Catalogus] Stellarum Variabilium* (9 Series, 1899–1941). [Charts and lists of about 24,000 comparison stars for 488 variables. The 5th Series (Berlin, 1906) is particularly useful, including all variables wholly observable with the naked eye or binoculars (minima brighter than mag 7). See also B.25.13.]
25.7　*Katalog und Ephemeriden Veränderlicher Sterne* (Vierteljahrsschrift der Astronomischen Gesellschaft, annually 1870–1926).
25.8　*Katalog und Ephemeriden Veränderlicher Sterne* (Berlin-Babelsberg, annually to 1941).
25.9　B. V. KUKARKIN *et al.*, *General Catalogue of Variable Stars*

* Primarily of historical value, B.25.9 now being the standard work.

(Moscow, 3rd edition, 1970). [Recognised by the I.A.U. as the standard work; data on 20,437 stars, extended by subsequent Supplements.]

25.10 R. PRAGER, *Geschichte und Literatur des Lichtwechsels der Veränderlichen Sterne* (Ferd. Dümmlers Verlagsbuchhandlung, Berlin; vol. 1, 1934; vol. 2, 1936). [See also B.25.11.]

25.11 R. PRAGER, History and Bibliography of the Light Variations of Variable Stars (*Harv. Ann.*, **111**, 1941). [Vol. 3 of B.25.10.]

25.12 Provisional Catalogue of Variable Stars* (*Harv. Ann.*, **48**, No. 3, 91, 1903). [1227 variables.]

25.13 J. STEIN & J. JUNKERS, [Index to the 9 Series of B.25.6.] (*Ricerche Astronomiche*, **4**; Specola Vaticana, 1941).

26. STAR CATALOGUES — BINARY STARS

26.1 R. G. AITKEN, *New General Catalogue of Double Stars within 120° of the North Pole* (Publ. Carneg. Instn., 417, 2 vols, 1932). [Includes all measures of 17,181 doubles prior to 1927; epochs 1900.0 and 1950.0. Abbrev: *ADS*.]

26.2 S. W. BURNHAM, *A New General Catalogue of Double Stars within 121° of the North Pole* (Publ. Carneg. Instan., 5, 2 vols, 1906). [Measures etc of 13,665 doubles; epochs 1880.0 and 1900.0. Part I, The Catalogue; Part II, Notes to the Catalogue. Abbrev: *BGC*.]

26.3 W. W. CAMPBELL & H. D. CURTIS, First Catalogue of Spectroscopic Binaries (*L.O.B.*, **3**, No. 79, 136). [Complete to 1905; 140 stars.]

26.4 W. W. CAMPBELL, Second Catalogue of Spectroscopic Binary Stars (*L.O.B.*, **6**, No. 181, 17). [To 1910; 306 stars. See also B. 26.3, 26.22−24.]

26.5 F. W. DYSON, *Catalogue of Double Stars from observations made at The Royal Observatory Greenwich with the 28-inch Refractor, 1893−1919* (H.M.S.O., 1921).

26.6 J. F. W. HERSCHEL, Descriptions and approximate Places of 321 new Double and Triple Stars (*Mem. R.A.S.*, **2**, No. 29, 459).

26.7 J. F. W. HERSCHEL, Approximate Places and Descriptions of 295 new Double and Triple Stars (*Mem. R.A.S.*, **3**, No. 3, 47).

26.8 J. F. W. HERSCHEL, Third Series of Observations . . . Catalogue of 384 new Double and Multiple Stars; completing a first 1000 of Those Objects (*Mem. R.A.S.*, **3**, No. 13, 1977).

26.9 J. F. W. HERSCHEL, Fourth Series of Observations . . . containing the Mean Places . . . of 1236 Double Stars [etc] (*Mem. R.A.S.*, **4**, No. 17, 331).

26.10 J. F. W. HERSCHEL, Fifth Catalogue of Double Stars . . . Places, Descriptions, and measured Angles of Position of 2007 of those objects [etc] (*Mem. R.A.S.*, **6**, No. 1, 1).

26.11 J. F. W. HERSCHEL, Sixth Catalogue of Double Stars . . . 286 of these objects (*Mem. R.A.S.*, **9**, No. 7, 193).

* Primarily of historical value, B.25.9 now being the standard work.

26.12 J. F. W. HERSCHEL, Seventh Catalogue of Double Stars (*Mem. R.A.S.*, **38**, 1870).

26.13 J. F. W. HERSCHEL (Ed. R. MAIN & C. PRITCHARD), Catalogue of 10,300 Multiple and Double Stars (*Mem. R.A.S.*, **40**, 1874).

26.14 W. HERSCHEL, Catalogue of Double Stars (*Phil. Trans.*, **72**, 112, 1782).

26.15 W. HERSCHEL, Catalogue of Double Stars (*Phil. Trans.*, **75**, 40, 1785).

26.16 W. HERSCHEL, On the places of 145 new Double Stars (*Mem. R.A.S.*, **1**, 166, 1821).

26.17 H. M. JEFFERS, W. H. VAN DEN BOS & F. M. GREEBY, *Index Catalogue of Visual Double Stars, 1961.0* Publ. Lick Obs., XXI, 1963). [Tabulates basic data for 64,247 double stars, but does not give measures of rapid binaries. Abbrev: *IDS*.]

26.18 R. JONCKHEERE, Catalogue and Measures of Double Stars discovered visually from 1905–1916 within 105° of the North Pole and under 5″ Separation (*Mem. R.A.S.*, **17**, 1917). [Virtually a Supplement to B.26.2.]

26.19 T. LEWIS, Measures of the Double Stars contained in the Mensurae Micrometricae of F. G. W. Struve (*Mem. R.A.S.*, **56**). [See B.26.26.]

26.20 J. MEEUS, Some Bright Visual Binary Stars, I (*Sky & Tel.*, **41**, No. 1, 21).

26.21 J. MEEUS, Some Bright Visual Binary Stars, II (*Sky & Tel.*, **41**, No. 2, 88).

26.22 J. H. MOORE, Third Catalogue of Spectroscopic Binary Stars (*L.O.B.*, **11**, No. 355, 141). [To 1024; 1954 stars. See also B.26.3, 4, 23, 24.]

26.23 J. H. MOORE, Fourth Catalogue of Spectroscopic Binary Stars (*L.O.B.*, **18**, No. 483, 1). [375 stars.]

26.24 J. H. MOORE & F. J. NEUBAUER, Fifth Catalogue of the Orbital Elements of Spectroscopic Binary Stars (*L.O.B.*, No. 521).

26.25 P. MULLER & C. MEYER, *Troisième Catalogue d'Ephémérides d'Etoiles Doubles* (Paris Observatory, 1969). [A standard work.]

26.26 J. SOUTH & J. F. W. HERSCHEL, *Observations of the Apparent Distances and Positions of 380 Double and Triple Stars, made in the Years 1821, 1822 and 1823* (Lond., 1825).

26.27 F. G. W. STRUVE, *Catalogus Novus Stellarum Duplicium et Multiplicium* (Dorpat, 1827). [The Dorpat Catalogue (Σ).]

26.28 F. G. W. STRUVE, *Stellarum Duplicium et Multiplicium Mensurae Micrometricae* (Petropoli, 1837). [See B.26.19.]

26.29 F. G. W. STRUVE, *Stellarum Fixarum imprimis Duplicium et Multiplicium Positiones Mediae pro Epocha 1830.0* (Petropoli, 1852).

26.30 O. STRUVE, *Revised Pulkova Catalogue* (Pulkova, 1850). [Abbrev: $O\Sigma$; Part II denoted by $O\Sigma\Sigma$.]

29. NEBULAE AND CLUSTERS — CATALOGUES
[See also under 'Star Atlases and Charts'.]

BIBLIOGRAPHY

27.1   S. I. BAILEY, Globular Clusters – A Provisional Catalogue (*Harv. Ann.*, **76**, No. 4, 43). [113 clusters.]

27.2   E. E. BARNARD, On the Dark Markings of the Sky, with a Catalogue of 182 Such Objects (*Ap. J.*, **49**, No. 1, 1). [See *Ap. J.*, **49**, No. 5, 360 for list of errata.]

27.3   H. D. CURTIS, Descriptions of 762 Nebulae and Clusters Photographed with the Crossley Reflector (*Publ. Lick Obs.*, **13**, No. 1, 9).

27.4   J. L. E. DREYER, New General Catalogue of Nebulae and Clusters of Stars (*Mem. R.A.S.*, **49**, 1888). [NGC. Based on B.27.5.]

J. L. E. DRYER, Index Catalogue (*Mem. R.A.S.*, **51**, 1895). [IC. Extension of NGC.]

J. L. E. DRYER, Second Index Catalogue (*Mem. R.A.S.*, **59**, 1908). [Extension of NGC.]

27.5   J. F. W. HERSCHEL, *General Catalogue of Nebulae and Clusters of Stars of the Epoch 1860* (Lond., 1864).

27.6   W. HERSCHEL, Catalogue of 1000 new Nebulae and Clusters of Stars (*Phil. Trans.*, **76**, 457, 1786).

27.7   W. HERSCHEL, Catalogue of a second 1000 Nebulae and Clusters of Stars (*Phil. Trans.*, **79**, 212, 1789).

27.8   W. HERSCHEL, Catalogue of 500 new Nebulae, Nebulous Stars, Planetary Nebulae, and Clusters of Stars (*Phil. Trans.*, 1802, 477).

27.9   J. HOLETSCHEK, Catalogue of Nebular Magnitudes (*Ann. K.K. Univ.-Stern.*, **20**, 114, Vienna, 1907).

27.10 P. J. MELOTTE, A Catalogue of Star Clusters shown on Franklin-Adams Chart Plates (*Mem. R.A.S.*, **60**, No. 5, 175). [245 objects.]

27.11 C. MESSIER, *Catalogue of 103 Nebulae and Clusters* (1771–84). [Abbrev: *M*. See also B.27.13.]

27.12 J. SULENTIC & W. TIFFT, *Revised New General Catalogue* (University of Arizona, 1975). [Updated version of B.27.4.]

27.13 H. SHAPLEY & H. DAVIES, Messier's 'Catalogue of Nebulae and Clusters' (*Obs.*, **41**, No. 529, 318). [Reprint of B.27.11, with corresponding NGC (B.27.4) numbers.]

27.14 G. DE VAUCOULEURS, *Reference Catalogue of Bright Galaxies* (University of Texas Press, 1964). [Major revision and enlargement of the 1932 Shapley-Ames catalogue; includes 2,599 galaxies, mostly brighter than mag 14.]

27.15 C. WIRTZ, Flächenhelligkeiten von 566 Nebelflecken und Sternhaufen (*Lund. Medd.*, **2**, No. 29, 1923).

APPENDIX

Table of Contents of 'Amateur Astronomer's Handbook'

APPENDIX

INDEX